Heterostructures on Silicon:
One Step Further with Silicon

NATO ASI Series

Advanced Science Institutes Series

A Series presenting the results of activities sponsored by the NATO Science Committee, which aims at the dissemination of advanced scientific and technological knowledge, with a view to strengthening links between scientific communities.

The Series is published by an international board of publishers in conjunction with the NATO Scientific Affairs Division

A	**Life Sciences**	Plenum Publishing Corporation
B	**Physics**	London and New York
C	**Mathematical and Physical Sciences**	Kluwer Academic Publishers
		Dordrecht, Boston and London
D	**Behavioural and Social Sciences**	
E	**Applied Sciences**	
F	**Computer and Systems Sciences**	Springer-Verlag
G	**Ecological Sciences**	Berlin, Heidelberg, New York, London,
H	**Cell Biology**	Paris and Tokyo

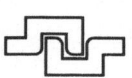

Series E: Applied Sciences - Vol. 160

Heterostructures on Silicon:
One Step Further with Silicon

edited by

Yves I. Nissim

Physics of Materials and Microstructures Division,
Bagneux Laboratory, C.N.E.T., Bagneux, France

and

Emmanuel Rosencher

Physics Division, Central Research Laboratory,
Thomson CSF, Orsay, France

Kluwer Academic Publishers

Dordrecht / Boston / London

Published in cooperation with NATO Scientific Affairs Division

Proceedings of the NATO Advanced Research Workshop on
Heterostructures on Silicon: One Step Further with Silicon
Cargèse, Corsica, France
May 15–20, 1988

Library of Congress Cataloging in Publication Data

Heterostructures on silicon.

  (NATO ASI series. Series E, Applied sciences ; 160)
  Includes index.
  1. Gallium arsenide semiconductors--Design and
construction--Congresses. 2. Epitaxy--Congresses.
3. Silicon--Congresses. I. Nissim, Yves I., 1953-
II. Rosencher, Emmanuel, 1952-   . III. Series.
TK7871.15.G3H48 1989   621.3815'2   88-27209

ISBN-13:978-94-010-6900-7   e-ISBN-13:978-94-009-0913-7
DOI: 10.1007/978-94-009-0913-7

Published by Kluwer Academic Publishers,
P.O. Box 17, 3300 AA Dordrecht, The Netherlands.

Kluwer Academic Publishers incorporates the publishing programmes of
D. Reidel, Martinus Nijhoff, Dr W. Junk and MTP Press.

Sold and distributed in the U.S.A. and Canada
by Kluwer Academic Publishers,
101 Philip Drive, Norwell, MA 02061, U.S.A.

In all other countries, sold and distributed
by Kluwer Academic Publishers Group,
P.O. Box 322, 3300 AH Dordrecht, The Netherlands.

printed on acid free paper

CONTENTS

OTHER III-V AND II-VI ON Si

SiGe HETEROSTRUCTURES

SUPERCONDUCTORS /Si HETEROSTRUCTURES

SILICON INSULATORS HETEROSTRUCTURES

* invited contribution

In the field of logic circuits in microelectronics, the leadership of silicon is now strongly established due to the achievement of its technology. Near unity yield of one million transistor chips on very large wafers (6 inches today, 8 inches tomorrow) are currently accomplished in industry.

The superiority of silicon over other material can be summarized as follow :

- The Si/SiO_2 interface is the most perfect passivating interface ever obtained (less than 10^8 eV^{-1} cm^{-2} interface state density)
- Silicon has a large thermal conductivity so that large crystals can be pulled.
- Silicon is a hard material so that large wafers can be handled safely.
- Silicon is thermally stable up to 1100°C so that numerous metallurgical operations (oxydation, diffusion, annealing...) can be achieved safely.
- There is profusion of silicon on earth so that the base silicon wafer is cheap.

Unfortunatly, there are fundamental limits that cannot be overcome in silicon due to material properties : laser action, infra-red detection, high mobility for instance.
The development of new technologies of deposition and growth has opened new possibilities for silicon based structures. The well known properties of silicon can now be extended and properly used in mixed structures for areas such as opto-electronics, high-speed devices. This has been pioneered by the integration of a GaAs light emitting diode on a silicon based structure by an MIT group in 1985.

As it will appear in this book, majority carrier devices (FET, Resonnant Tunneling diode...) realized with GaAs on silicon have now characteristics equivalent to those obtained on bulk GaAs. Minority carrier devices (lasers, detectors...) which are much more demanding on material quality begin to appear but there is still a long way to go. Some important fields of research with their potential application for future devices can be briefly listed as follow :

- GaAs on silicon for intrachip communication.
- InGaAsP on silicon for fiber telecommunication.
- HgCdTe, InSb, PbSnTe on silicon for far-infrared detection.
- Polymers on silicon for chemical FETs or thermal imaging.

- Insulator (CaF_2, SiO_2) for 3D-integration or lattice matching with other semiconductors.

The Figure shown below is a example of what could be considered as the "total integration concept" where the silicon chip could communicate optically, chemically etc... with the outside world.

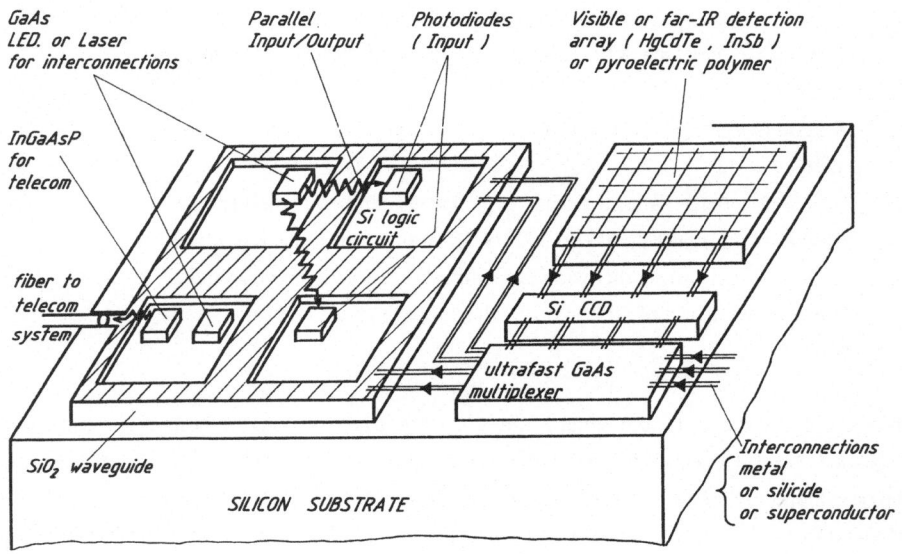

Fig. 1 : an artist's view of tomorrow's integration : a multiple material heterostructure where the logic and CCD applications are fabricated on silicon while all other functions are designed on the optimum choice of material epitaxially grown on Silicon.

The other challenge of heteroepitaxy on Silicon is the realization of new class of devices :

- Silicide/Si heterostructures for Metal Base Transistors
- Si/Ge or SiC/Si heterostructures for Heterojunction Bipolar Transistors
- Superconductor/Si heterostructures for Superconducting Transistors
  etc...

The crossroad of all these topics and applications lies in the understanding of growth mechanisms dominated by the interface aspects.

All these subjects have been treated by numerous scientists coming from ATT, Philips, France Telecom, Tokyo Institute of Technology, Thomson-CSF etc...

In their presentation, the authors strongly emphasized their motivations to study a particular structure. The net result of this Advanced Research workshop that is well reflected in this book is that it is possible to forsee the future for device structures and integration. This would not have been possible only few years ago, and two years from now a similar Workshop will definitely validate these new concepts. For this reason, the editors would like to thank all the authors for their valuable contribution which made this workshop a successful enterprise.

Emmanuel Rosencher, Yves Nissim

ACKNOWLEDGEMENTS

The editors would like to acknowledge the efficient contribution of Dr. P. Siffert and Dr. K. Ploog co-directors of the workshop.

This NATO Workshop has benifitted from the co-sponsorship of the following :
. The European Material Research Society
. C.N.E.T.
. D.R.E.T.
. Riber
. C.G.E (Laboratoire de Marcousis)
. C.I.T. Alcatel

The organizers are grateful to all these co-sponsors for their support.

We were particularly fortunate to have a series of high quality plenary and invited lectures to set the scene for each session. For these we are grateful to :

K. Ploog	J.C. Bean	J.M. Bureau
K. Woodbridge	T. Pearsall	J. Janata
D. Shaw	R.C. Frye	D. Bensahel
E. Kasper	A.W. Kleinsasser	H. Matsunami
D. Ankri	J.M. Phillips	T. Asano
J.C. Pfister		

For maintaining good timing in each session and orchestrate discussions and debates we would like to acknowledge the efficiency of the session chairmen :

K. Ploog	E. Rosencher	T. Asano
J.C. Bean	J. Phillips	J. Janata
J.C. Pfister	D. Bensahel	

It would have been impossible to organise successfully this NATO Workshop without the experienced contribution of Dr. Josselyne de Montlaur from DRET who directed our organising committee.

Dr. D. Bensahel provided us his skillfull organizing experience on the site of the NATO Workshop.Many thanks go to him.

Mme Marie France Hanseler has managed the local arrangements. The organizers and the participants are unanimously grateful for her valuable contribution.

Finally, particular thanks go to Dr. C. Sinclair, director of the NATO Advanced Research Workshop Program for his encouragement and assistance.

MBE GROWTH OF GaAs AND III-V QUANTUM WELLS ON Si

K. WOODBRIDGE

Philips Research Laboratories, Redhill, Surrey RH1 5HA, England

1.INTRODUCTION

The extensive research currently in progress on GaAs on Si reflects the important role this work could have in future utilisation of III-V material properties in combination with the well established silicon device technology. The problems inherent in this heteroepitaxy have now been well documented in the literature and are related mainly to the large lattice mismatch between the materials and the polar-nonpolar nature of the interface. The resulting generation of high densities of structural faults such as threading dislocations and anti-phase domains (APD's) are of immediate concern because of their possible effect on device performance especially in the case of thin layer structures such as multiple quantum wells (MQW's) and superlattices. In this work we will report on growth of both GaAs and GaAs/AlGaAs quantum wells on Si with special reference to the effect of strained layer superlattices and annealing on material quality. In-situ reflection high energy electron diffraction (RHEED) data has been correlated with ex-situ TEM and XRD structural data. A systematic study of photoluminescence properties of MQW's on Si has also been carried out.

2.EXPERIMENTAL

Early layers were grown for this work in a laboratory-built ion pumped MBE system but more recent material has been grown in a commercial Varian Modular GEN II MBE machine equipped with 3" non-bonded substrate capability and in-situ Auger analysis. Si wafers were thermally cleaned in a separate chamber prior to loading into the growth chamber to avoid the possibility of contamination. Substrate temperatures were monitored with both a thermocouple mounted behind the rotating wafer holder and an infrared pyrometer external to the system. Substrate rotation was employed during all growths to ensure lateral uniformity.

TEM(110) cross-sectional specimens were prepared by mechanical polishing and ion beam milling and were examined in a Philips EM400(T) electron microscope. The XRD studies were made on a powder diffractometer using a 002 reflection and CuKα radiation. Room temperature photoluminescence and photoluminescence decay were measured using a mode locked Kr$^+$ laser.

3.GaAs GROWTH

Prior to growth of MQW material we have examined the growth of GaAs layers in order to obtain good quality single domain surfaces on which to commence MQW growth. The so-called "two step" growth nucleation procedure in which a thin layer of GaAs is deposited at a low temperature before the substrate is raised to the growth temperature has been widely reported in the literature for both MBE and MOCVD growth. We have used a similar procedure (1) consisting of thermal cleaning at 800°C and deposition of

1

Y. I. Nissim and E. Rosencher (eds.), Heterostructures on Silicon: One Step Further with Silicon, 1–6.

500Å of GaAs at about 300°C before raising the substrate to the growth temperature. We have used a full multiple step etch (2) for most of these growths but more recently we have found that the use of an HF dip only followed by a passivation etch produces similar results. Auger analysis shows a similar clean up behaviour at about 800-900°C for both types of etch. The RHEED pattern following this cleaning procedure usually showed a strong 2x2 reconstruction. Inadequate thermal cleaning resulted in the formation of a large number of microtwins at the substrate epilayer interface as observed by RHEED when growth was started and subsequently confirmed by TEM.

The use of substrates misoriented from (001) has been reported in the literature to result in the reduction of threading dislocations and the suppression of APD's (3-5) due to step doubling and a possible mechanism for this has recently been proposed by Aspnes and Ihm (6). We have found that substrates tilted towards [100] showed RHEED evidence of formation of high densities of APD's whereas these were mainly absent on substrates tilted towards [110]. This is consistent with observations of growth morphology on spherical substrates which showed milky texture in the [010] and [001] axes (7). In the case of GaAs grown on substrates tilted 3° towards [110] no APD's were detected in cross-sectional TEM indicating that any present are of less than about 0.5μm lateral dimension and confined very close to the interface.

These initial studies showed it was possible to get smooth single domain GaAs after about 2-3μm growth on the substrate misoriented 3° off (001) towards [110] if the correct cleaning procedure was used. We have therefore subsequently grown a series of MQW structures to assess the structural and optical quality for a range of buffer layers.

4.MQW GROWTH

The MQW structures grown in this work were identical to structures routinely grown for optical assessment of GaAs/AlGaAs quantum well material on GaAs (8). The 60 period MQW had nominal well and barrier thicknesses of 55Å and 175Å respectively with the barrier Al content being 40%. The MQW region was cladded with 1400Å of $Al_{0.4}Ga_{0.6}As$ and the substrate temperature for the MQW growth was about 680°C. All buffer layers prior to the quantum well structure were grown at 550°C following the initial low temperature nucleation stage. Some initial layers were grown using only a GaAs buffer for structural assessment and examined in cross-sectional TEM and by X-ray diffraction profiling.

Fig. 1 shows a TEM cross-sectional image of a structure grown on a 3° off orientation substrate with a 3%m GaAs buffer layer. It can be seen that the uniformity of the MQW appears quite good and is better than that observed for MQW's grown on substrates closer to orientation (1). This is confirmed by comparing an X-ray diffraction profile of this layer with ones we have previously obtained using substrates close to orientation (1). This sample shows a marked improvement in number and sharpness of satellite peaks (Fig. 2) and the profile illustrated is very similar to that obtained from high structural quality MQW's grown on GaAs.

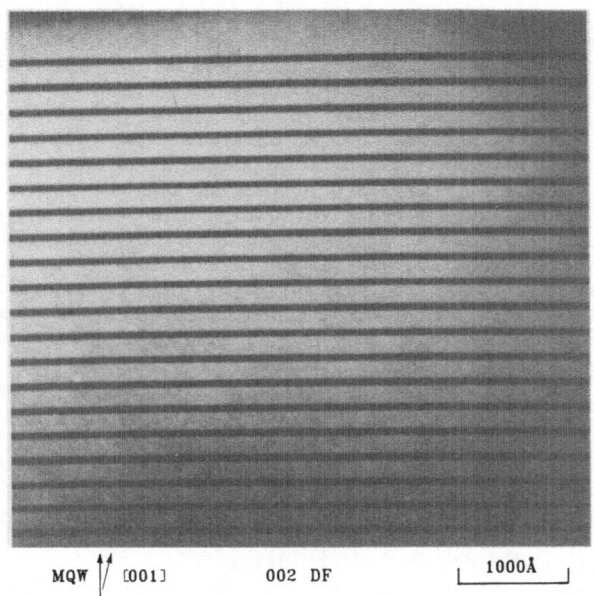

MQW ↕↑ [001]   002 DF |⎿ 1000Å ⏌|

FIGURE 1. Cross-section TEM micrograph of an MQW structure grown on a Si(001) substrate misoriented 3° towards [110].

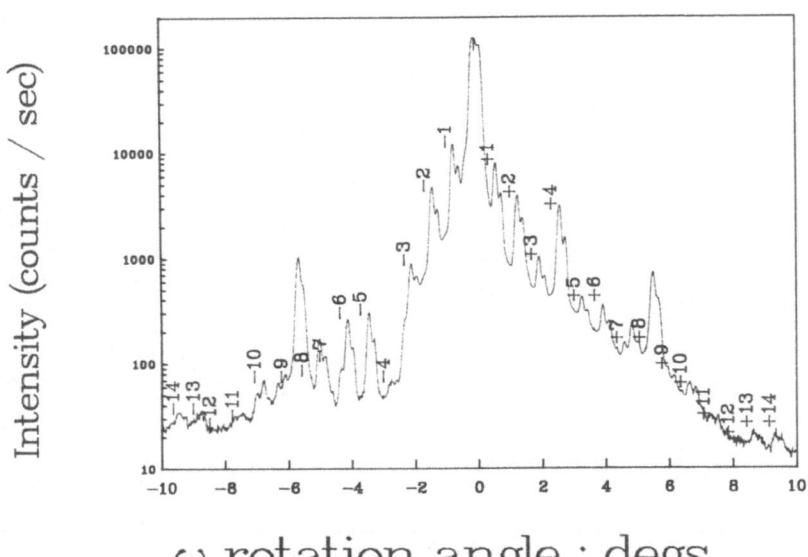

FIGURE 2. X-ray diffraction profile of an MQW structure grown on Si(001) misoriented 3° towards [110].

There have been several reports of the use of GaInAs/GaAs strained layer superlattices as a means of deflecting the threading dislocations and thus leading to a reduction in defect density in subsequent layers (3,4). The mechanism for this process is not clear but may be related to the influence of the strain field of the SLS. We have therefore used a GaAs/InAs SLS in order to maximise the strain field and indeed found that a single GaAs/InAs SLS in the buffer region of an MQW sample appeared to deflect many thread-ing dislocations which were propagating upward through the layers (1). Careful control of layer thickness is required in this case however since a large number of misfit dislocations are generated by the SLS itself if the InAs layer is more than about 15Å thick. The defect density at the surface in these layers was still fairly high ($\sim 10^8$ cm^{-2}) and it was not clear that the single SLS had drastically reduced defect densities since some dislocations may only be deflected for a short distance before resuming their upward path. We have therefore examined the effect of multiple 10Å/ 10Å GaAs/InAs SLS layers spaced about 0.5μm apart to see if a cummulative defect reduction can be achieved. In addition it has been reported that thermal annealing (9-11) shows some effect in reducing dislocation densities and we have combined SLS buffer layers and thermal annealing and examined the photoluminescence efficiency of these layers. The structure of the samples grown for this study are given in Table 1.

TABLE 1. Parameters of MQW sample buffer layers

Sample number	Buffer type	Annealing
M28	3μm GaAs	None
M31	3μm GaAs with 4xSLS at about 0.5μm intervals starting 1μm from the substrate	None
M35	As M31 + 3μm GaAs grown prior to above	3 steps of 10 min at 700°C after each 1μm during initial GaAs
M40	As M31	Steps of 10 min at 700°C at 0.1μm after each SLS.
M43	As M35	As in M35 and M40

Following growth the samples were examined in photoluminescence at 300K. The results of these measurements are shown schematically in Fig. 3.

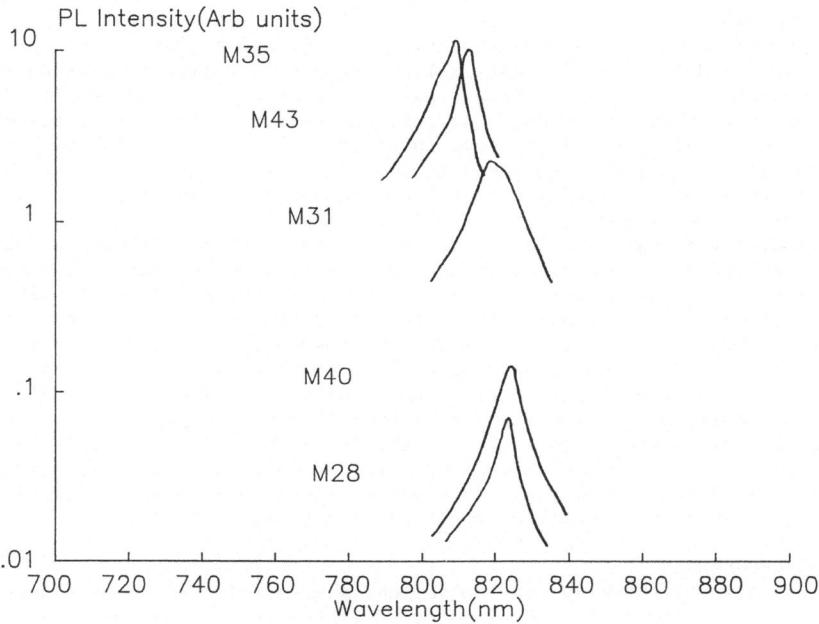

Fig. 3. Room temperature photoluminescence from the series of MQW samples
showing the improvement with SLS incorporation and thermal annealing.

It can be seen that the introduction of the multiple SLS layers results
in a significant improvement in PL efficiency. Annealing steps during the
initial GaAs seems to cause a further improvement although annealing after
each SLS seems to decrease the PL efficiency. The best result appears to be
obtained with a combination of the annealed GaAs buffer and SLS buffer. The
PL results from samples M31, 35 and 43 are comparable with MQW structures
grown on GaAs and show that a high optical quality of quantum well material
has been achieved. The spectra of these samples is not as sharp as that
generally found with MQW's on GaAs but this may be related to strain in the
layer which shifts the light and heavy hole exciton peaks in opposite
directions (12).

TEM studies of samples M31 and M35 reveal that a fairly large number of
threading dislocations are however still present, the measured figures at
$4\mu m$ from the substrate being $3x10^8$ cm^{-2} for M31 and $1x10^8$ cm^{-2} for M35.
The correlation between these results and the photoluminescence efficiency
measurements on these samples shows that the optical efficiency of the
material is, as expected, strongly influenced by the dislocation density.
Despite these high defect densities the optical efficiencies of M35 and M43
were high enough for photoluminescence decay measurements to be made and
carrier lifetimes of around 3ns were measured for these samples. This is
close to the values we have found for laser quality material and indicates
that optical device quality MQW material has been obtained.

5.CONCLUSIONS

The growth of GaAs on Si is now showing considerable promise for the production of device quality material. The use of substrates oriented several degrees off (001) towards [110] allows growth of good quality MQW material which shows structural characteristics comparable with similar structures on GaAs. The improvement in structural quality of MQW material grown on these substrates compared to substrates close to orientation seems to be related to the elimination of APD's rather than reduction in dislocation densities which are still quite high.

The use of InAs/GaAs SLS layers in combination with thermal annealing cycles has resulted in a marked increase in photoluminescence efficiency. We have been able to measure carrier lifetimes in these samples similar to those achieved on GaAs substrates indicating that a high optical quality of material has now been achieved. This improvement appears to be linked with the reduction in the number of dislocations which act as non-radiative recombination centres. The fairly high density of dislocations still present in this material is still of concern however and may be a serious problem in the case of high injection devices such as lasers. Nevertheless it is clear that this quantum well material could be suitable for many device applications and work is in progress to explore both the optical and electrical device possibilities of these structures.

6.ACKNOWLEDGEMENTS

I would like to thank my many colleagues who contributed to this work especially J.P. Gowers for TEM studies, P.F. Fewster for XRD data, P. Dawson and K.J. Moore for photoluminescence measurements and C. Roberts for assistance with MBE growth.

REFERENCES

1. K. Woodbridge, J.P. Gowers, P.F. Fewster, C. Curling, P. Dawson and K.J. Moore: Les Editions de Physique, Vol. XVl, 329, 1987.
2. A. Ishizaka and Y. Shiraki: J. Electrochem. Soc. 133, 666, 1986.
3. R. Fischer, D. Neuman, H. Zabel, H. Morkoc, C. Choi and N. Otsuka: Appl. Phys. Lett. 48, 1223, 1986.
4. R. Fischer, H. Morkoc, D.A. Neumann, H. Zabel, C. Choi, N. Otsuka, M. Longerbone and L.P. Erickson: J. Appl. Phys. 60, 1640, 1986.
5. H. Kroemer: J. Cryst. Growth 81, 193, 1987.
6. D.E. Aspnes and J. Ihm: Phys. Rev. Lett. 57, 3054, 1986.
7. T. Ueda, S. Nishi, Y. Kawarada, M. Akiyama and K. Kaminishi: Jap. J. Appl. Phys. 25, L789, 1986.
8. P. Dawson, G. Duggan, H.I. Ralph and K. Woodbridge: Superlattices and Microstructures 1, 173, 1985.
9. H.L. Tsai and J.W. Lee: Appl. Phys. Lett. 51, 130, 1987.
10. R.A. Lum, J.K. Klingert, B.A. Davidson and M.G. Lamont: Appl. Phys. Lett. 51, 36, 1987.
11. H. Okamoto, Y Watanabe, Y. Kadota and Y. Ohmachi: Jap. J. Appl. Phys. 26, L1950, 1987.
12. C. Jagannath, S. Zemon, P. Norris and B.S. Elman: Appl. Phys. Lett. 51, 1268, 1987.

EPITAXY OF GaAs ON PATTERNED Si SUBSTRATES BY MBE.

M.N. Charasse, B. Bartenlian, J.P. Hirtz, A. Peugnet, J. Chazelas
THOMSON-CSF - Domaine de Corbeville - B.P. 10 -91401 ORSAY - France

H. Blank
THOMSON THM/DAG - B.P. 48 - 91401 ORSAY - France

I. INTRODUCTION.

In recent years a great deal of activity has taken place to obtain good quality GaAs on Si. Two main reasons have motivated this work. The first-one is the Si substrate which is larger, mechanically stronger, with a better thermal conductivity and less expensive than GaAs. The second-one is the possibility of integration of high speed or optical GaAs devices to the standard Si integrated circuit technology. Another possibility which has been emerging very recently is the fabrication of devices consisting of layers in Si and layers in GaAs. A GaAs-Si heterojunction bipolar transistor has already been proposed /1/. The growth of GaAs on patterned Si substrates is essential for these two last applications. As the Si processes are made at a higher temperature than the GaAs-ones, they have to be made first. Afterwards the Si devices are covered with a pattern made of SiO_2 or Si_3N_4 before growing the GaAs layers in the unmasked areas. Then after a lift off, GaAs and Si devices are found close together and can be interconnected.

Another feature of the localized epitaxy of GaAs on Si concerns the reduction of wafer bending. The difference in the thermal expansion coefficients of GaAs and Si produces a wafer bending which is undesired during the device fabrication process. For example,the curvature radius of a 2 µm thick GaAs on Si wafer is about 15 m and it decreases with increasing GaAs thickness /2/. If GaAs is grown on small areas of Si, wafer bending would be reduced.

However, only a few papers have been published on the growth of GaAs on patterned Si substrates. Soga et al /3/ have achieved it on a pattern with holes of 0.5mm x 0.5mm by MOCVD (Metallorganic Chemical Vapor Deposition). Their paper deals mainly with the ridge growth observed at the substrate/mask interface. Matyi et al /4/ have grown by MBE (Molecular Beam Epitaxy) GaAs on a mask with holes of 1 mm x 1 mm and made structural characterizations with SEM (Scanning Electron Microscopy) and TEM (Transmission Electron Microscopy). Lee et al /5/ have also grown by MBE GaAs on a mask with 2 mm long stripes with widths ranging from 10 to 100 µm. They have found a higher photoluminescence intensity in the GaAs grown on the smallest area and suggest they have less defects than in larger areas. In this paper, we report the MBE growth of a 3000 Å thick GaAs layer on a pattern with features as small as 3 µm in one direction, which is smaller than any published result. This study applies directly to the growth of GaAs-Si devices with thin layers of GaAs. The epilayer is characterized by SEM and STEM (Scanning Transmission Electron Microscopy). We also give results on a complete selective growth of GaAs on Si.

7

Y. I. Nissim and E. Rosencher (eds.), Heterostructures on Silicon: One Step Further with Silicon, 7–12.
© *1989 by Kluwer Academic Publishers.*

II. MBE GROWTH.

The (001) 4° off towards $[110]$ Si substrates are covered with a 3000Å thick SiO_2 layer. Photolithographic etching is made using a "test mask" with various patterns oriented along the $[110]$ and $[1\bar{1}0]$ directions. The smallest features are 3 µm wide. Before MBE growth, the substrates are degreased and successively oxidized, etched and oxidized with the following procedure :

- boiling HNO_3, 5 min
- (HF : H_2O) (1:15) , 30 sec
- boiling HNO_3, 5min

According to our calibration, the SiO_2 mask must have been etched away on about 100Å during the HF solution step.

The substrate is then loaded in a Riber 2300 MBE system and considered as a usual Si substrate. It is outgassed at 500°C in the buffer chamber, and heated up to 950°C for 15 min in the growth chamber to remove the native oxide. Then the temperature is lowered down to 350°C. The Arsenic cell is opened during this cooling when the temperature is around 700°C. At this stage, the RHEED pattern consists of the superposition of a very intense background (due to the amorphous SiO_2) and very thin streaks (due to Si).

When the Ga shutter is opened, the thin streaks become long thin spots and discontinuous rings appear on the intense background. The spots come from the monocrystalline GaAs which is growing on Si and are much thinner than in the case of the growth of GaAs on a normal substrate. We believe that this comes from the smaller diffracting areas. The discontinuous rings are due to the polycrystalline GaAs growing on SiO_2. The mono or polycrystallinity of GaAs is confirmed by STEM (Scanning Transmission Electron Microscopy) observations (see IV).

The sample we describe here consists of three successive 1000Å thick GaAs layers grown respectively at 400,450 and 500°C, as measured by thermocouple. For the last value, the measurement by the pyrometer indicates 570°C. After growth, the sample is annealed at 600°C (pyrometer measurement) for 15 min. During the growth, we observe long thin spots, almost continuous, and discontinuous rings. After growth, the GaAs grown in the unmasked regions appears shiny and metallic in colour, whereas the GaAs grown on SiO_2 is dull and milky, as shown on figure 1.

III. SEM.

The observation of the sample in SEM shows a very good definition of the pattern as shown on figure 1. In the unmasked regions, the surface is similar to that of GaAs on non-patterned Si substrate, ie with a slight orange peel morphology. The GaAs grown on SiO_2 is very rough. We can see a small fracture between these two regions, as shown on figure 3, and small holes in the corners of the squared areas.

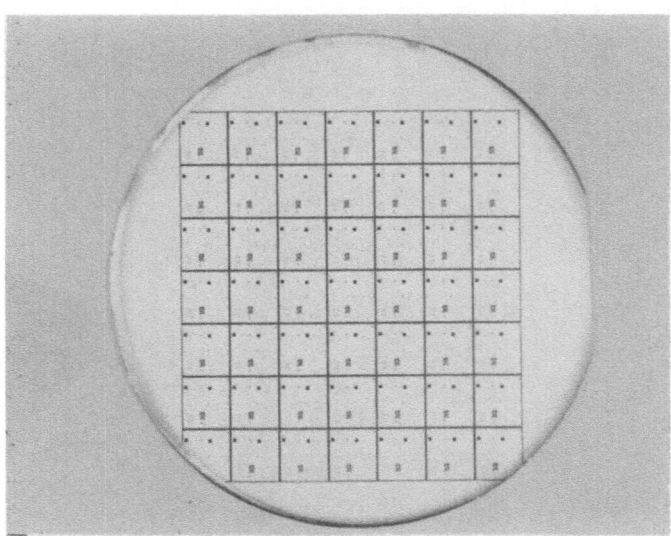

Figure 1. Picture of the surface of GaAs grown on a patterned 2 inches Si substrate. The monocrystalline and polycrystalline GaAs appear respectively metallic and milky.

IV. STEM.

Cross sections of the sample have been observed in STEM in order to characterize the defects and to examine the transition from the monocrystalline to the polycrystalline GaAs. It is found that the defect density of the GaAs grown on the unmasked regions is about $5 \times 10^7 \, cm^{-2}$, which is comparable to what we observe for the growth on a non-patterned substrate. Figure 4 shows a picture taken in the zone of the transition mask/substrate. Using the diffraction mode, evidence is given that the GaAs grown on SiO_2 is polycrystalline and the GaAs on the unmasked regions is monocrystalline.

As shown on the picture, it is found that the Silicon substrate has been underetched during the window opening process, leading to a (311) and (111) Si facets on the edge. These unexpected Si facets are responsible of the small disoriented GaAs crystal situated between the polycrystalline GaAs and the (001) GaAs. The free space between this crystal and the polycrystalline GaAs growing on the SiO_2 inclined wall must correspond to the small fracture identified in SEM.

It is often suggested that in the growth on a patterned substrate the polycrystalline zone will tend to extend over the monocrystalline-one during the crystal growth. Matyi et al /4/ have reported that they did not observe this effect in the case of GaAs on Si. We have not grown thick enough GaAs to have a definitive answer to this problem, but on this 3000A thick sample, the polycrystalline GaAs does not extend over the monocrystalline-one, as shown on picture 4 and on the other zones we have observed.

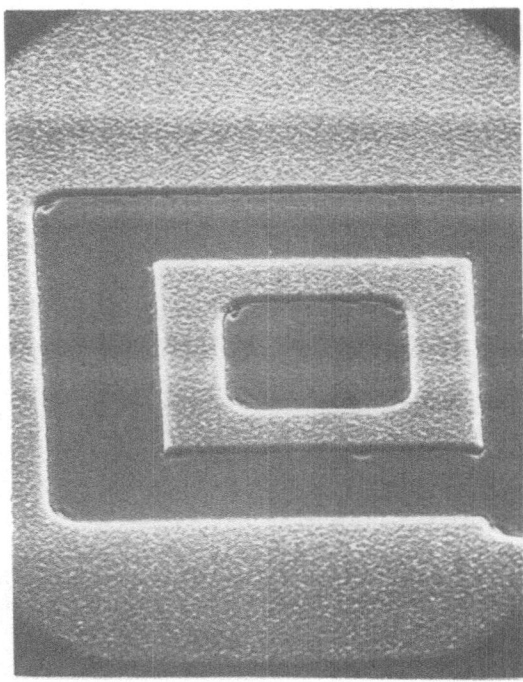

Figure 2. SEM micrograph
of GaAs   grown on a
patterned Si substrate
showing   the   good
definition   of   the
pattern.

5 μm

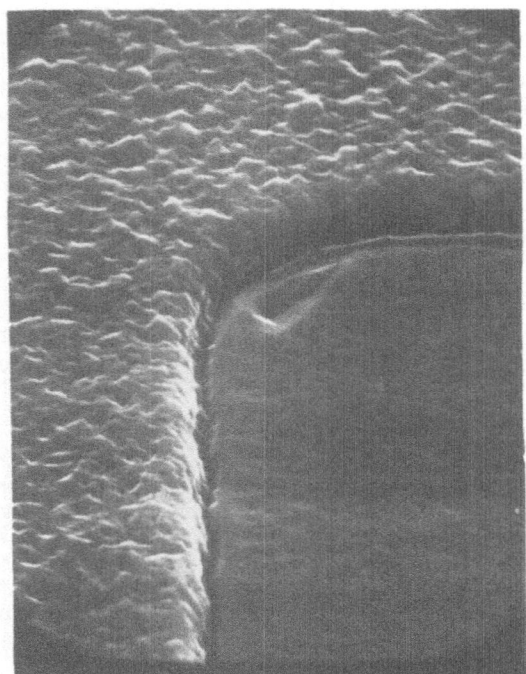

Figure 3. SEM micrograph
    of GaAs grown on a
patterned Si
substrate showing the
corner of a squared
area.

1 μm

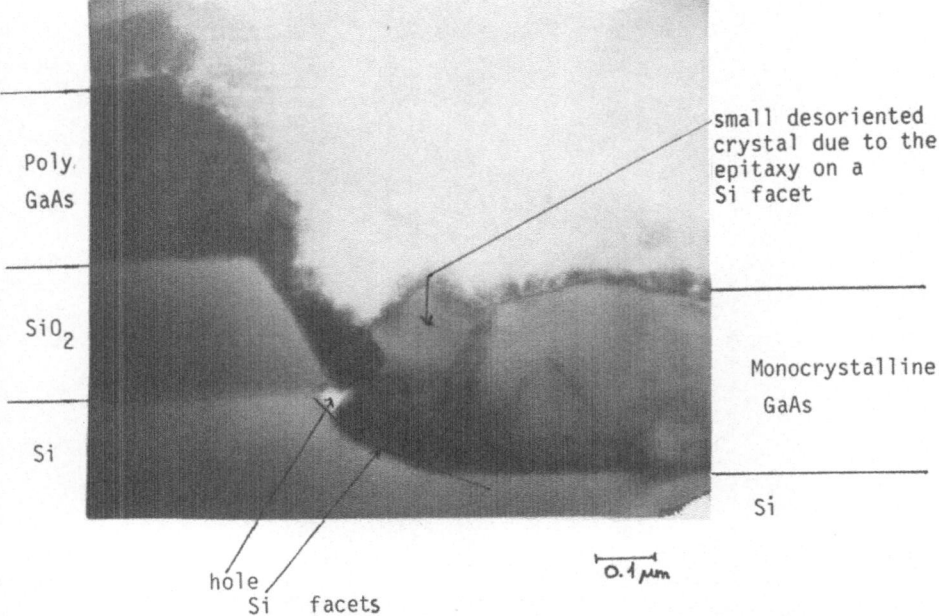

Poly.
GaAs

SiO$_2$

Si

small desoriented
crystal due to the
epitaxy on a
Si facet

Monocrystalline
GaAs

Si

hole
Si facets

0.1 μm

Figure 4. Transition between polycrystalline and monocrystalline GaAs observed by STEM. Note the Si facets due to unexpected underetching.

V. SELECTIVE GROWTH.

In many applications, it is necessary to remove the polycrystalline GaAs. From our experience, SiO$_2$ lift off in acid solution is difficult and takes a long time. Selective growth (no growth of SiO$_2$) is interesting to solve this difficulty. Okamoto et al /6/ have obtained it on GaAs subs-trate. We have applied their work to GaAs on Si and grown GaAs on a pat-terned Si substrate with a growth temperature of 710°C. At such a high temperature, the sticking coefficient of GaAs is zero on SiO$_2$ and not on Si or GaAs. Taking into account this high growth temperature, the Arsenic flux had been increased significantly (we did not have any ion gauge working to measure it). However, GaAs RHEED pattern was bad and SEM shows a very rough surface (figure 5). An optimization of growth conditions may improve the GaAs quality.

VI.CONCLUSION.

We have shown that the epitaxy of a 3000A thick GaAs layer on a patterned Si substrate with features as small as 3 μm can be obtained by MBE with a very good definition of the pattern. The transition mono-polycristalline GaAs is abrupt and the formation of desoriented crystals should be avoided on non-underetched substrates. This result is very important for the future of GaAs-Si devices. It is also quite promising for the integration of GaAs and Si devices where the GaAs thicknesses envolved are larger, a few microns instead of a few thousand Angströms.

12

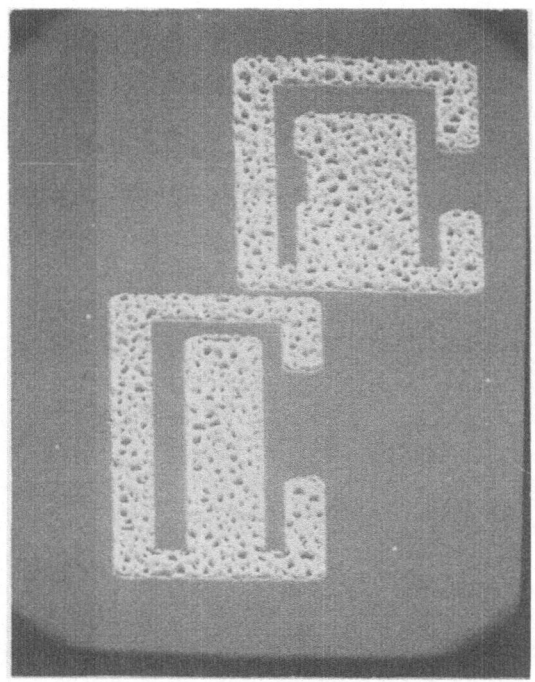

Figure 5. SEM micrograph of GaAs grown selectively on the unmasked areas of a patterned Si substrate.The bad morphology comes from the high growth temperature.

VII.ACKNOWLEDGEMENTS.

The authors thank G. Amendola of the ESIEE (Ecole Supérieure d'Ingénieurs en Electrotechnique et Electronique) for preparing the patterned Si substrates and M. Peschang and M. Magis for technical assistance on cross—sections preparation and STEM observations respectively.

REFERENCES.

1. J. Chen, T. Won, M.S. Unlü and H. Morkoç, D. Verret, Appl. Phys. Lett. 52 (10),822 (1988)
2. T. Soga, S. Hattori, S. Sakai and M. Umeno, J. Cryst. Growth, 77, 498 (1986).
3. T. Soga, S. Sakai, M. Umeno and S. Hattori, Jpn. J. Appl. Phys. 26(2), 252 (1987).
4. R. J. Matyi, H. Shichijo, T.M. Moore and H-L. Tsai, Appl. Phys. Lett. 51 (1), 18 (1987).
5. H.P. Lee and S. Wang, Y-H. Huang and P. Yu, Appl. Phys. Lett. 52 (3), 215 (1988).
6. A. Okamoto and K. Ohata, Appl. Phys. Lett. 51 (19), 1512 (1987).

EMBEDDED MOLECULAR BEAM EPITAXY FOR A COPLANAR GALLIUM - ARSENIDE ON SILICON TECHNOLOGY

J. DE BOECK, J.B. LIANG, J. VANHELLEMONT, G. BORGHS

IMEC v.z.w., Kapeldreef 75, B-3030 Leuven, BELGIUM

1. INTRODUCTION

Heteroepitaxy of GaAs on Si has gained interest over the past years because of the promising combination of optoelectronic and high speed properties of GaAs with the well established high density silicon technology. Devices with performances comparable to those of their GaAs on GaAs counterparts are frequently reported ([1,2,3]). Although the high density of threading dislocations together with the remaining stress in the device layers are limiting factors for the lifetime and reliability of the minority carrier devices, some successes in the monolithic integration of e.g. Si MOSFET's and double heterostructures LED's are achieved ([4]). A serious problem that has not been adressed properly is the strong non-planarity of the surface after growth and definition of the GaAs device islands. This leads to severe problems during the following processing steps. A coplanar surface is desirable from a process point of view. One way to achieve such a coplanar surface is to recess wells several microns deep into the Si substrate and refill them with the epitaxial layer, such that the GaAs surface levels with the Si surface after the GaAs growth. Since the III-V devices are to be fabricated in the windows the embedded layer quality has to be at least comparable with the large area GaAs on Si. The possible introduction of surface roughness and the loss of the intentional misorientation after recessing the Si can lead to dramatic degradation of the morphology and the introduction of antiphase domains. It has been anticipated that the exposure of different crystallographic orientations and the more severe geometrical constraints for the embedded GaAs film may disturb the single crystal quality of the GaAs at the well edges ([5]). Furthermore voids between the GaAs-filling and the Si-substrate are a hazard for interconnection and must be avoided carefully. In this paper we present an approach for etching the wells that takes care of the above mentioned problems and stress some of the important structural results and growth phenomena. A GaAs on Si coplanar technology by MBE which is suitable for fine line metallization with good step coverage is demonstrated.

2. WAFER PREPARATION AND EXPERIMENTAL PROCEDURE.

2.1. Wafer preparation.

Light p-doped 10 ohm-cm 3" (001)Si wafers 4^0 off towards [110] were used. The preparation before growth involves a standard cleaning of the Si wafer, forming and patterning a 400nm thick SiO_2 layer and exposing the Si where it is to be etched. The mask consists of 4x4mm^2 squares containing micro-windows with different widths(20-150μm) and spacings (5-15μm). Consequently the holes are etched 2-3μm deep in the

13

Y. I. Nissim and E. Rosencher (eds.), Heterostructures on Silicon: One Step Further with Silicon, 13–18.
© *1989 by Kluwer Academic Publishers.*

Si substrate. Different solutions of HNO_3 and HF diluted in H_2O or CH_3COOH were tested to form a featureless flat bottom and to preserve the intentional misorientation. Especially the curvature in the bottom corners was investigated because there the inclination of the surface showed the largest deviation from the 4^0 off. An isotropic solution of $HF:HNO_3$ (1:19) was found in this work to meet the smoothness requirements mentioned above as can be apreciated from Fig. 1. The etching process tends to be diffusion limited with a moderate and precisely controllable rate and the maximun effective angle at the bottom corner led to a deviation from the intentional surface misorientation of plus-minus less than one degree ([6]). After etching holes the SiO_2 overhang appears because of the effect of undercutting, as can be seen on Fig. 1. Embedded MBE growth on such a structure generates voids between the epitaxial GaAs layer and the Si sidewall which are undesirable for interconnection of monolithically integrated devices in both materials. The overhang is carefully removed in buffered HF, by etching for more than half the time needed to remove the oxide layer completely. During the final treatment, similar to the last step in the Ishizaka ([7]) cleaning procedure, a thin oxide layer was formed in $HCL:H_2O_2:H_2O$ that is readily decomposed upon heating in the growth chamber.

2.2. Experimental procedure.

After the preparation the wafer was indium-free mounted on a molybdenum block and loaded into a Riber 2300 MBE system. Two heat treatments of 300^0C and 850^0C were used to remove the moisture and gases and the thin surface oxide respectively. The oxide decomposition could be monitored by the reflection high energy electron diffraction image obtained from one half of the wafer where the oxide was stripped after Si etching. Besides the advantage of the RHEED imaging this oxide removal from one half of the wafer makes it possible to compare the crystal quality of the GaAs grown on the large area not etched Si in between the macro-squares, with the quality of the film embedded in the wells. Typically 2μ thick films are grown to fill the wells. For the growth schedule either the standard two step growth technique or thermally strained layers as reported by Lee ([8]) in the first 300nm from the interface were adapted.

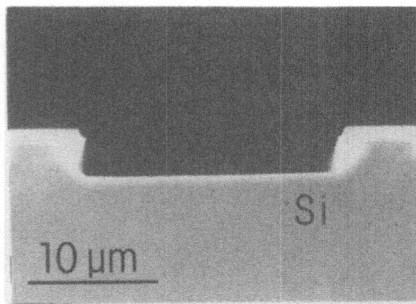  

Fig. 1. An SEM cross-section photo of the etched silicon wafer. The SiO_2 overhang has to be removed to avoid shadow formation during growth.

Fig. 2. Nomarski contrast photo's of the identical GaAs on Si morphology outside (a) and inside (b) the etched well.

A nomarski optical microscope and a JEOL JXA 840 scanning microanalizer were used to investigate the surface morphology and the crystallinity. Transmission electron microscopy was performed on a JEOL 1250 TEM and a JEOL 200CX for high voltage (HVEM) and high resolution (HREM) images respectively. This techniques allow the cross-section imaging of the dislocation network and special growth features (HVEM) and the study of the epitaxial relation on the GaAs/Si interface. The preparation of the samples is descibed in detail elsewhere ([9]).

3. STRUCTURAL PROPERTIES.
3.1. Embedded window growth vs. large area GaAs on Si.

Different growth schedules give different surface texture, especially when the temperature and growth rate in the initial stage is changed. For a given wafer however the GaAs morphologhy was identical for large area GaAs on Si grown on the not etched part where the Si-wafer was stripped before growth and the embedded GaAs on Si layers. When the initial temperature was higher a grain-like morphology was found. The features were oriented perpendicular to the tilt axis. When using the thermally strained layers a very shiny surface was obtained. Figure 2 shows a pair of Nomarski contrast photographs taken from the large area film (Fig. 2(a)) and the region of embedded growth (Fig. 2(b)) for a layer with a TSL-buffer. Both clearly exhibit the same morphology indicating the nucleation behaviour and the resulting morphology to be the same for the two different Si environments. In experiments on growth behaviour morphology degradation is observed when growing in a well were the effective curvature angle at the bottom corner was large.

The crystallinity is investigated with the electron channeling patterning technique on the JEOL JXA 840 scanning microanalizer. The channeling contrast observed by collecting the secondary electrons emerging from the sample was identical for GaAs layers in and outside the windows. Also the photoluminescent behaviour for the embedded layers was identical to that found for the large area GaAs on Si. These results indicate that equal crystalline and optical quality can be achieved for both Si-environments and are discussed in more detail elsewhere ([10]).

3.2. TEM study of embedded layers.

Heteroepitaxy is feasible for monolithic device applications when the threading dislocation density can be reduced to about $10^4 cm^{-2}$. This is very important for minority carrier devices such as bipolar transistors and optoelectronic devices. Another material defect possibly causing problems is the presence of antiphase domain boundaries, they can act as electrically active charge sheets and thus be detrimental for device operation. Growing on tilted (001) substrates 2^0 to 4^0 off towards [110], after a short heat treatment to form double atomic steps, has shown to lead to a reduction of threading dislocations and single phase epitaxial films ([11,12]). The results of our HREM-studies confirmed that the beneficial off orientation is preserved after the subtractive wet etch used to form the recess. Furthermore, surface atomic flatness after etching is comparable with that obtained after Ishizaka cleaning ([7]) thus providing identical Si surfaces for the identical growth behaviour found on both etched and non-etched surfaces ([10]).

High voltage transmission electron micrographs show the dislocation network present in the single crystal material. Typically the defect density is very large within a $0.5\mu m$ region near the interface and is reduced towards the top layers. The threading dislocations are grown in defects due to the lattice mismatch of both materials and are of the meander type. The dislocations introduced by glide during cool down appear as very straight lines in the cross-section images ([12]). Figure 3 shows the HVEM cross-section of an embedded GaAs-layer. The flatness of the well bottom and the high dislocation density near the GaAs/Si interface are clearly observable. This sample was grown in a standard two-step fashion and no additional measures, such as the incorporation of strained layer superlattices (see e.g. [11,13,14]), were taken to suppres the dislocations from penetrating the top layers.

When the SiO_2 overhang is properly removed shadow formation during growth is avoided. This leads to complete epitaxial filling of the wells when the thickness of the layer made equal to the well depth. We studied this filling in detail with HVEM and HREM. As a first important conclusion the GaAs/Si relation on the interface is found to consist of coherently grown regions separated by amorphous looking spots. These spots can be identified as SiO_2 nuclei interrupting the crystallinity on the interface. A small misorientation between the GaAs crystal and the Si substrate is observed, so that the GaAs $< 001 >$ lies in between the surface normal and the Si $< 001 >$. Dislocations with Burgers vectors inclined approximately 45° to the interface are found to generate a tilt in this direction ([15]). Yao et.al.([16]) have mentioned the possible presence of low-angle grain boundaries as found in the GaP on Si system([17]) to be responsible for the observed tilt.

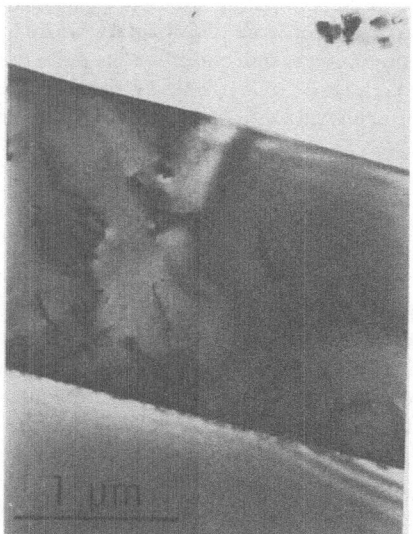

Fig. 3. The dislocation network in the embedded GaAs-film on the bottom of the etched well by HVEM.

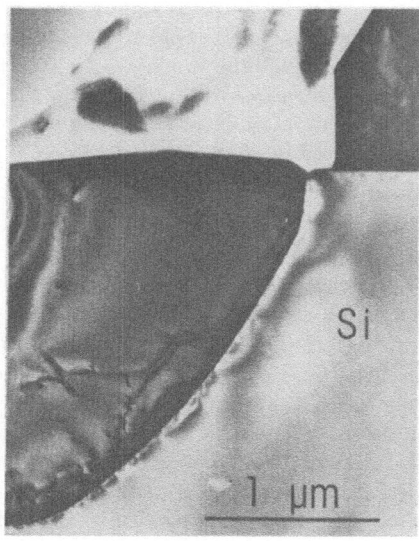

Fig. 4. A HVEM image showing the cross-section of the embedded layer at the sidewall.

An HVEM cross-section of the dislocation dislocation network at the sidewall is shown in Fig. 4. As obvious from the micrograph, most of the meander type dislocations encapsulated during growth are restricted to a narrow region parallel to the slope. This prevents them from dramatically threading the top layer and reduces the chance on possible side effects with implemented devices. The reduction of the threading dislocation density is possibly due to a different distribution of the coherency strain field and a more favourable annihilating interaction of dislocations during growth on the sidewall. Sometimes glide on (111) planes induces straight crystal defects during cool down from the growth temperature.

Excessive etching of the SiO_2 during the overhang removal mentioned above, so that too much Si is exposed, leads to the formation of a single-crystal GaAs grain at the rim of the platform. On Fig. 4 one can see the shape of such a GaAs mono-crystal that can cause serious problems during the lift off process discussed below. Furthermore the non-growth behaviour on the horizontal edge of the platform, as can be observed on Fig. 4, is an interesting feature. This is due to a diffusion of the growth controlling Ga atoms and leads to a perfect vertical (110) plane, rising on the platform. A detailed study on the growth mechanism of this non-planar heteroepitaxy is under way.

4. COPLANARITY AND GAAS-TO-SI INTERCONNECTION.

The GaAs devices are to be fabricated in the embedded single crystal layers in the wells. They are surrounded by the polycrystalline deposit on the masking dielectric. The realization of the coplanar GaAs on Si surface requires precise control of growth rate in order to equalize the thickness of the deposit and the depth of the well and to obtain complete filling at the edge. Using a photoresist mask to protect the monocrystalline GaAs, the polycrystalline GaAs can be removed by etching the underlying oxide layer. After this lift off, the GaAs surface step height is $0.7\mu m$ for a $2.3\mu m$ thick layer on the GaAs to Si border. A metallization test is performed over the border to test the step coverage of the metal line. An Au-evaporation followed by lift off to define the metal wires is performed. The thickness of the Au coverage is 150nm with varying width and spacing from $10\mu m$ to $1.25\mu m$. The interconnection from GaAs to Si is achieved without loss of continuity and no supplementary smoothening layer such as e.g. polyimide was used ([6]).

5. CONCLUSIONS AND SUMMARY.

In conclusion, the recessed windows are accomplished by a controlled solution of HF and HNO_3 and etched several microns deep in the Si substrate by an amount equal to the thickness of the epitaxial layer. The etched surface meets the requirements of flatness and retaining the intentional misorientation of several degrees from [110]. Morphology, crystallinity and optical behaviour for the GaAs deposit inside and outside the wells were found to be identical for a given sample whatever growth scheme was used. This is an important result because the findings of the research on quality improvement of large area GaAs on Si can directly be adapted for embedded growth. The dislocations at the sidewalls are restricted to a narrow region near the GaAs/Si interface. When the growth is precisely controlled the GaAs surface step height and sharpness obtained with this process do not cause metallization problems.

The interconnection over the bi-material border without loss of continuity indicates the feasibility of the monolithic integration of GaAs on Si by embedded MBE.

6. ACKNOWLEDGEMENTS

The authors wish to thank the ESAT-laboratory of the Catholic University Leuven for the photolithography on the Si wafers. We also thank P. Roussel for SEM, W. De Raedt and M. Van Hove for processing support and W. Van de Graaf for technical assistance during MBE-growths.

REFERENCES
1. R. Fishcher, N. Chand, W. Kopp, C.K. Penn, H. Morkoc, K.R. Gleason, and D. Scheitlin, IEEE Trans. Eletron. Devices, ED-33 , p.206 (1986).
2. L.T. Tran, R.J. Matyi, H.T. Yuan, and J.W. Lee, IEEE Electron.Dev.Lett, EDL-8, p.50 (1987).
3. H. Shichiji, J.W. Lee, W.V. Mclevige, and A.H. Taddiken, IEEE Electron.Dev.Lett. EDL-8 , p.121 (1987).
4. H.K. Choi, G.W. Turner, T.H. Windhorn, and B.Y. Tsaur, IEEE Electron.Dev.Lett, EDL-7 , p.500 (1986).
5. D. Shaw, in "Heterostructures on Silicon II", in *MRS Symposia Proceedings No.91,* edited by J.C.C. Fan, J.M. Phillips, and B-Y. Tsaur (Materials Research Society, Pittsburg, PA, 1987), p.15.
6. J.B. Liang, J. De Boeck,and G. Borghs, submitted to J.Vac.Sci.Tech.
7. A. Ishizaka and Y. Shiraki, J.Electrochem.Soc. 133 (4), p.666 (1986).
8. J.W. Lee, in "GaAs and Related Compounds", edited by W.T. Lindley (Institute of Physics, Bristol, 1986) p.111.
9. J. Vanhellemont, H. Bender, and L. Rousou, to be published in the proceedings of symposium W of the MRS Fall Meeting, Boston (1987).
10. J. De Boeck, J.B. Liang, J. Vanhellemont, K. deneffe, D.J. Arent, C. Van Hoof, G. Borghs, and R. Mertens, submitted to Appl.Pys.Lett.
11. R. Fisher, D. Neumann, H. Zabel, H. Morkoc, C. Choi, and N. Otsuka, Appl.Phys. Lett. 48 , p. 1223 (1986).
12. J.S. Harris,Jr. , S.M. Koch, and S.J. Rosner, in "Heterostructures on Silicon II", in *MRS Symposia Proceedings No.91,* edited by J.C.C. Fan, J.M. Phillips, and B-Y. Tsaur (Materials Research Society, Pittsburg, PA, 1987), p. 3.
13. J.W. Lee, in "Heteroepitaxy on Silicon", in *MRS Symposia Proceedings No.67,* edited by J.C.C. Fan, J.M. Poate (Materials Research Society, Pittsburg, PA, 1986), p. 29.
14. P.N. Uppal, and H. Kroemer, J. Vac. Sci. Tech. B4 (2), p.641 (1986).
15. R.J. Matyi, J.W. Lee, and H.F. Schaake, J.Electron.Mater. 17 (1), p. 87 (1988).
16. T. Yao, Y Okada, H. Kawanami, S. Matsui, A. Imagawa, and K. Ishida, in "Heterostructures on Silicon II", in *MRS Symposia Proceedings No.91,* edited by J.C.C. Fan, J.M. Phillips, and B-Y. Tsaur (Materials Research Society, Pittsburg, PA, 1987), p. 63.
17. O. Igarashi, Jpn.J.Appl.Phys, 15 ,1435 (1976).

SUPPRESSION OF DEFECT PROPAGATION IN HETEROEPITAXIAL
STRUCTURES BY STRAINED LAYER SUPERLATTICES

Zuzanna Liliental-Weber,[1] E.R. Weber,[1,2] J. Washburn,[1,2] T.Y. Liu[3], and
H. Kroemer[3]

[1]Materials and Chemical Sciences Division, Lawrence Berkeley Laboratory,
 Berkeley, CA 94720, USA
[2]Department of Materials Science and Mineral Engineering, University of
 California, Berkeley, CA 94720, USA
[3]Department of Electrical and Computer Engineering, University of
 California, Santa Barbara, CA 93106, USA

Abstract
  Defects present in GaAs on Si(211) heteroepitaxial layers grown by MBE
have been analyzed in detail by TEM. Efficient reduction of dislocation
propagation by strained layer superlattices was found. The mechanisms of
defect reduction were suggested based on Burgers vector analysis. It was
shown that additional threading dislocations can glide into the epilayer
during cooling process and that misfit dislocations at the interface can
be forced to dissociate on a (111) plane inclined to the interface leaving
one ·partial dislocation at the interface and forming extended stacking
faults.

1. INTRODUCTION
  The heteroepitaxy of GaAs thin films on Si substrates (GaAs on Si) has
attracted considerable interest in recent years [1-4], mainly for two
reasons: (1) the possibility of fabricating existing GaAs-based devices
on large, low cost Si substrates, and (2) the exciting potential of
monolithic integration of GaAs-based electronic and optoelectronic devices
with Si integrated circuits. However, the density of structural defects
such as dislocations, stacking faults, and microtwins in GaAs on Si
heteroepitaxy is still too high for many applications. These defects are
formed because of the different lattice constants and thermal expansion
coefficients in the substrate and epilayers. As a result of these
mismatches, defects in the epilayer are formed initially during the growth
process or during postgrowth cooling by propagation into the epilayer.
Even for GaAs grown on Si(001), where a large fraction of dislocations
formed at the interface have Burgers vector's in the interfacial
plane [5], dislocation densities in the range of 10^6 to 10^7 cm^{-2} are
usually found. This is over three orders of magnitude greater than the
dislocation density for GaAs films grown directly on GaAs substrates.
Another difficulty arising in the growth of a polar on a nonpolar crystal
is the presence of antiphase disorder and the formation of a very large
intrinsic electric charge, which can act as a sheet of very high doping.
One solution to these problems may be the use of (211)Si substrates to
grow GaAs [4]. However, many misfit dislocations in the GaAs grown on
(211)Si have Burgers vectors inclined to the interface [6,7], making them
susceptible to dissociations and gliding into the GaAs layer. In order to
obtain device-quality epitaxial GaAs material, a reliable method for
suppressing defect propagation in the epilayer is necessary. One
promising method is to use strained layer superlattices (SLSL's).

Y. I. Nissim and E. Rosencher (eds.), Heterostructures on Silicon: One Step Further with Silicon, 19–26.
© 1989 by Kluwer Academic Publishers.

In this paper, we compare, by transmission electron microscopy (TEM), the effectiveness of SLSL's (InGaAs/GaAs) in controlling dislocation propagation into the GaAs epilayer grown on Si(2$\bar{1}$1) substrates. The influence of furnace annealing and rapid thermal annealing (RTA) was investigated as well.

2. EXPERIMENTAL

GaAs crystal growth on a Si(2$\bar{1}$1) substrate was conducted in a molecular beam epitaxy (MBE) system (Varian-360). A two step cleaning procedure was applied for the Si wafers. The first step involved a procedure described by Ishizaka [8] where four major steps were involved: degreasing, acidic oxidation, alkaline oxidation, and boiling in HCl:H$_2$O:H$_2$O$_2$ (3:1:1) for 5-7 min followed by DI water rinse. After this, the Si wafers were mounted on a molybdenum block with In, and the Si wafer was dried with filtered nitrogen. The second step involved Ga reduction in the MBE chamber. The Si sample temperature was raised to 800°C, and a beam of Ga was simultaneously impinged on the sample surface. This "Ga-reduction" procedure lasted for ~10 min. Then the Ga furnace shutter was closed, but the sample temperature was still kept at 800°C for one more minute to eliminate the excess Ga on the surface. After this procedure the surface was considered oxide free, and the specific layers were grown.

Four different kinds of structures were investigated: (a) sample "23", with 50 layers of GaAs/InGaAs (5 nm thick each) grown directly on the Si surface followed by an 0.5-μm-thick GaAs epilayer, (b) sample "60", with a 50-nm-thick GaAs buffer layer followed by 10 layers of GaAs/InGaAs (5 nm thick each) and a 1-μm-thick GaAs epilayer, (c) sample "72", with three sets of SLSL's, and (d) sample "62", with only one InGaAs layer (30 nm thick), followed by 1μm of GaAs. Detailed informations about these structures were described previously [6]. Those structures were grown to investigate the influence of the presence of a GaAs buffer layer and the thickness of the SLSL sequence on defect density. The influence of annealing (furnace and RTA) was investigated for sample 60 and compared with similar annealing for GaAs grown on (001)Si.

All structures shown were investigated by using a JEOL JEM 200CX electron microscope with a point-to-point resolution of a 2.4 Å and by the Atomic Resolution Microscope (ARM) at Berkeley with its 1.7 Å point-to-point resolution. All samples were investigated in cross sections prepared along Si[111] and Si[011] parallel to the electron beam.

3. RESULTS

Before discussion of the experimental results of the dislocation study, it is useful to consider the possible geometries of misfit and threading dislocations in the GaAs/Si heterostructures. Complete dislocations have a translation vector of the crystal lattice as Burgers vector (b). In the diamond structure the shortest translation vector is of the type $\frac{a}{2}$ <011>. The dislocations found in GaAs/Si heteroepitaxial layers can be distinguished with respect to their Burgers vectors in the following way: GaAs/Si(001): Type I -- $\frac{a}{2}$ [$\bar{1}$10] and $\frac{a}{2}$ [110]; type II -- $\frac{a}{2}$ [011], $\frac{a}{2}$ [$\bar{1}$01], $\frac{a}{2}$ [0$\bar{1}$1], and $\frac{a}{2}$ [101]; GaAs/Si (2$\bar{1}$1): type I -- $\frac{a}{2}$ [011]; type II -- $\frac{a}{2}$ [110] and $\frac{a}{2}$ [$\bar{1}$01]; type III -- $\frac{a}{2}$ [101] and $\frac{a}{2}$ [0$\bar{1}$1]; type IV -- $\frac{a}{2}$ [$\bar{1}$10]. With this classification type I dislocations have Burgers vectors

in the hetero-interface. Type II dislocations have Burgers vectors inclined to the heterointerface. For GaAs/Si (001) these dislocations can
glide on {111} planes, which are inclined 55° to the heterointerface.
Type II dislocation in GaAs/Si (2$\bar{1}$1) can glide on the (1$\bar{1}$1) plane, which
is inclined only 19.5° to the interface. Type III dislocations in GaAs/Si
(2$\bar{1}$1) can glide on the (11$\bar{1}$) plane, which is perpendicular to the (2$\bar{1}$1)
interface plane. Type III as well as type IV dislocations can glide on
(111) or ($\bar{1}$11), both of which are 61.9° from (2$\bar{1}$1).

3.1. As grown epilayers

A cross-section TEM study (Fig. 1) of sample "23" shows that the use of
SLSL's can reduce the dislocation density by about two orders of
magnitude. A very important feature of the blocking of dislocation
propagation in these samples is that it occurs almost entirely at the
uppermost interface between the strained layers and the final GaAs layer.
Therefore, the reduction of dislocation density is only weakly dependent
on the thickness of whole set of strained-layer superlattices. Many
stacking faults occur through out SLSL's (visible as straight lines in
Fig. 1). These stacking faults are formed primarily on (11$\bar{1}$) planes
perpendicular to the interface. Many dislocations interact with each
other in the epilayer. Imaging with two-beam approximation and using
$\mathbf{g.b}$ = 0 (invisibility conditions) for particular dislocations allowed
one to determine Burgers vector of the threading dislocations.
Dislocations A in Fig. 1 vertical to the interface are type III.
Dislocations C parallel to the interface are type II. Dislocations B and
D are always visible for all low-index diffraction vectors in the (011)
plane. In this sample areas with very low dislocation densities have been
found. However, on the average the dislocation density in the area
directly above the SLSL's was in the 10^8 cm^{-2} range. Close to the
surface the dislocation density was around 5×10^7 cm^{-2}.

This study shows that the suppression of defect propagation depends only
weakly on the combined thickness of all the SLSL's. Samples "60" are
prepared with only 10 layers of SLSL's grown on the 5-nm-thick buffer
layer. These samples, with a buffer layer grown at 505°C, turned out to
have large dislocation-free areas in the GaAs, and the average dislocation

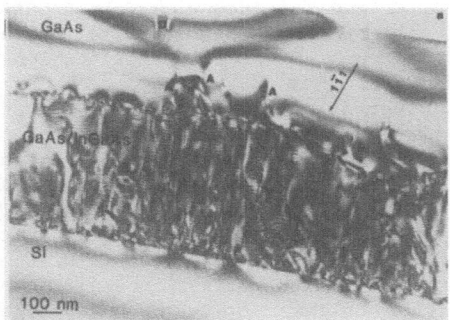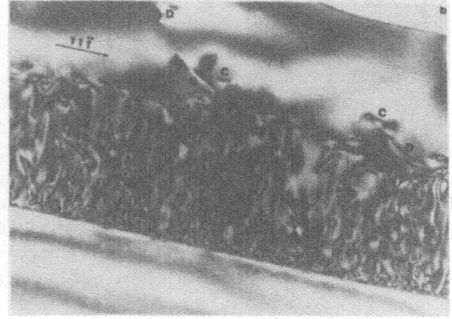

FIGURE 1. TEM cross-section micrograph of sample "23". Note that the
large number of stacking faults on the (11$\bar{1}$) plane was suppressed by the
SLSL's of InGaAs/GaAs and a low defect density was observed in the GaAs
epi-layers; a and b show that disappearance of particular dislocations for
different diffraction conditions (e.g., $\mathbf{g}$ = [11$\bar{1}$] and [1$\bar{1}$1]).

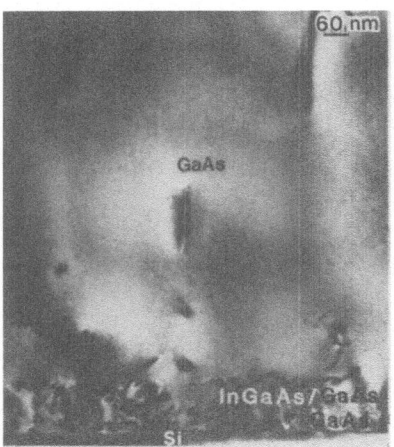

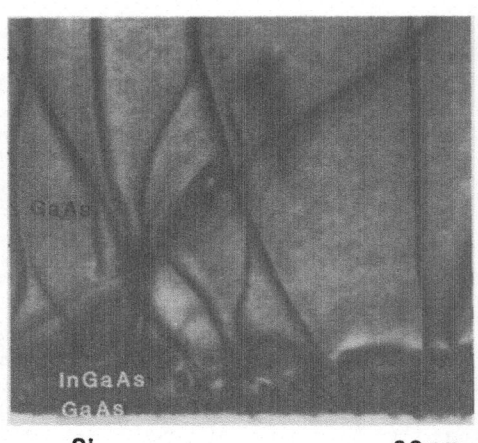

FIGURE 2. TEM micrograph of sample "60". Note the SLSL grown on the GaAs buffer layer. Many areas with low dislocation densities were found.

FIGURE 4. Dislocation bending by only one 30-nm-thick layer of InGaAs (sample "62").

density was in the range of 10^7 cm^{-2} at 150 nm from the Si interface (Fig. 2). Similar kinds of threading dislocations were observed in the GaAs epilayer above the SLSL's: arc dislocations (type IV) resulting from the interaction between these dislocations in the SLSL's, dislocations inclined to the interface with SLSL's of (type II), and dislocations that were always visible for all three <220> directions in the (111) plane. Stacking faults and microtwins were formed in the buffer layer on both (11$\bar{1}$) and (1$\bar{1}$1) planes. Study of these samples shows that formation of stacking faults and microtwins is often associated with the presence of some impurities on the interface (Fig. 3).

It was interesting to observe that only one strained layer of InGaAs (sample 62) was enough to bend many dislocations, but many of these dislocations still propagated into the GaAs. Bending of dislocations again occurred on the upper interface with GaAs (Fig. 4).

Because of the observation that the upper interface of SLSL's is most efficient in bending of dislocations, three sets of SLSL's were grown on the Si separated by 50 nm of the GaAs buffer layer (sample "72"). Each set was expected to reduce the dislocation density on the upper GaAs layer. Indeed, each set of SLSL caused additional dislocation bending, and there were many dislocation free areas (3–5 μm long) (Fig. 5). But there were also areas where additional dislocations were formed at the lower interface between the buffer layer and the SLSL. Therefore, in some areas the dislocation density was slightly higher; however, an average dislocation density in this sample was in the ~2x10^7 cm^{-2} range.

In all samples investigated, the interface with Si was contaminated less by impurities than described in earlier reports of samples grown on Si(001) [9]. These results show that cleaning by "Ga reduction" is

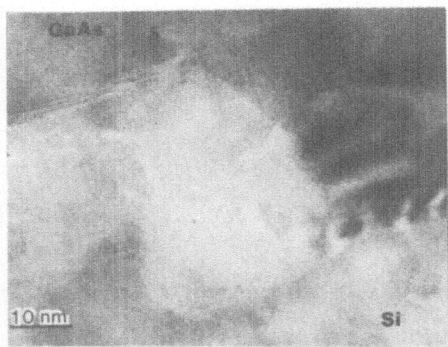

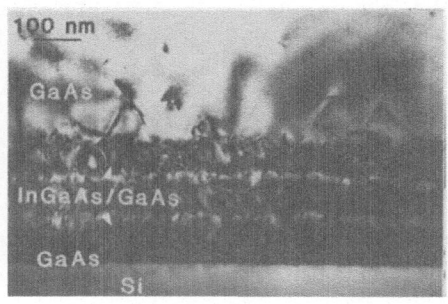

FIGURE 3. Impurities present on the Si surface cause additional defects in the form of protrusions. Such areas are an additional source for the formation of stacking faults.

FIGURE 5. TEM micrograph of sample "72". Note bending of dislocations by three sets of SLSL's.

clearly an improvement over earlier procedures [8]. However, wherever Si surface contamination was present, origination of stacking faults and microtwins was observed. Burgers circuits for the perfect misfit dislocations observed on (011) projections show that the Burgers vector lies in the slightly inclined (1Ī1) plane. For the interfaces viewed in [111] projection, two types of Burgers vectors were observed: lying on the (2Ī1) interface plane and lying on the (1Ī0) plane.

3.2. Annealed epilayers

Furnace annealing of the samples "60" at 800°C for 10 min change the defect rearrangement very slightly, (Fig. 6). The dislocation density remains in the same range as for "as-grown samples." Arc and vertical dislocations (type IV) (A and C in Fig. 6, respectively), and bowed dislocations (type II) were observed in the GaAs above the uppermost interface with SLSL. As in "as-grown samples," there were many dislocations for which invisibility conditions were not found (for low index planes). Some improvement of GaAs quality was observed (Fig. 7) when the samples 60 were annealed by RTA for 10 seconds in a commercial Heatpulse furnace by the capless close-proximity method. However, as in all previously investigated samples, cracking of the GaAs epilayers was observed in these samples. The main change in RTA-annealed samples was the disappearance of the stacking faults on the (1Ī1) planes perpendicular to the (2Ī1) interface. The stacking faults on the (1Ī1) planes inclined to the interface were still present (Fig. 7). The TEM study of GaAs grown on 4°-off (001)Si annealed under the same conditions did not reveal stacking faults formed at the interface [10]. Type II and IV threading dislocations were observed after RTA. As in all previously observed cases, there were dislocations for which invisibility conditions were not found.

FIGURE 6. TEM micrograph of sample "60" after furnace annealing.

FIGURE 7. Sample "60" after RTA. Note the stacking faults on the (111) plane and cracking of the GaAs.

4. DISCUSSION

The formation of threading dislocations and stacking faults in GaAs/Si heteroepitaxial growth can be caused by defect formation during growth and/or during cooling after the growth. During growth, the formation of misfit dislocations and their subsequent glide to the hetero-interface can result in threading dislocations in the epitaxial layer, as suggested by Matthews[11]. In GaAs/Si ($2\bar{1}1$), these dislocations are type I or II. At the end of the growth process, the epitaxial layer has very little residual strain if the misfit dislocation density at the heterointerface corresponds to the lattice mismatch of GaAs (a = 5.675 Å at 600°C) and Si (a = 5.439 Å at 600°C). However, the difference in thermal expansion coefficient (α_{GaAs} = 6.8 10^{-6}/°C, α_{Si} = 2.6 10^{-6}/°C) will produce a new strain during cooling from the growth temperature. In GaAs/Si, this strain is of the opposite sign to the lattice mismatch strain, and photoluminescence measurements of GaAs/Si structures indeed confirm the presence of tensile strain after growth [12]. The tensile strain observed experimentally is considerably lower than the expected value of 2.4×10^{-3}, indicating strain relief by plastic flow. Cooling from 600°C to 400°C is sufficient to generate a biaxial tensile stress far above the experimentally determined critical resolved shear stress of 15 MPa at 400°C [13], which can result in the glide of additional threading dislocations of various types into the epilayer. In addition, misfit dislocations at the interface can be forced to glide back into the epilayer or to dissociate on a {111} plane inclined to the interface, leaving one partial dislocation at the interface and forming an extended stacking fault. The formation of extended stacking faults by glide processes was first found in plastically deformed semiconductors cooled under high stress [14,15].

Type I misfit dislocations with an edge component cannot move into the epilayer by glide, as their glide plane is the interface plane. Only screw dislocations at the interface with a type I Burgers vector can either glide as complete dislocations or dissociated into two partial dislocations, leaving a vertical stacking fault in GaAs/Si ($2\bar{1}1$), as frequently observed (Fig. 1). A second source for vertical stacking faults can be dissociated type III dislocations. Type II dislocations have a glide plane inclined to the interface, allowing for each dislocation gliding as a complete dislocation or dissociated into two partial dislocations, forming a stacking fault. In GaAs (001), type II dislocations can dissociate in this way, e.g., a dislocation along [$\bar{1}$10] with a Burgers vector $\frac{a}{2}$ [011] → $\frac{a}{6}$ [$1\bar{2}\bar{1}$] + $\frac{a}{6}$ [112], leaving a partial dislocation with $\underline{b} = \frac{a}{6}$ [12$\bar{1}$] at the heterointerface, which still relieves all of the misfit strain of the total dislocation. In GaAs/Si (2$\bar{1}$1), the dissociation reaction is for type II dislocations along [011], with $\underline{b} = \frac{a}{2}$ [110] → $\frac{a}{6}$ [21$\bar{1}$] + $\frac{a}{6}$ [121], in which case the partial dislocation with $\underline{b} = \frac{a}{6}$ [121] still relieves all the misfit compared to the complete dislocation.

Thermal annealing at temperatures above the original growth temperature reverses the sign of the stress due to the difference in thermal expansion coefficient, providing a driving force in the opposite direction as compared to the original cooling process. This stress may force the reversal of the glide processes discussed above, including the removal of stacking faults by recombination of the partial dislocations. The removal of stacking faults in GaAs/Si (001) and (2$\bar{1}$1) and the disappearance of type III dislocations in GaAs/Si (2$\bar{1}$1) seem indeed to be the most clearly observed annealing effects. It is not surprising that furnace annealing is less efficient in the reduction of threading defects, compared to rapid thermal annealing. The subsequent slow cooling period after furnace annealing might reverse the beneficial effects of the high-temperature treatment, whereas the rapid quenching after RTA more likely preserves those effects.

Bending of dislocations at a strained interface provides an opportunity for dislocations to react with each other, which can result in annihilation of the threading parts or in the formation of immobile stair-rod dislocations. Thus, the total density of threading dislocations can be reduced substantially by SLSL's.

Early studies clearly showed bending of the threading dislocations at each interface of a SLSL [11]. Our work clearly shows that in superlattices consisting of periods of 10 nm or less, the dislocation bending effect and reduction of dislocation densities by several orders of magnitude is confined to the first and last interface of the SLSL (see Figs. 1-2, 4-7). This result can be easily explained by the fact that the dislocation strain field is a far-ranging 1/r field. If the thickness of the individual strained layer is too small, the total energy of a dislocation moving through a SLSL is only slightly modulated, depending on its position within the SLSL (in an area of positive or negative strain). Only a dislocation entering or leaving a SLSL or a dislocation moving through a SLSL consisting of thick layers experiences a strong change in total energy in each layer and thus a force bending it into the interface. However, if the individual layers of the SLSL exceed the critical thickness for strained structures, new misfit dislocations will be formed. Therefore, a careful optimization of the growth parameters of SLSL's is necessary.

5. CONCLUSION

This study shows that the density of threading defects in lattice mismatched heteroepitaxy can be substantially influenced by post-growth annealing treatments and by the insertion of strained layer superlattices. A detailed analysis of the character of the observed dislocations and stacking faults has allowed us to suggest mechanisms for the observed density reduction. It is obvious that the current growth technology is not optimum. We have shown that it is possible to grow low-defect density lattice mismatched structures with a controlled network of misfit dislocations at the heterointerface by removing the misfit strain. A method for reliably eliminating threading dislocations has not been found yet. Our studies show occasional ~1-μm^2-wide areas of GaAs/Si structures free of all threading defects. It can be expected that careful optimization of the design of strained layer superlattices and of annealing sequences during and after growth will help to reach the final goal of device-quality GaAs/Si and other mismatched structures.

ACKNOWLEDGMENT
This work was supported by the Materials Science Division of the U.S. Department of Energy under Contract No. DE-AC03-76SF00098.

Use of the electron microscopes at the National Center for Electron Microscopy of Lawrence Berkeley Laboratory is gratefully acknowledged.

REFERENCES
1. Fischer, R., Henderson, T., Klem, J., Masselink, W.T., Kopp, W., Morkoc, H., and Litton, C.W.: Electronics Letters **20**, 945, 1984.
2. Uppal, P.N. and Kroemer, H.: J. Appl. Phys. **58**, 2195, 1985.
3. Fischer, R., Kopp, W.F., Gedymin, J.S., and Morkoc, H.,: IEEE Transactions on Electron Devices **ED33**, 1407, 1986.
4. Kroemer, H.: Mat. Res. Soc. Symp. **67**, 3, 1986.
5. Otsuka, N., Choi, C., Nakamura, Y., Nakagawa, S., Fischer, F., Peng, C.K., and Morkoc, H.: Mat. Res. Soc. Symp. Proc. **67**, 85, 1986.
6. Liliental-Weber, Z., Weber, E.R., Washburn, J., Liu, T.V., and Kroemer, H.: Mat. Res. Soc. Symp. **91**, 91, 1980.
7. Ahearn, J.S., Uppal, P., Liu, T.K., and Kroemer, H.: J. Vac. Sci. Techn. **B5**, 1156, 1987.
8. Ishizaka, A., Nakagawa, N., and Shiraki, Y., in **Proceedings of the 2nd International Symposium on Molecular Beam Epitaxy**, Tokyo, p. 183, 1982, Jpn. Soc. of Appl. Phys.
9. Lo, Y.H., Charasse, M.N., Lee, H., Vakhshoori, D., Huang, Y., Yu, P., Liliental-Weber, Z., Werner, M., and Wang, S.: Mat. Res. Soc. Symp. **91**, 149, 1987.
10. Lee, H.P., Huang, Yi-He, Liu, X., Lin, H., Smith, J.S., Weber, E.R., Yu, P., Wang, S., and Liliental-Weber, Z.: Mat. Res. Soc. Symp. **116** (in press).
11. Matthews, J.W.: in **Epitaxial Growth**, edited by J.W. Matthews (Academic, New York, 1975), Part B, Chapter 8.
12. Bugajski, M., Nauka, K., Rosner, S.J., and Mars, D.: Mat. Res. Soc. Symp. 116 (in press).
13. Bourret, E.D., Tabache, M.G., Beeman, J.W., Elliot, A.G., and Scott, M.: J. Crystal Growth **85**, 275, 1987.
14. Wessel, K. and Alexander, M.: Phil. Mag. A35, **6**, 1523, 1977.
15. Kuesters, K.H., DeCooman, B.C., and Carter, C.B.: Phil. Mag. **A53**, 1, 141, 1986.

GROWTH OF GaAs and GaAlAs DOUBLE HETEROSTRUCTURES ON SILICON BY MOCVD

R. AZOULAY, E.V.K. RAO, B. SERMAGE, G. LEROUX, L. DUGRAND, N. DRAIDIA

Centre National d'Eudes des Télécommunications
Laboratoire de Bagneux
196 avenue Henri Ravera - 92220 BAGNEUX - FRANCE

INTRODUCTION

There has been a considerable interest in recent years on the heteroepitaxial growth of GaAs and its associated devices on silicon substrates. Some majority carrier devices have already been realized with these layers and have been found to yield performances comparables to their equivalents on GaAs substrates (1), (2), (3). On the other hand, minority carrier based devices such as laser diodes have consistently exhibited performances inferior to their counterpart on GaAs substrates, at least as regard to device operation life time (4), (5). In this work, after assessing the quality of GaAs grown on silicon (surface morphology, crystalline quality, photoluminescence properties), we have studied the influence of some growth conditions on the optical properties of GaAs layers sandwiched between two GaAlAs layers.

1. EXPERIMENTAL

The growths have been performed in a vertical atmospheric pressure MOCVD reactor described previously (6). Trimethylgallium, trimethylaluminium, pure arsine have been employed as starting gases. The substrate were placed on a rotating, graphite pedestal designed to accommodate 2 inch wafers. Boron doped p-type silicon substrates, disoriented 2° and 4° off (100) towards direction <110> have been employed. The following procedure for substrate preparation have been used : they were degreased in trichloretylène, acétone, methanol, desoxied in HF, rinsed in D.I. and blown dry with N_2 before loading into the reactor via a glove box. Before the growth, the substrates were annealed at 1000 C in a H_2+AsH_3 atmosphere, then the temperature of the substrate was lowered to the growth temperature. The "two steps" method described by Akyama (7) has been used. The first prelayer was grown at 450C and was 100A thick. As discussed below, different buffer layers have been grown prior to realizing the final GaAs layer or GaAs/GaAlAs double heterostructure at 750C.

2. CHARACTERIZATION

The surface morphology of the layers has been examined by optical micrography and ALPHA step measurements. Since the surface roughness was found to increase with the thickness of the layer, the ratio of the roughness over layer thickness is taken as a measure of the smoothness of the layers. Double X-ray diffraction, photoluminescence measurement at 300K and 77K photoluminescence imaging and non radiative life time measurements have been performed.

Y. I. Nissim and E. Rosencher (eds.), Heterostructures on Silicon: One Step Further with Silicon, 27–36.
© 1989 by Kluwer Academic Publishers.

3. RESULTS

It has been reported that a GaAs/AlAs superlattice may have some beneficial effect in blocking the dislocations propagating from the substrate into the layers (8). We have used 3 different types of buffer layers :

Type I : A 100Å thick prelayer deposited at 450°C
Type II : A 100Å GaAs prelayer grown at 450°C and a GaAs/AlAs superlattice, with a periodicity of 150A grown at 750°C (SL 1)
Type III : type II buffer layer, and another GaAs/AlAs superlattice (SL 2) with a periodicity of 300A, grown at 1μ or 0.7μ from SL 1

3.1. GaAs single layers

. Morphology

The grown layers have a roughness of 150Å to 1000Å depending on the total thickness and the growth conditions. The ratio roughness/thickness ranges from 0.006 to 0.04.

. X-ray diffraction

The X-ray single diffraction was measured along the <001> direction. Figure 1 shows a typical single crystal X-ray rocking curve recorded along <001> direction of a 0.4μ GaAs layer grown over type II buffer layer. These measurements revealed always a small disorientation in the ranges of 0.02° to 0.04° between the planes of the substrate and the epilayer. Measurements performed on direct (100) substrate show no such disorientation. In accordance with the observation by Nagai (9), this effect is due to the lattice mismatch between the two crystals and the lattice deformation of the epitaxial layers along the atomic steps of the surface. Single X-ray diffraction has also been employed to assess the quality of the superlattice in type II and type III buffer layers. The satellite peaks of SL 1, near the interface Si/GaAs are weak and broad, and those related to SL2 located away from the interface are more intense. This is indicative of a strong Al-Ga intermixing induced by defect propagation and silicon outdiffusion in SL 1.

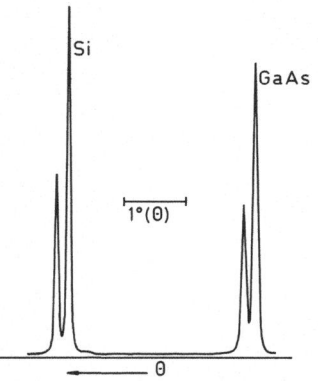

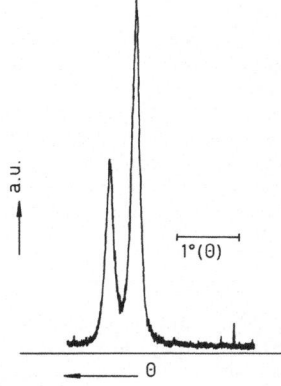

Fig. 1 : Single X-ray rocking curve of a GaAs layer on silicon 004 reflection.

Fig. 2 : 006 reflection of a single X-ray rocking curve of a GaAs layer on silicon.

As reported earlier (10), the lattice parameters of GaAs measured along the growth direction is 5.644A, and the difference with the parameter of GaAs is due to the strain of the layer. The layer is under biaxial tensile strain because of the difference in thermal expansion between the two materials (11). The OO6 reflection has a strong intensity, which shows that the layers is raisonnably free from antiphase domaines (12) (cf Fig. 2).

The half width of the 004 reflection of GaAs is a criterion of the crystalline quality of the material. For layers thicker than 3μ, a half width ranging from 180 arcsec to 230 arcsec has been measured by double X-ray rocking curve. The lowest half width (180 arcsec) has been measured on a 4μ GaAs layer grown with type II buffer layer (total thickness : 4.6μ). For a given thickness d > 3.5μ, the influence of the type of the buffer layer on the half width of the double X-ray racking curve is very weak.

. Photoluminescence

The 300K photoluminescence spectrum of a GaAs layer grown on silicon is presented in Fig. 3. Excepting the location at a small lower energy (8 meV to 10 meV) recorded on the layer grown on silicon, the band gap emission exhibits a comparable intensity and a half width (about 40 meV) with a similar spectrum recorded on a layer grown on GaAs. Similar results have been obtained using type II or type III buffer layers.

In the low energy region of the spectrum a band centered around 1eV is often observed on layers grown on silicon. Its intensity is comparable or superior to the band gap emission, depending on the layer thickness. Its intensity relative to the band gap emission decreases with increasing layer thickness. This band is often attributed to complexes involving native defects and silicon atoms (13).

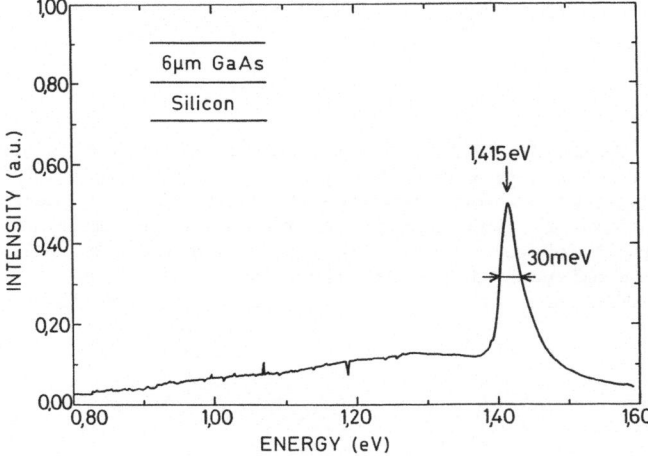

Fig. 3 : 300 K photoluminescence spectra of a GaAs layer grown on silicon with type I buffer layer.

Figure 4 shows a comparison of the 77K spectra of a GaAs/GaAs layer and a GaAs/Si layer. As reported earlier (14), the influence of the biaxial strain of the epitaxial layer on silicon is clearly seen in this figure, as the shrinkage of the band emission together with the splitting in the valence band. The magnitude of the splitting (11 meV) is consistent with the calculation of the strain in the layer (15).

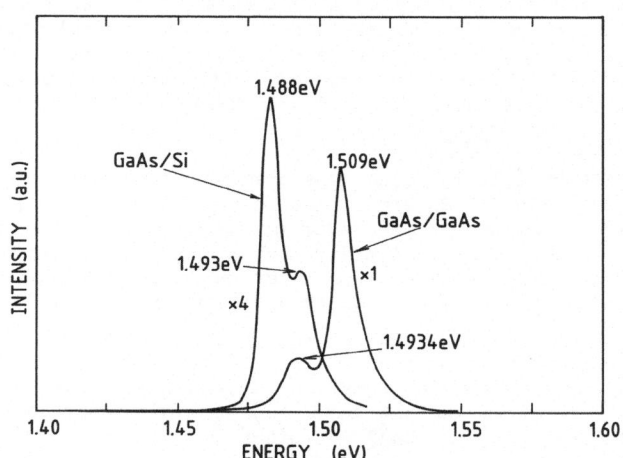

Fig. 4 : 77 K photoluminescence spectra of GaAs grown on silicon and GaAs grown on GaAs.

Figure 5 shows the 300K PL topography of the GaAs-Si heterointerface exposed by a chemically etched low angle level made in a 3µ thick layer grown over type III buffer layer. This topography is recorded by exciting PL with an Ar laser (5145 A line) and realizing the image with the band gap emission. Close to the interface, a high density of well structured regions can be observed, indicating the presence of a high defect density. Even though this density decreases away from the interface, it remains very high.

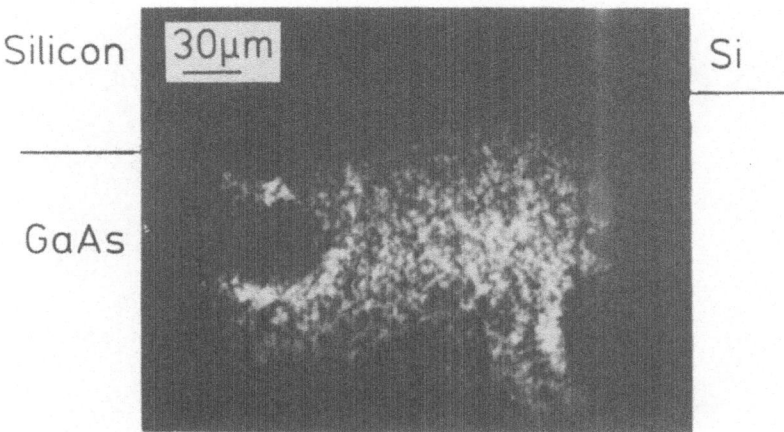

Fig. 5 : 300 K photoluminescence topography of a GaAs-Si heterointerface.
The topography is recorded by exciting the luminescence with
an Ar laser (5145 A line) and realizing the image with the
bandgap emission.

. Electrical measurement

C(V) measurements have been performed with a mercury probe on these
layers, and a residual level of 4 to $8x10^{16}$ cm^{-3} has been measured. As the
residual carrier concentration of GaAs grown with the same condition is n
= $4x10^{15}$ cm^{-3}, it appears that a small contamination from the substrate
till exists in thick layers. This is consistent with the 300K PL spectra
reported earlier.

. Post growth annealing

Post growth annealing has been performed in the MOCVD reactor with an
arsine partial pressure of 10^{-2} Torr at 750°C during 30 min. A decrease
from 200" to 150" of the half width of the peak of the 004 reflection of
the X-ray racking curve has been observed. Neither change in PL spectra
nor extra outdiffusion of silicon has occured, as is inferred by SIMS
measurements. Increasing the temperature of anneal to 850°C did not lead
to further improvement of the crystalline quality of the material.

3.2. GaAs/GaAlAs double heterostructures

To study the minority carrier based properties of GaAs, we have grown
the following undoped double-hétérostructures : 1μ
$Ga_{0.7}Al_{0.3}As$/0.15μGaAs/0.5μ $Ga_{0.7}Al_{0.3}As$ with the three different buffer
layers defined earlier.

In Fig. 6 is presented the optical micrograph of a DHS grown with type III buffer layer. The lowest ratio roughness/thickness is 0.006, with a roughness of 200A. The crystalline properties of the DHS are similar to those of GaAs single layers, as measured by X-ray diffraction.

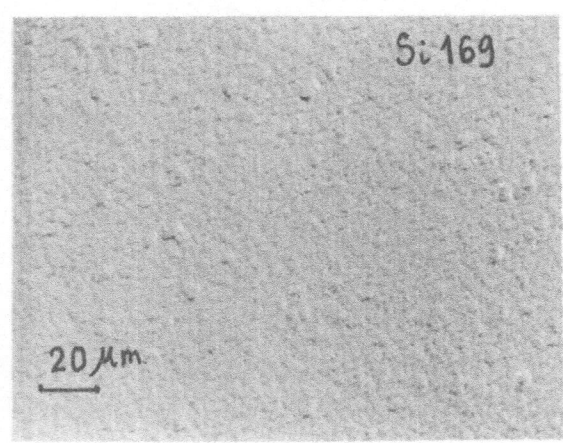

Fig. 6 : Optical micrograph of a GaAs/GaAlAs DHS grown on silicon
with buffer layer type III.

Photoluminescence

In table I, is presented the intensity of the band gap emission of the GaAs confined layer and of the GaAlAs upper layers relatively to the intensity measured on a similar DHS grown on GaAs. We observed a considerable improvement of the luminescence properties of the confined GaAs confined layer on the structure grown with type III buffer layer in comparison with structures grown on type I and type II buffer layers. On the other hand, type I and type II buffer layers show similar results. In a few experiments with type III buffer layer, the location of SL2 has been changed, but no difference in the recorded PL intensities has been observed.

As an example, the PL spectra of a DHS grown on silicon with type III buffer layers and of a DHS grown on GaAs are shown in fig. 6. Under similar exitation conditions, the intensity of the confined GaAs layer grown on silicon is only about 10 times lower than the intensity of the similar DHS layer grown on GaAs. Besides, the higher intensity of GaAs compared to those of the GaAlAs upper layer (10 times for the DHS grown on silicon, 800 times for the DHS grown on GaAs) is a clear indication of the good quality of the heterointerface. Indeed, in the present experiment, the photocarriers are generated within the GaAlAs layer and diffuse into the GaAs layer underneath. The PL intensity of the GaAs confined layer is consequently a fonction of the carriers that have diffused into the GaAs layer. This is indicative of a lower non radiative recombination and thus a good quality of the heterointerface. But the heterointerface of the DHS grown on silicon is not as "clean" as that of the DHS grown on GaAs.

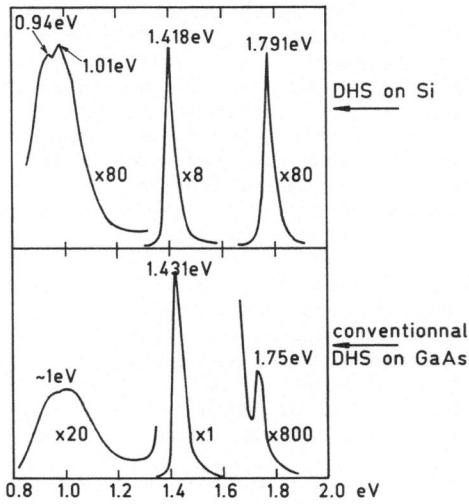

Fig. 7 : 300 K photoluminescence spectra of a GaAs/GaAlAs DHS grown
on silicon with buffer layer type III and the same DHS grown
on GaAs.

300K PL topograph examination on a chemically bevelled DHS grown on
silicon reveals the presence of a large number of defects, as was shown
previously in the case of single layer. An example is shown in Fig. 8.

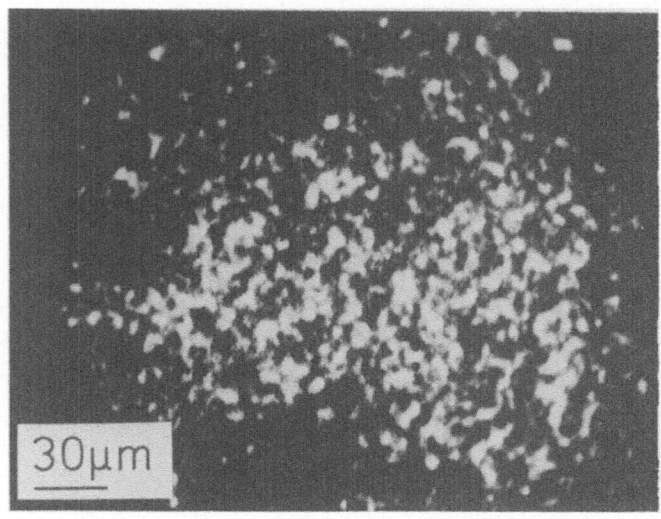

Fig. 8 : 300 K photoluminescence topography of the GaAs confined
layer on the DHS grown on silicon with type III buffer layer.

The radiative carrier lifetime measurements have been made on the different DHS at room temperature by recording the decay of the luminescence efficiency of the sample after excitation by an ultra short laser pulse (16). On the DHS grown on GaAs, the non radiative lifetime is 43 ns, whereas it is 0.33 ns on the structure grown with type I buffer layer and 0.55 ns with type III buffer layer. By comparing this results with the PL topograph, it seems that this low value is most likely due to bulk recombination than interface recombination : a high density of defects is observed in the GaAs layer.

4. DISCUSSION

A notable improvement of the optical properties of the DHS grown with type III buffer layer has been observed. The same results have been obtained by replacing the buffer layer by a superlattice. We have shown by X-ray diffraction measurements, that SL1 is strongly mixed because of the outdiffusion of silicon from the substrate and the propagation of defects. Auger spectroscopy measurements performed on very thin layers containing only SL1 show that the Al-Ga intermixing occurs at the beginning of the growth.

This explains why another SL grown at 0.7μ or 1μ away from the fig. 7 SL is necessary to obtain a real improvement of the PL intensity of the confined GaAs layer in the DHS. As has been measured by X-ray diffraction, the intermixing in the SL2 is very weak or not existent.

As the optical properties of the confined GaAs layer are strongly related to the defects density, we have shown that, accordingly with the conclusions reported recently (17), a GaAs/AlAs superlattice has some efficiency in "bending" the misfit dislocations threading from the interface. But the low values of non radiative carrier lifetime as well as the PL topographs show that further improvement is needed to obtain DHS compatible with the need of high performance laser diodes.

The improvement due to the second SL in the buffer layer has not been observed on GaAs single layers neither by X ray diffraction nor by PL measurement. This shows that a DHS is a more interesting structure to assess the quality of the material.

5. CONCLUSION

High quality GaAs monolayers have been grown on silicon and characterized by PL and X-ray diffraction. Three types of buffer layers, including 1 or 2 GaAs/AlAs superlattices have been used for growing GaAs monolayers and GaAs/GaAlAs double heterostructure on silicon. By using the buffer layers including two SL, one being 1μ away from the interface, photoluminescence measurements exhibit a considerable improvement by comparison with the other buffer layers.

This shows the efficiency of a GaAs/AlAs superlattice to decrease the density of defects in the confined layer of the DHS, as long as this SL does not present a strong Ga-Al intermixing. But the low values of non radiative lifetime measured on DHS grown on silicon show that more work is needed to improve the optical quality of the material.

Some work is in progress on thermal cycling and the study of the influence of the periodicity of the SL.

TABLE I

PL MEASUREMENTS ON GaAs/GaAlAs DHS GROWN ON SILICON

SL 1	SL 2	PL	I_{GaAs}	I_{GaAlAs}
no	no	1.8eV 1.431eV	$3\ 10^{-2}$ $\tau_{nr} = 0.33$ ns	10
yes	no	1.78eV 1.415eV	$1.6\ 10^{-2}$	0.6
yes	yes	1.77eV 1.418eV	5.10^{-1} $\tau_{nr} = 0.44$ ns	5
yes	yes	1.745eV 1.418eV	2.10^{-1} $\tau_{nr} = 0.55$ ns	2.4

The intensities are relative to intensities measured on a DHS grown on GaAs.

References

1 - L.T. Traon, J.W. Lee, IEEE Electron Device Lett. 8, 2, 1987, p 50.
2 - T. Nonaka, M. Akiyamu, Y. Kawarada, K. Kaminishi, J.J.A.P., 23, 1984, p L 919.
3 - M.I. Aksun H. Morkoc, L.F. Lester, K.H.G. Duh, P.M. Smith, P.C. Chao, M. Lorqueborne, L.P. Erickson, APL, 49(24), 1986, p 1654.
4 - N. Chand, J.P. Van der Ziel, R.D. Dupuis, A.M. Sergent, Optoelectronics, 2(2), 1987, p 329.
5 - R.D. Dupuis, J.P. Van der Ziel, R.A. Logan, J.M. Brown, C.J. Pinzone, APL, 50(7), 1987, p 407.
6 - R. Azoulay, B. Jusserand, G. Leroux, P. Ossart, L. Dugrand, J. of Cryst. Growth, 77 (1986), p 546.
7 - M. Akiyama, Y. Kawavada, K. Kaminishi, 68 (1984), p 21.
8 - M. Shinohara, APL, 52(7), 1988, p 543.
9 - N. Nagai, JAP 45(9), p 3789 (1974).
10 - T. Soga, S. Hattori, S. Sakai, M. Umeno, J. of Cryst. Growth 77, (1988), p 428.
11 - K. Ishida, M. Akiyama, S. Nishi, J.J.A.P., 26(5), (1987), L 530.
12 - D.A. Neumann, X. Zhu, H. Zabel, T. Henderson, R. Fisher, W.T. Masserlink, J. Klen, C.K. Peng, H. Morkoc, J. Vac. Sci. Technol. B4(2), 1986, p 692.
13 - B.A. Wilson, C.E. Bouner, R.C. Miller, S.K. Sputz, T.D. Harris, M.G. Lamont, R.D. Dupuis, J. of Electronic Mat., 17(2), 1988, p 115.
14 - Y. Huang, P.Y. Hu, H. Lee, S. Wang, APL 52(7), 1988, p 579.
15 - W. Stolz, F.E.G. Guimaraes, K. Ploog, JAP 63(2), p 492 (1988).
16 - B. Sermage, EMRS Meeting, June 1987.

DEVELOPEMENT OF MOLECULAR BEAM EPITAXY FOR LOW TEMPERATURE AND LATTICE-MISMATCHED SYSTEMS GROWTH OF III-V COMPOUNDS

Luisa González, Ana Ruiz, Yolanda González, Angel Mazuelas and Fernando Briones.

Centro Nacional de Microelectrónica, CSIC. Serrano, 144.
28006 Madrid, Spain.

INTRODUCTION

Two major material problems should be solved to achieve an actual integration of III-V semiconductors with silicon IC'S: 1) the difference of thermal properties between Si substrate and epitaxial layer and 2) the difference in lattice parameters.

In connection with lattice-mismatch, the main problem is the high density of threading dislocations still present in the epilayers, even for layer thicknesses of several microns. Progress in improving crystalline quality has been achieved by growing strained layer superlattices between substrate and epilayer. However, taking into account that the mechanism for lattice relaxation in mismatched systems should depend on growth kinetics, it is worth to investigate modified growth processes capable of improving, in principle, the confinement of strain relaxation centers close to the interface.

Related to point 1, due to the difference in thermal expansion coefficients of Si and GaAs, even a completely relaxed film with an appropiate network of decoupling dislocations at the interface will develop a residual stress (in plane-tensile) by cooling from the growth or annealing temperature. This effect will cause device degradation and a very disturbing warpage of the Si substrate. Of course, the best way to mitigate this problem is to lower growth temperature by MBE as far as possible. Another possible approach is to grow the GaAs device layer on an alloy buffer layer (or short period superlattice) slightly mismatched but still congruent to GaAs. If adequately designed (GaP_xAs_{1-x} is a good candidate) it will generate a mismatch induced compressive stress on the active GaAs layer exactly compensating the cooling induced tensile strain.

We present a new development of MBE, denominated Atomic Layer Molecular Beam Epitaxy (ALMBE) that can be instrumental to the solution of the above mentioned problems. This new technique is based on a cyclic and generalized perturbation of the growth front, at atomic layer level, by periodically pulsing or interrupting the individual molecular beams in adequate phase relative to the monolayer by monolayer growth sequence characteristic of MBE[1].

The main effect of this cyclic perturbation seems to be an

37

Y. I. Nissim and E. Rosencher (eds.), Heterostructures on Silicon: One Step Further with Silicon, 37–44.
© *1989 by Kluwer Academic Publishers.*

enhancement of the 2D nucleation process, approaching in some cases, but not reproducing, an Atomic Layer Epitaxy (ALE) mechanism[2] and separating substantially from the proposed mechanism for MEE (Migration Enhanced Epitaxy)[3].

In the following we present growth experiments for GaAs and InAs, analyzed by means of RHEED intensity recordings under a wide range of growth conditions, in order to illustrate the modified growth mechanism. Results are also reported on the application of ALMBE to the growth of various III-V compounds, showing advantages such as excellent flat morphology even for large misfit with the substrate, elimination of oval defects, low temperature growth and production of sharp interfaces. Furthermore, the possibility of growing Short Period Superlattices containing not only various group III elements (Ga, In, Al) but also two group V elements (As_4 and P_4) is demonstrated. This unique capability of ALMBE solves the problem of As_4 versus P_4 competition for incorporation by supplying these species as separate pulses from specially designed cells. Enhanced reactivity of P_4 over the group III atomic layers contributes also to reduce significantly the phosphorus load in the MBE system.

For the present work, demonstrative of the capabilities of ALMBE, GaAs substrates are used, but results can be of general applicability to highly mismatched systems and, in particular to the GaAs on Si problem.

EXPERIMENTAL

Samples are grown on $(001)^{\pm} 0.5^{\circ}$ off GaAs substrates etched, passivated and In mounted on Mo holders in the conventional way. After the removal of the oxide, they are properly oriented to have (110) or (100) azimuth, and RHEED conditions are set to observe a maximum contrast for the specular beam (<2 μA, 10 keV e-beam, 25 mrad). RHEED pattern is obtained on an Al back-coated fluorescent screen followed with a high sensitivity CCD camera and a video recording system.

Growth mode consists on a periodic modulation of the impinging species on the sample by interrupting or pulsing the group V flux, or alternating group III and V beams. Growth rates are typically 0.5-1.5 ml/s. Shutter operation is controlled by a desk-top computer with fast response time: 0.1 seconds for the magnetically operated shutters of Ga, In and Si cells and even less for the As_4, P_4 and Al cells, which are pneumatically operated.

Special As_4 and P_4 cells have been designed to provide short pulses (down to 0,1 s.) of group V molecules, and a large ON/OFF flux ratio of typically $\emptyset_{ON}/O\emptyset_{FF} = 150$.

Specular beam intensity oscillations are recorded during growth with a low noise blue sensitive photodiode facing the (00) spot on the TV screen. Intensity distributions along selected lines on the RHEED diagram are obtained by mechanical scanning the photodiode over a single stationary image. Time evolution of this distribution is obtained by repeating the scanning image by image on the recorded video. In this way, measuring the distance between appropiate streaks, on each scan, we can follow the evolution of the in-plane lattice parameter caused by lattice relaxation across interfaces through the growth process. A calibration of the camera length is provided by the GaAs substrate (00) and (10) rods distance.

Structural characterization of the layers is performed after growth by X-ray diffactometry using CuK_α radiation. We obtain the superlattice period from the (002) or (004) reflections in the ordinary way.

Surface morphology of the samples is checked with a Nomarsky interference microscope.

GROWTH PROCESS

Two different ways of growing by ALMBE are used:
a) Group III and group V fluxes are alternated.
b) Group III flux is supplied continiously while group V flux is modulated periodically.

Figure 1 shows shutter operation for both methods for InAs samples grown at Ts= 350°C, as well as the corresponding RHEED intensity oscillations (I_{00} beam). In the first case, (fig. 1 a) the time interval during which In cell is open, t_{InON}, is set to coincide with the period of RHEED intensity oscillations during a previous standard MBE growth. t_{AsON} is usually set approximately the same as t_{InON} although, as we will see, it is not critical. The effective growth rate is consequently halved compared with MBE. In case b) (fig.1b) group V pulses are programmed to occur once per monolayer, so that its periodicity is also determined by the growth rate. The length of the pulse is also not relevant, provided the integral flux of group V beam is enough to complete one monolayer.

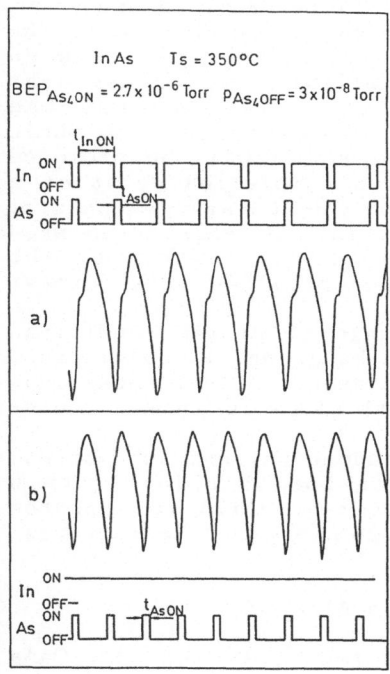

Fig 1: Specular beam intensity during ALMBE growth of InAs at T_s = 350°C by
a) alternating In and As beams and
b) pulsed As_4 supply while In is supplied continiously.
Shutter operation is also shown

Conventional MBE growth process is a competition of various mechanisms such as steps propagation, 2D growth and, at low temperatures, 3D islands formation. Over an optimized range of temperatures , V/III flux ratios and low densities of surface steps, a 2D growth mechanism predominates[4]. In general, that competition of growth processes can be described also by a local and time dependent incorporation kinetics[5] so that the 2D growth rate is a quantity that can be considered statistically fluctuating with time and position over the sample[1]. Surface migration of adsorbed species provides the mechanism of mass exchange among areas over the sample with different incorporation kinetics. Different local growth kinetics occur because different areas might be at different moments of the monolayer formation cycle (half completed monolayers are more reactive than flat or completed surfaces). Besides, normal surfaces are not perfectly homogeneous, showing regions of increased density of steps, defects or dislocations which will modify the growth kinetics at those areas. Locally enhanced growth kinetics is, for example, the origin of the self flattening effect during MBE under optimum growth conditions, but is also the cause of hillocks and oval defects formation around substrate defective points the and origin of rough morphology under non-optimal growth conditions or presence of large mismatch. Reduced surface migration at low temperatures mitigates this effect but will promote 3D growth mode.

ALMBE improvement over MBE growth process is determined by the cyclic perturbation of the growth front caused by either interrupting, alternating or pulsing the impinging beams in synchronism with the monolayer by monolayer growth sequence.

Every time that one of the beams is shuttered ON or OFF, the whole surface will be suddenly and simultaneously separated from equilibrium (T_S and flux dependent). The consequent temporal separation from stoichiometry will enhance the surface chemical reactivity simultaneously over the whole surface. In this way, formation of 2D nuclei will be promoted even on those areas that being at a completion point of a monolayer are flat and less reactive. The detailed mechanism by which, under cyclic repetition of this process (even when the period is not exactly a multiple of the inverse of the growth rate) is too complex to be described in this context and will be published elsewhere.

This broadens considerably the range of growth conditions under which good epitaxial layers can be grown. In particular, uniform 2D nucleation will reduce the need for long range mass exchange or surface migration and will allow to grow at lower substrate temperatures.

We think that this mechanism of enhanced uniform nucleation, forcing 2D growth, is also the reason for excellent growth morphology and confinement of lattice relaxation to a narrow interfacial region in the case of large mismatched structures.

RESULTS

1.- Low temperature growth of GaAs and AlAs/GaAs heterostructures.

Atomic Layer MBE has been used to grow GaAs and AlAs/GaAs

heterostructures at substrate temperatures Ts < 400° C, with very good control of doping and interfaces. Using this growth mode, highly Si-doped GaAs has been grown at a substrate temperature of 350°C. Hall densities n=2x10^{19} cm^{-3} together with excellent morphology (defect free surface) epilayers are obtained.

GaAs quantum wells (QW) and GaAs/AlAs superlattices have also been grown by Atomic Layer MBE. Photoluminescence characterization of these structures[6] shows that good optical quality can be obtained for growth temperatures as low as 400°C. In particular ALMBE mode seems to reduce well width fluctuations, being comparable to those observed in similar samples grown by conventional MBE under optimized conditions, at higher temperatures.

Modulated doped GaAs/AlAs short period superlattices (SPSs) are another evidence of the capabilities of ALMBE for growing at low substrate temperatures with control in doping and composition at atomic level. For example, Hall measurements on a SPS (GaAs)$_4$/(AlAs)$_4$ grown at Ts=380°C, in which only the two central atomic planes in GaAs wells were doped with Si, show a carrier density of n= 2x10^{19} cm^{-3} with good mobility ($\mu \sim 10^3$ cm^2/Vs) and low thermal activation energy for the donors (E$_a \sim$ 3meV)

2.- Growth of highly mismatched systems.

As we have mentioned before, the relaxation of strained lattices is one of the problems in heteroepitaxy of lattice-mismatched systems. It is well stablished that layers with large lattice mismatch can be grown without generation of misfit dislocations as long as their thicknesses are kept below a critical value[7] which depends on the elastic properties of the materials involved. However, the actual critical thickness appears to be dependent also on growth conditions. In order to know the influence of Atomic Layer MBE growth mode in the generation of misfit dislocations, we have chosen to grow InAs and AlAs/InAs superlattices on (001) GaAs substrates as highly mismatched systems (up to 7%). For practical convenience, these layers were grown by pulsing only the arsenic beam at Ts=400°C and r$_g \simeq$1 ml/s. First layer is InAs directly on the substrate without any buffer layer of GaAs.

The width of RHEED streaks increases during the first layers of InAs, demonstrating surface disorder. The diffraction streaks become narrower and sharper as deposition proceeds but no evidence of spotty RHEED diagram or 3D growth is seen during the growth of the interfacial layer.

On figure 2 we represent the evolution of surface lattice constant (a$_{//}$) as a function of InAs thickness evaluated by measuring the streak separation in the RHEED pattern at the onset of growth. We observe that the fully-unrelaxed situation only holds up to one monolayer of InAs. This thickness is much smaller than the 7 ml value estimated by elastic calculations[5]. Fast lattice relaxation occurs in a transition region of about 6 ml. So, we can conclude that a thickness of 7 ml of InAs on the GaAs substrate is enough to reach complete lattice relaxation.

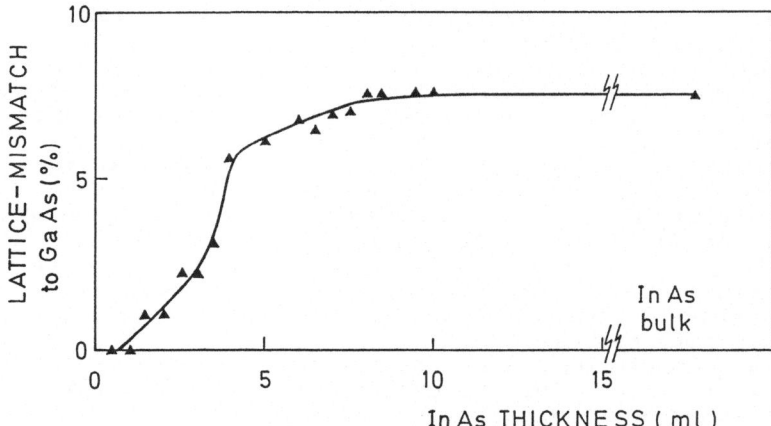

Fig. 2: Lattice-mismatch to GaAs as a function of thickness of AlMBE grown InAs on GaAs substrate.

X-ray diffraction of as grown InAs layers with thicknesses in the range of 0.25-2 μm exhibits peaks with FWHMs 1.3 times those of GaAs substrate.

A fast lattice relaxation process is also observed for AlAs/InAs superlattices grown by AlMBE at Ts=400°C. Figure 3 shows the evolution of the surface lattice parameter during growth of a 0.3 μm thick $(AlAs)_{10}/(InAs)_{10}$ superlattice grown on 0.15 μm $Al_{0.46}In_{0.54}As$. The alloy buffer lattice parameter is also depicted in the same figure, showing that unstrained alloy can be grown after the first 15 ml. It is also shown that lattice parameter of the superlattice remains constant

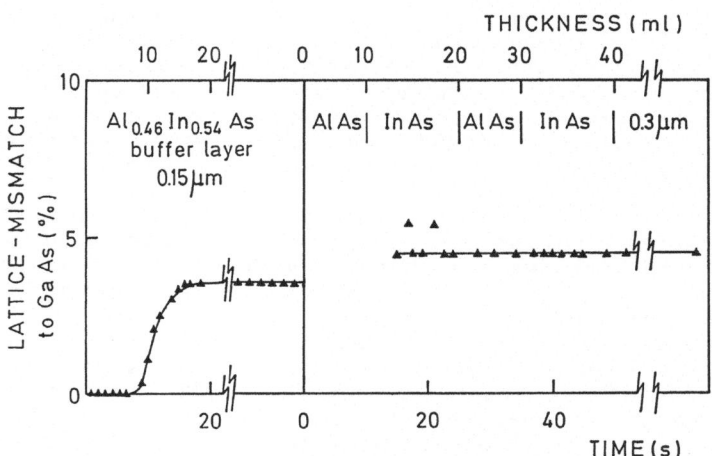

Fig. 3: Lattice mismatch to GaAs as a function of thickness for AlMBE grown $(InAs)_{10}/(AlAs)_{10}$ on a 0.15 μm $Al_xIn_{1-x}As$ buffer layer. Evolution of lattice parameter of the alloy buffer layer during growth on GaAs substrate is also shown.

from the 1st period. It is interesting to note the fact that once a critical thickness is reached, no relaxation of the in-plane lattice parameter between AlAs and InAs is observed even for individual layer thicknesses larger (> 10 ml) than the critical for a single interface. Finally, we find a superlattice in-plane parameter which differs from that of InAs by a 2%.

We want to remark that morphology of the InAs bulk layers, as well as that of the superlattices, even for the thicker ones, is absolutely flat or specular.

X-ray diffraction measurements of an $(InAs)_{10}/(AlAs)_{13}$ superlattice (see figure 4) shows the good crystalline quality of the SL layers and gives a mean $a_{\perp} = 5.95$ A.

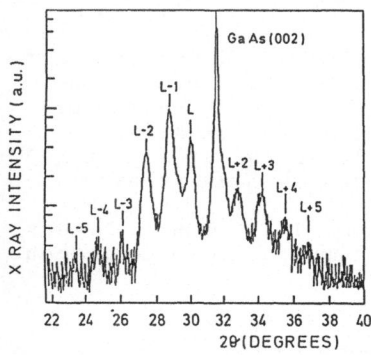

Fig. 4: X-ray intensity distribution around the (002)GaAs substrate reflection for a 0.3 μm thick $(InAs)_{10}/(AlAs)_{13}$ superlattice grown directly on GaAs.

3.- Short period superlattices with various III and V elements.

We have already mentioned that production of sharp interfaces is one of the advantages shown by Atomic Layer MBE. In this paragraph we present results demonstrating that ALMBE is a good method to grow short period superlattices containing not only group III elements (Ga, In, Al) but also different group V elements, arsenic and phosphorous. This is extremely interesting for the growth of phosphorous containing compounds as dislocations propagation barriers in heteroepitaxy of GaAs on Si.

In this context, two types of superlattices are presented as examples: $(GaAs)_m/(GaP)_n$ and $(Al_xIn_{1-x}As)_m/(Al_xIn_{1-x}P)_n$.

Both types of structures were grown at Ts=400°C by opening continuously group III furnaces, and pulsing As_4 and P_4 beams. Figures 5 and 6 show x-ray diffractograms from respectively 0.3 μm thick $(GaAs)_7/(GaP)_2$ and $(Al_{0.25}In_{0.75}As)_3/(Al_{0.25}In_{0.75}P)_5$ superlattices grown on GaAs substrates.

The appearance of satellite peaks typical of superlattices demonstrate the full incorporation of phosphorous. It is remarkable to note that the problem of competition of As_4 versus P_4 for incorporation during MBE growth of alloys is solved by ALMBE by supplying these species as separate pulses from specially designed cells during the growth of an equivalent short period SL. Moreover, lower phosphorous beam equivalent pressures are necessary due to the high adsorption

44

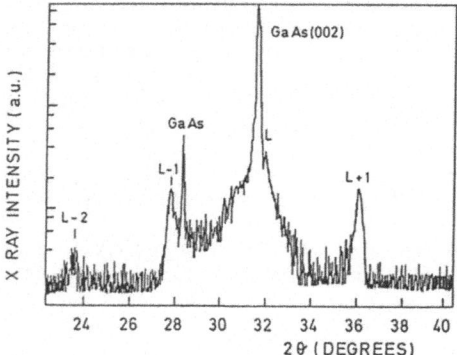

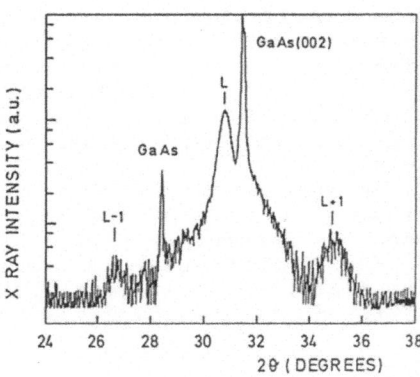

Fig. 5: X-ray intensity distribution around the (002)GaAs reflection of a 0.27 μm thick $(GaAs)_7/(GaP)_2$ SL grown on GaAs substrate.

Fig. 6: X-ray intensity distribution around the (002)GaAs reflection of a 0.15 μm thick $(Al_x In_{1-x}As)_3/(Al_x In_{1-x}P)_5$ grown on GaAs substrate.

probability of this element on group III atomic layers, contributing to a reduction of the amount of phosphorous inside the MBE system.

CONCLUSIONS

Atomic Layer MBE, besides the already known advantages of low temperature growth, production of sharp interfaces and elimination of oval defects, is also very useful for the growth of highly mismatched III-V heteroepitaxial layers and SL with excellent crystallinity and morphology. Its ability to concentrate mismatch relaxation in a narrow region near the interfaces is a consequence also of the forced 2D growth mode. A further important possibility offered by the new technique, in connection with the growth of dislocation propagation barrier layers for GaAs on Si, is based on the controlled growth of phosphorous containing short period SLs with a mean lattice parameter between those of Si and GaAs.

REFERENCES

1.- F. Briones, D. Golmayo, L. Gonzàlez, A. Ruiz. J. Crystal Growth **81** (1987) 19.
2.- C.H.L. Goodman and M.V. Pessa, J.Appl.Phys. **60** (1986) R65.
3.- Y. Horikoshi, M. Kawashima and H. Yamaguchi. Jpn. J. Appl. Phys. **27** (1988) 169.
4.- J. H. Neave, P. J. Dobson, B. A. Joyce and J. Zhang. Appl. Phys. Lett. **47** (1985) 400.
5.- S. V. Ghaisas and A. Madhukar. J. Vac. Sci. Technol. **B3** (1985) 540.
6.- F. Briones, L. Gonzàlez, M. Recio and M. Vàzquez. Jpn. J. Appl. Phys. **26** (1987) L1125.
7.- J. W. Matthews and A. E. Blakeslee, J. Cryst. Growth **27** (1974) 118.

MOMBE AND PEMOCVD GROWTH OF GaAs ON Si (100) SUBSTRATES

M. Kamp, J. Leiber, J. Musolf, A. Brauers, M. Weyers,
H. Heinecke*, H. Lüth and P. Balk

Institute of Semiconductor Electronics and II. Physikalisches
Institut, Technical University Aachen, 5100 Aachen, FRG
*present address: Siemens Research Laboratories
8000 München, FRG

1. Introduction

The concentrated study of the epitaxial growth of GaAs layers
on Si substrates in recent years has lead to the preparation
of device quality material /1,2/. Both MOCVD (metal organic
chemical vapor deposition) and MBE (molecular beam epitaxy)
have been used successfully. In order to deal with the problems
of lattice mismatch and differential thermal contraction often
the so-called "two step method" /3,4/ is being used. This
growth sequence starts with the deposition of a thin (10-100
nm) buffer layer at reduced substrate temperature (usually
below 650K) before depositing the top layer at temperatures
conventionally used in GaAs homoepitaxy. GaAs growth on Si
substrates appears to require large V/III ratios for the
starting materials in the gas phase, indicating that an excess
of the group V component is an essential condition for high
quality heteroepitaxial growth. For the growth of InP on Si
using the MOCVD approach the same situation holds true /5/.
Since it is likely that the availability of the group V element
rather than that of the undissociated group V source compound
is an essential requirement the use of technological approaches
providing large amounts of elemental As appears to be
indicated.

For this reason we have studied two methods meeting this
criterion: MOMBE (metal organic MBE) and PEMOCVD (plasma
enhanced MOCVD). In the first approach arsine AsH_3 is being
cracked at a hot Ta filament before being injected into the UHV
chamber to react with a metalorganic Ga compound (for example,
$Ga(CH_3)_3$, trimethyl Ga or $Ga(C_2H_5)_3$), triethyl Ga) at the
substrate surface. The second method uses the same reactants,
but in a carrier gas. Here the AsH_3 is being cracked in a
plasma before injecting it in the deposition zone. Thus, in
both methods a large As concentration at the surface is
obtained for modest ratios of the injected amounts of the group
V and group III compounds. PEMOCVD has the additional advantage
of permitting growth at rather low temperatures for preparing
the buffer layer.

2. Experimental

The deposition systems were described in detail in /6/ and /7/.
In our experiments in MOMBE growth we used TEG directly from
the container. In PEMOCVD both TEG and TMG were used as Ga
sources; these compounds were transported by the carrier gas

Y. I. Nissim and E. Rosencher (eds.), Heterostructures on Silicon: One Step Further with Silicon, 45–50.
© 1989 by Kluwer Academic Publishers.

hydrogen. In both approaches undiluted AsH₃ was employed which was precracked at a hot Ta filament in MOMBE and in a DC plasma discharge (canal ray configuration) in PEMOCVD. Substrates were n or p-type (100) Si of o.1 Ω cm resistivity. After degreasing the substrates in organic solvents they were dipped in ammonia fluoride (NH_4F) for 10 sec. and allowed to oxidize in deionized water. For MOMBE growth samples were attached to the molybdenum substrate holder with indium, as usual in MBE, and subsequently introduced into the growth chamber. The growth cycle was started with a 1000K anneal for 1h in MOMBE, a 1200K anneal for 15 min. in PEMOCVD in a flux of uncracked arsine in both cases. This treatment sufficed to remove the SiO_2 to within the limit of detection by physical methods (XPS, RHEED). These temperatures were the maximum ones attainable in the respective system. It is to be expected that the use of higher annealing temperatures before growth, which is presently under study, will strongly reduce the occurence of antiphase domains. After the annealing procedure, when the substrate had reached the deposition temperature, growth was initiated by opening the shutter in front of the Ga capillary in MOMBE or switching the flux of TMG or TEG into the MOCVD reactor.

3. Results and discussion
a) Low temperature buffer layer
In both MOMBE and MOCVD low temperature growth is governed by the pyrolysis of the Ga alkyl, as long as cracked arsine is supp lied in excess. Transfer of energy from the arsine plasma to the Ga alkyls changes this situation to some extent in PEMOCVD /8/. In the latter system best results regarding the morphology of the buffer layer (20-50nm) and the subsequently grown upper layers were obtained at growth temperatures of 640K (TEG) and 750K (TMG) at total pressures of 300-500 Pa using an AsH₃ plasma. The plasma allowed a reduction of the V/III ratio from above 100 to less than 25 in both cases.

In MOMBE growth kinetics poses a lower temperature limit of 630K for the deposition of the buffer. Below this temperature with fluxes yielding a growth rate of 0.7 μm /h in the

1 μm

Fig.1:SEM micrograph of MOMBE GaAs buffer layer (0.4 μm) grown on Si(100) at 740 K

injection limited growth regime at 850K in 1h deposition was not observed using Auger Electron Spectroscopy (AES). RHEED studies on buffer layers grown at higher temperatures yielded diffuse diffraction patterns which turned spotty at temperatures above 720K. Fig.1 shows the SEM micrograph of a 400 nm thick buffer layer deposited at 740K. This layer looked mirrorlike to the naked eye but under higher magnification roughness on a small scale was still visible. Double crystal X-ray diffraction (DCXD) rocking curves of this buffer layer using the GaAs (115) plane and CuK radiation revealed rather broad diffraction peaks with a FWHM of 1200 arcs. Such buffer

layers permitted the growth of mirrorlike top layers. Layers
grown on buffers deposited below 720K were inhomogeneous with
hazy areas and a GaAs peak was not detected in DCXD. Raising
the deposition temperature for the buffer to above 750K
resulted in rough surfaces of the buffers and of the layers
grown on top of them. Thus when using TEG in MOMBE a deposition
temperature of 720 to 740K for the buffer layer appears to be
optimum.

b) Top layer
In contrast to InP GaAs is rather sensitive to radiation damage
/8,9/. For this reason in PEMOCVD the top layer was deposited
without plasma stimulation at 10^4 Pa total pressure. Like for
the buffer the optimum temperature for TMG (920K) was about
100K higher than that for TEG (840K) due to the higher thermal
stability of this compound. With TMG (fig. 2b) smoother
surfaces were obtained than with TEG (fig. 2a) and additionally

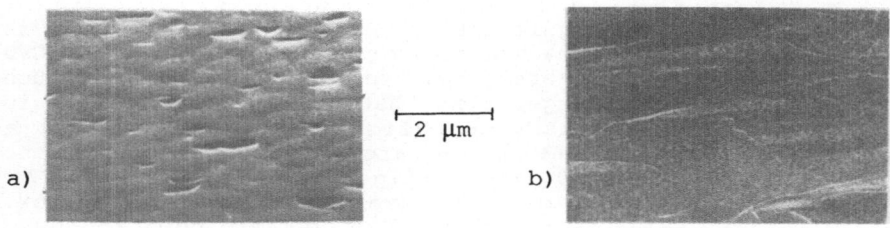

a)   b)

Fig.2: SEM micrograph of PEMOCVD GaAs epilayer grown from
a) TEG at 840K with V/III ratio of 25 on buffer deposited at 640K
b) TMG at 920K with V/III ratio of 15 on buffer grown at 730K

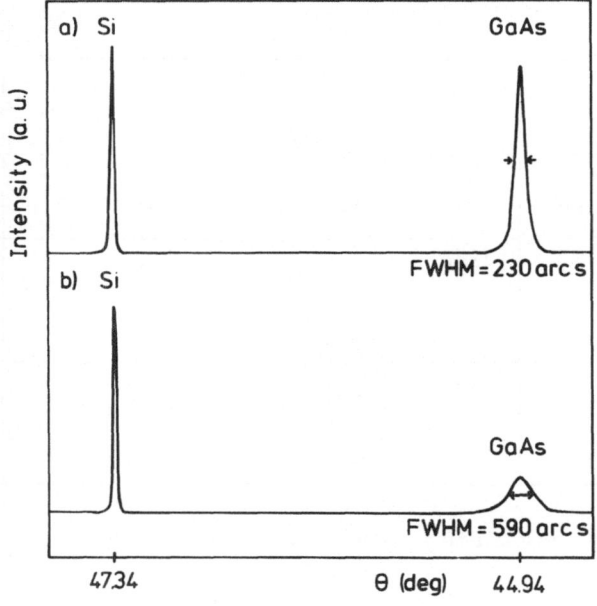

Fig.3: Double
crystal X-ray
diffraction
spectra of GaAs
layers
g r o w n   i n
PEMOCVD
a) from TMG
b) from TEG

the crystalline properties were improved: While the FWHM in the DCXD spectrum of a layer grown from TMG was only 230 arcs (fig. 3a) and compares favorably with the best results obtained by MBE/4/ and MOCVD/10/, the value is more than twice as large (590 arcs) for layers grown from TEG (fig. 3b).

In MOMBE top layers were grown on mirrorlike buffers by raising the temperature from 740K to the growth temperature without growth interruption; this temperature grading took approx. 60 min. While the deposition temperature for the buffer layer is a very critical parameter the effect of the growth temperature of the top layer on the crystalline and morphological properties was not very pronounced. However, using TEG best results were obtained at a growth temperature of 830K which is about 50K lower than the optimum temperature for GaAs homoepitaxy. Layers deposited at these conditions are mirrorlike but show considerable roughness at higher magnification (fig.4). The roughness is somewhat greater than that of the PEMOCVD layers grown from TEG (fig. 2a) and much larger than that of TMG grown layers in this method (fig. 2b). DCXD curves show a FWHM of 750 arcs. Further reduction to 600 arcs (fig. 5) was possible by employing a lower cooling rate of approx. 5K/min instead of 20K/min down to 500K. This value of 600 arcs is comparable with the 590 arcs obtained in PEMOCVD. The parameters used for obtaining the best layers in both systems are listed in table 1.

1 μm

Fig.4:SEM micrograph of a MOMBE epilayer (T=830K,d=2μm) grown on low temperature bufferlayer (T=740K)

	MOMBE	PEMOCVD	
	TEG	TEG	TMG
anneal	1000K AsH$_3$(uncracked)	1200K AsH$_3$ (without plasma)	
buffer	740K	640K	750K
	V/III=2 d=400 nm	300-500 Pa, V/III=24 with plasma, d=40 nm	
top layer	830K V/III=2	840K V/III=25	920K V/III=15
		10^4 Pa, without plasma	

table 1: Conditions for obtaining layers with optimized properties

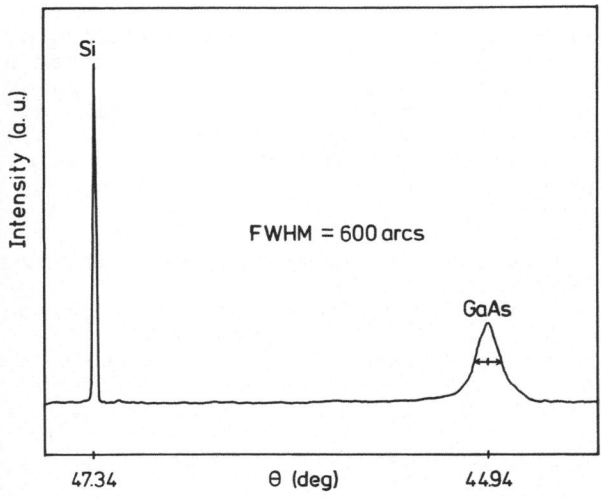

Fig.5: DCXD plot of MOMBE GaAs epilayer grown at 820K from TEG

c) Optical properties
MOMBE layers grown at the parameter of table 1 show a 2K photoluminescence peak which is shifted to approx. 1.495 eV and has a FWHM of 9 meV (fig.6a). The shift of the PL response indicates the presence of stress and is often observed in GaAs/Si heteroepitaxy /11/. In PEMOCVD TEG grown layers show an unshifted but broadened peak (FWHM =15.2 meV) at 1.513 eV together with an acceptor related peak (1.493 eV) of comparable broadness (FWHM 12 meV). The presence of an unshifted peak indicates that stress in these layers has been relieved by an as yet ununderstood mechanism. Like the crystalline properties, the optical properties of the PEMOCVD layers are improved when using TMG as the Ga precursor. The PL response is more intense and much sharper peaks are observed (fig. 6b). The band edge luminescence shows a FWHM of 5.1 meV and the acceptor related transition a FWHM of 6.3 meV, values that are comparable to the best reported in the literature /12,13/. Again the peaks are not shifted.

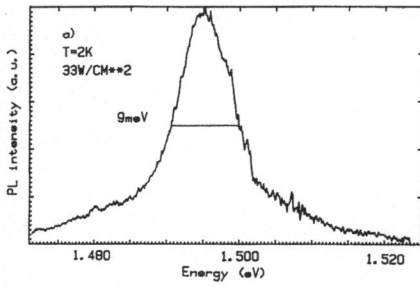
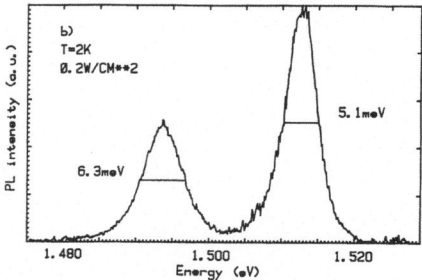

Fig.6: Photoluminesence spetrum of GaAs epilayer grown
       a) by MOMBE from TEG   b) by PEMOCVD from TMG

4.Conclusions
The results of this first study of the MOMBE and the PEMOCVD of
GaAs on Si show that both methods are in principle suitable
for this purpose. They permit growth at modest V/III input
ratios and in the case of PEMOCVD from TMG the crystalline and
optical properties of the layers (FWHM of DCXD and PL peak) are
among the best reported up to now. Compared to this the use of
TEG leads in both systems to somewhat inferior properties.

References
1. J.C.C. Fan and J.M. Poate eds.: Heteroepitaxy on Silicon.
   MRS Proc. Vol. 67 (MRS, Pittsburgh, 1986)
2. J.C.C. Fan, J.M. Phillips and B.Y. Tsaur: Heteroepitaxy on
   Silicon II. MRS Proc. Vol. 91 (MRS, Pittsburgh, 1986)
3. M. Akiyama, Y. Kawarada and M. Yamaguchi: Jpn. J. Appl.
   Phys. 23 , L843, 1984
4. S. M.Koch, S.J. Rosner, D. Sisnon, J.S. Harris, see ref. 1
   p.37
5. A.Yamamoto, N. Uchida and, M. Yamaguchi: Optoelectronics -
   Devices and Technology 1, 41, 1986
6. H.Heinecke, A. Brauers, H. Lüth and P. Balk: J. Crystal
   Growth 77 241, 1986
7. N. Pütz, E. Veuhoff, H. Heinecke, M. Heyen, H. Lüth and
   P.Balk: J. Vac. Sci. Technol. 83, 671, 1986
8. A.Brauers, F. Grafahrend, H. Heinecke, H. Lüth and P. Balk:
   Proc. of Advanced Materials for Telecommunications (EMRS,
   Strassbourg, 1986) XIII, les editions des physique, p. 231
9. H.Heinecke, F. Grafahrend, A. Brauers, H. Lüth and P. Balk:
   MRS Proc. Vol. 75 (MRS, Boston, 1987) p. 747
10.S.J. Pearton, C.R. Abernatky, R. Caruso, S.M. Vernon, K.T.
   Short, J. M. Brown, S. N. G. Chu, M. Starola and V.E. Haven:
   J.Appl.Phys.63, 775, 1988
11 T. Soga, T. Imori, M. Umeno and S. Hattori: Jpn. J. Phys.
   26, L536, 1987
12 R.M.Lum, J.K. Klingert, B.A. Davidson and M.G. Lamont:
   Appl. Phys. Lett. 51, 36, 1987
13.W.I. Wang: J.Vac.Sci.Techn. 83, 552, 1985

CORRELATION BETWEEN STRUCTURAL AND OPTICAL PROPERTIES OF
GaAs-on-Si GROWN BY MOLECULAR BEAM EPITAXY

K. Ploog, F.E.G. Guimaraes, and W. Stolz

Max-Planck-Institut für Festkörperforschung
D-7000 Stuttgart-80, Federal Republic of Germany

1. INTRODUCTION

The current activities in the heteroepitaxial growth of polar III-V compounds on nonpolar Si substrates are stimulated by the technological importance awaited from the integration of GaAs- and Si-based devices on a common
Si substrate (1). Although the previous major obstacles hampering the progress in this field, i.e. the formation of antiphase boundaries, the large
lattice mismatch of 4.1%, and the difference in the thermal expansion coefficient, have now been overcome to a certain extent, the understanding of the
microstructural growth process and its influence on the electronic properties of GaAs-on-Si is still not adequate. In this paper we describe the correlation between the structural and the optical properties of GaAs layers of
thickness ranging from 0.05 - 4.0 μm which are grown by molecular beam epitaxy (MBE) on (100)Si substrates oriented 2^0 off in [011] direction. High-
resolution double-crystal X-ray diffraction (XRD) is used to evaluate the
crystal quality and to determine the strain existing in the epilayer, while
photoluminescence (PL) as well as photoluminescence excitation (PLE) spectroscopy are applied for the characterization of the optical properties. A
comparison with GaAs-on-GaAs grown under similar conditions is also made.

2. EXPERIMENTAL

The recent achievements of accurate growth control down to monolayer dimensions during MBE of III-V compounds (2) has been exploited to the deposition of GaAs with alternating planes of As and Ga on vicinal (100) Si substrates. The (100) Si substrates oriented 2^0 off in [011] direction are
cleaned and prepared by the procedure described in Ref. (3) but without the
final acidic oxidation step. Before deposition of a prelayer of arsenic the
substrate is heated to 950 ^{0}C for 30 min in the MBE growth chamber to evaporate the thin oxide layer. After oxide removal a reconstruction of the
(100) Si surface evolves at this elevated temperature which creates step
heights of an even number of atomic planes, e.g. double steps, and a (2x1)
orientation of the Si dimers parallel to the steps (4). This reconstruction
is probably responsible for the absence of antiphase domain boundaries in
the subsequently deposited GaAs film. First, a thin ($\sim$ 250 Å) GaAs layer is
deposited on the arsenic prelayer using a low substrate temperature of 300
^{0}C and a low growth rate of 0.1 - 0.2 μm/hr, in order to form a uniform
GaAs film on the Si substrate. The substrate temperature is then increased
to 600 ^{0}C while the Ga shutter is closed (suspended growth). After this heat
cycling, which may be repeated several times, the "active" GaAs layer is
grown at a substrate temperature of 550 - 630 ^{0}C using the conventional MBE
process (i.e. Ga and As_4 are impinging continuously onto the substrate) with
a growth rate of 0.7 - 1.0 μm/hr. Attainment of the optimum growth conditions
is indicated by the appearance of the As-stabilized (2x4) surface reconstruction in the RHEED pattern.

51

Y. I. Nissim and E. Rosencher (eds.), Heterostructures on Silicon: One Step Further with Silicon, 51–59.
© 1989 by Kluwer Academic Publishers.

In Sect. 3 we show that the difference in the thermal expansion coefficient between GaAs and Si requires a substantial reduction of the growth temperature of the "active" GaAs layer. A reduction of the substrate temperature to T_s = 300 °C can be achieved by depositing the constituent Ga and As layers of the (100) GaAs crystal monolayer by monolayer in a controlled manner (5). The constituent molecular beams are thus not impinging continuously onto the substrate surface. Instead, they are modulated by interrupting the fluxes synchronously to the deposition of the respective monolayer. Growth rates are typically one monolayer in one or two seconds. The real-time control of monolayer growth is accomplished by monitoring the RHEED intensity oscillations.

Details of the computer controlled high-resolution double-crystal X-ray diffractometer used for the present experiments have been described previously (6) and will not be repeated here. Different laser lines of a Kr^+ ion laser are used to excite the luminescence of the samples kept in a variable temperature cryostate. Special attention is paid to the adjustment of the power density impinging onto the sample surface. The excitation density has a marked influence on the peak energy and intensity, particularly of the extrinsic luminescence features, and in the past this has led to considerable confusion in the assignment of the various PL lines obtained from GaAs-on-Si.

3. RESULTS AND DISCUSSION

Three distinct phenomena determine the structural and the optical properties of GaAs-on-Si, (i) the formation of antiphase domain boundaries due to the growth of a polar semiconductor on a nonpolar substrate, (ii) the large lattice mismatch (a_{Si} = 5.43072 Å, a_{GaAs} = 5.65315 Å), and (iii) the large difference in the thermal expansion coefficient of the two components (Si: 2.3 x 10^{-6} K^{-1}, GaAs: 6.0 x 10^{-6} K^{-1}). The 4.1 % lattice parameter difference gives rise to the formation of a network of misfit dislocations at the substrate-epilayer interface and to an angular misorientation, or tilt, between the epilayer and the substrate. As a consequence of the Si

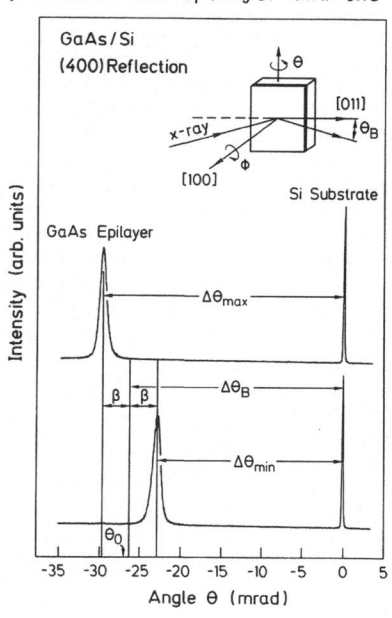

FIGURE 1. Double crystal XRD patterns of 3-μm GaAs-on-Si obtained at two azimuthal rotation angles ϕ. Θ_0 indicates the angular position of unstrained GaAs, and β the misorientation (tilt) angle of the epilayer with respect to the Si substrate.

substrate misorientation by a few degrees, which provides a higher-index plane necessary for the suppression of antiphase domains in the epilayer, the GaAs/Si heterointerface is no longer coincident with the (100) planes of either component.

In order to determine the lattice constants normal and parallel to the substrate surface and the angular misorientation of the MBE grown GaAs-on-Si layers, the high resolution XRD patterns are recorded in the vicinity of the symmetric (400) and the asymmetric (511) and($\bar{5}$11) reflections. Figure 1 shows characteristic XRD patterns around the (400) reflection obtained at two different azimuthal rotation angles ϕ around the [100] direction. From the shift of the GaAs peak position $\Delta\Theta_B$ with respect to the Si (400) reflection we can determine the lattice constant $a^\perp$ of the GaAs layer normal to the (100) surface. The value of $a^\perp$ (see Table 1) for GaAs-on-Si is smaller than the unstrained lattice constant of GaAs by about 0.09%. The XRD linewidth of GaAs-on-Si (FWHM) observed for 3 - 4 μm thick layers amounts to about 1 mrad ($\cong$200 s of arc) and is thus about one order of magnitude larger than that of GaAs-on-GaAs due to the rather high defect density.

TABLE 1. Structural properties of GaAs-on-Si layers of different thickness grown on vicinal (100) Si oriented 2° off towards [011] as determined by high-resolution XRD using CuKα_1 radiation for the symmetric (400) and the asymmetric (511) and ($\bar{5}$11) reflections.

Thickness (μm)	1.0	2.0	4.0
ß (sec)	404	669	727
Direction	(011)	(011)	(011)
$\measuredangle$ (sec)	113	113	113
$a_\perp$ (Å)	5.6491+0.0007	5.6486+0.0007	5.6483+0.0007
$a_\parallel$ (Å)	5.6609+0.0026	5.6605+0.0026	5.6602+0.0026
$\varepsilon_\perp$	-7.30×10^{-4}	-8.2×10^{-4}	-8.70×10^{-4}
$\varepsilon_\parallel$	$+13.5\times10^{-4}$	$+12.8\times10^{-4}$	$+12.2\times10^{-4}$

Inspection of Figs. 1 and 2 reveals that the angular separation $\Delta\Theta_B$ of the GaAs and Si (400) reflections systematically changes as a function of azimuthal rotation around the sample surface normal. This shift during azimuthal rotation is caused by the misorientation, or tilt, between the substrate and the epilayer. The sense of the tilt is such that the GaAs [100] lattice vector is directed between the surface normal and the [100] direction of the Si substrate. The angular separation $\Delta\Theta$ between the (400) reflections of Si and GaAs is thus composed of the difference in Bragg

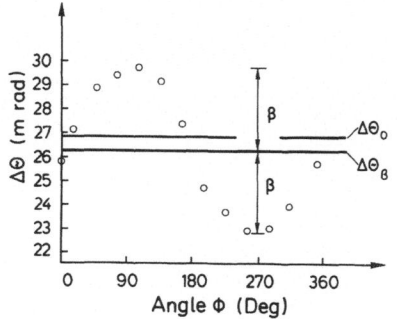

FIGURE 2. Variation of the misorientation angle ß between GaAs epilayer and Si substrate as a function of azimuthal rotation ϕ around the [100] direction (see Fig. 1 for geometry of the epilayer with respect to the X-ray beam).

angles $\Delta\Theta_B$ and the misorientation angle β, as per $\Delta\Theta = \Delta\Theta_B + \beta \cos\phi$. As shown by the data of Table 1, the misorientation angle β determined by means of this equation increases with increasing layer thickness, and it also increases when the substrate misorientation from the [100] direction is increased. Growth on a substrate with lower symmetry can result in the generation of a relative tilt between crystallographic planes that are nominally parallel to the interface. The origin of this tilt arises from dislocations with inclined Burgers vectors or from bond stiffening in specific crystallographic directions. For the evaluation of the structural properties of GaAs-on-Si it is thus important to distinguish the effects of unit cell distortion from those arising from misorientation.

The GaAs lattice constant $a^{||}$ parallel to the (100) growth surface is obtained by recording the diffraction patterns of the asymmetric (511) and $(\bar{5}11)$ reflections and by taking into account the measured value of $a^{\perp}$. At room temperature the value of $a^{||}$ is found to be <u>expanded</u> by about 0.13% as compared to bulk GaAs. This expansion exists although the lattice parameter of bulk GaAs is 4.1% <u>larger</u> than the Si substrate lattice parameter. The data compiled in Table 1 indicate that within the given error bars the lattice constants of the GaAs-on-Si layers do not depend on the thickness for layers thicker than 0.5 μm, provided that the layers are grown at the same growth temperature. Also included in Table 1 are the strain values at room temperature with respect to bulk GaAs obtained from the measured lattice constants (7). The in-plane tensile strain observed at room temperature follows from the fact that the thermal expansion of $a^{||}$ is forced to follow the thermal expansion of the Si substrate. The GaAs-on-Si layers are therefore under biaxial tensile strain at room temperature, and this strain results from the superposition of coherency strain (existing at the growth temperature) and thermal strain imposed upon cooling of the sample to room temperature.

In GaAs-on-Si layers of thickness below 0.1 μm a cross-over between in-plane contraction and expansion can occur, and the elastic strain from coherency effects and the thermal strain may even balance. In very thin (< 0.05 μm) layers the lattice parameter $a^{||}$ is found to be contracted and $a^{\perp}$ to be expanded, in contrast to the case of thicker (> 0.1 μm) layers. A systematic study of the structural properties of these very thin GaAs-on-Si layers by XRD is difficult because of the high defect density, and it is beyond the

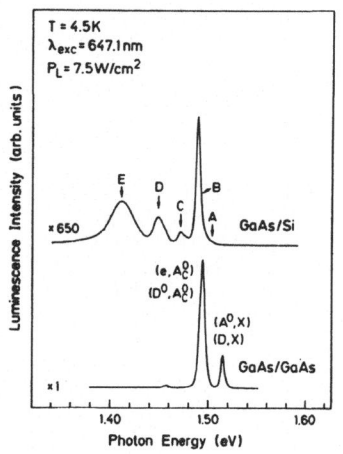

FIGURE 3. Low-temperature PL spectra of nominally undoped 4μm thick GaAs layers grown on GaAs substrate (bottom) and on Si substrate (top).

scope of this paper.

In addition to the translational incommensurability we also observe a certain rotational incommensurability in GaAs-on-Si layers, i.e. the GaAs in-plane [010] direction does not exactly align with the Si [010] axis but deviates by a few degrees. This rotation of the GaAs layer can amount to 5° and it takes probably place after the growth of a few monolayers. As yet, however, we have not established a systematic variation of this oriental relation as a function of layer thickness, substrate orientation, and growth conditions, and this phenomenon is the subject of further investigations.

The biaxial tensile strain existing in GaAs-on-Si layers leads to a pronounced low-energy shift of the bandgap and to a splitting of the heavy- and light-hole valence band (7). In Fig. 3 we compare the low-temperature PL spectra of nominally undoped GaAs layers grown on GaAs and Si substrates under comparable growth conditions. The homoepitaxial GaAs exhibits the characteristic luminescence properties, where the peak at 1.513 eV is due to excitons bound to shallow acceptor and donor states, while the dominant PL feature below 1.50 eV is due to band(donor)-to-carbon acceptor recombinations. The PL spectrum of the GaAs-on-Si layer shows the characteristic five luminescence lines which are labeled A to E according to the literature. The peak energy of the dominant feature at 1.489 eV does not depend on the excitation density, whereas the lines C, D, and E shift to higher energy with increasing excitation density. This behaviour is characteristic for extrinsic recombination processes. Only the line D saturates at high excitation densities. The high-energy shoulder of the main luminescence line at 1.504 eV (peak A) is only observed at the medium excitation level of 7.5 W/cm^2. At lower excitation densities in the mW/cm^2 range, the peak A disappears, and the linewidth of peak B becomes narrower. We obtain a linewidth of 2.9 meV (FWHM), which to our knowledge is the narrowest ever reported for GaAs-on-Si grown without any superlattice buffer layer. Because of the energy shift of the extrinsic lines C, D, and E it is important to note that the respective peak energies given in Table 2 are valid only for the applied excitation density of 7.5 W/cm^2.

When the GaAs layers are intentionally doped with Si, the measured n-type doping level is comparable for growth on GaAs or Si substrates. Most of the Si atoms are thus incorporated on Ga sites also in GaAs-on-Si. In Si-doped n-type GaAs-on-Si we observe a marked increase of the PL intensity and a broadening of the luminescence lines. For doping concentrations up to 5×10^{16} cm^{-3} the same five-peak structure as in undoped samples is observed. Only the line B appears at slightly lower energy. At doping concentrations beyond 10^{17} cm^{-3} the extrinsic line C dominates the spectrum and line B is only a high-energy shoulder (7).

The assignment of the excitonic PL features and the lifting of the valence-band degeneracy follows directly from the PLE spectrum displayed in Fig. 4, where the recombination line C is chosen to monitor the luminescence.

TABLE 2. Assignment of PL features obtained from undoped GaAs-on-Si layers using the 6471 Å laser line at an excitation density of 7.5 W/cm^2

Peak	Energy (eV)	Assignment
A	1.504	Recombination of free excitons and/or exciton bound to shallow acceptor and donor levels.
B	1.489	
C	1.470	Band-to-carbon acceptor recombination.
D	1.449	?
E	1.441	Defect related recombination

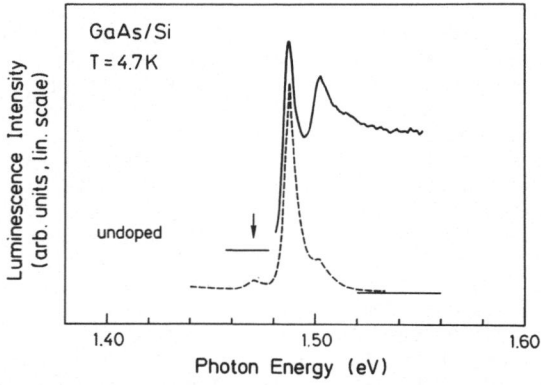

FIGURE 4. Low-temperature PL and PLE spectra of undoped GaAs-on-Si. The PL spectrum is excited with the 7800 Å laser line. The arrow indicates the detection wavelength for the PLE spectrum.

The energy position of the excitonic resonances in the PLE spectrum at 1.504 and 1.489 eV is identical to the position of the peaks A and B, resp., observed in the PL spectrum. The excitonic resonance becomes even more pronounced under the same excitation conditions when we record the PLE spectrum with line B as detection process. These findings demonstrate the _intrinsic_ excitonic nature of the luminescence lines A and B. They result from a recombination of free excitons and/or excitons bound to shallow acceptor and donor levels. In order to analyse the valence-band splitting and the bandgap shrinkage quantitatively, we have recently compared the experimentally detected values with a theoretical estimate for a biaxially tensile strained GaAs layer (7). As discussed before, this strain is caused by the different thermal expansion coefficients of GaAs and Si. The biaxial tensile strain can be divided into a hydrostatic component, leading to a reduction of the band gap, and into an uniaxial component, which lifts the valence-band degeneracy. Taking into account the exciton binding energy, we deduce a band gap of 1.494 eV associated with the heavy-hole valence band and a gap of 1.507 eV associated with the light-hole valence band from the PLE data. At 4.5 K the valence band splitting thus amounts to 13 meV consistent with a strain of $\varepsilon = 1.7 \times 10^{-3}$. This result indicates the increase of the strain level of GaAs-on-Si with decreasing temperature, which is expected from the different temperature behaviour of the linear thermal expansion of the materials (7).

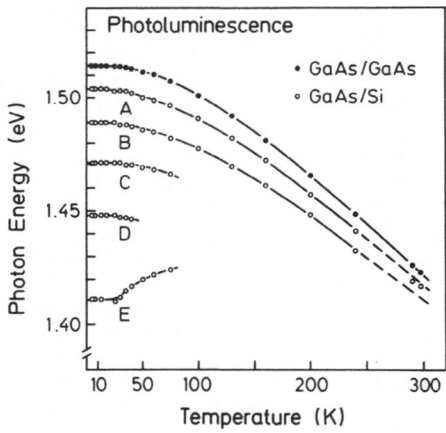

FIGURE 5. Variation of PL peak energy as a function of sample temperature for GaAs-on-Si (open circles) and for GaAs-on-GaAs (full circles) observed under the same excitation conditions.

We next perform temperature-dependent luminescence measurements, in order to analyse the reduction of strain with increasing temperature in the range 4.5 to 300 K and to assign the extrinsic luminescence lines C, D, and E. The data of Fig. 5 clearly show that the valence-band splitting (lines A and B) and the difference of the band gap between GaAs-on-Si and GaAs-on-GaAs decrease with increasing temperature. For temperatures up to 80 K the energy separation of the lines A and B remains nearly constant while for higher temperatures the valence-band splitting decreases and reaches a value of about 9 meV at room temperature. This splitting is consistent with a strain value of $\varepsilon \sim 1.2 \times 10^{-3}$ at room temperature, in agreement with the results of the XRD measurements shown in Table 1. The important result is that the strain in GaAs-on-Si increases when we cool the sample from 300 to 80 K and it then remains constant upon further cooling down to 4.5 K. This finding is a direct consequence of the different temperature dependence of the linear thermal expansion of GaAs and Si.

In the whole temperature range 4.5 to 300 K the electron-to-heavy-hole transition remains the dominant recombination process. The luminescence lines A and B are no longer separated at room temperature because of the temperature-broadened linewidth (31-meV FWHM at 300 K). Referring to the temperature dependence of the extrinsic recombination processes (lines C, D, E) depicted in Fig. 5, the first line to drop in luminescence intensity is line D which vanishes at about 40 K, whereas the lines C and E persist up to about 80 K. While the peak energies of lines C and D exhibit a similar temperature dependence as the electron-to-heavy-hole band-to-band transition, we observe an interesting high-energy shift for the recombination process leading to line E. The luminescence line C at 1.470 eV (4.5 K) is assigned to the band-to-carbon acceptor (e, A_C^0) recombination, because its peak energy coincides with the energy separation of the conduction band to the strain-split carbon acceptor level associated with the heavy-hole valence band. Under low excitation density also the donor-to-carbon acceptor (D^0, A_C^0) recombination at 1.465 eV (4.5 K) can be detected. The extrinsic luminescence line D at 1.449 eV (4.5 K) shows rather a peculiar temperature dependence. Despite the large energy difference of 45 meV with respect to the heavy-hole-related band gap, this luminescence process becomes ionized at relatively low temperature. A rough estimate of the related binding

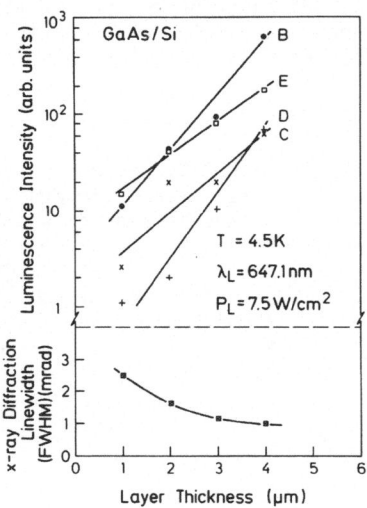

FIGURE 6. Comparison of PL peak intensities and of XRD linewidths [FWHM of (400) reflection] of GaAs-on-Si as a function of epilayer thickness. The solid lines serve as a guide to the eyes.

energy gives a value of about 10 meV. A tentative explanation for this luminescence process is the recombination of an exciton bound to a deep charged defect center. However, additional experiments are necessary for a clear assignment of this luminescence line D. The luminescence line E at 1.411 eV (4.5 K), finally, is attributed to a defect-to-carbon acceptor (d, A_C^0) recombination. This assignment is based on the comparable decrease in integrated luminescence intensity as the (e, A_C^0) recombination and on the high-energy shift of this line with temperature from 40 to 80 K. The total high-energy shift of 25 meV with respect to the conduction band-heavy-hole valence band gap is almost identical to the carbon acceptor binding energy of 27 meV. This shift to higher energy is caused by the increasing amount of defect-to-heavy-hole band recombination. The resulting defect energy level follows to be located 56 meV below the conduction band edge.

The defect center responsible for the luminescence line E is located in the GaAs/Si interface region. This assumption is confirmed by the application of different laser excitation wavelengths and by the investigation of samples with different epilayer thicknesses. When we use a laser wavelength of 4762 Å (≅ 2.60 eV) with smaller penetration depth for excitation, which is absorbed near the GaAs surface, the defect-related luminescence line E is strongly reduced as compared to the exciton-related near-band-gap luminescence. The data of Fig. 6 clearly show that in 0.5 μm thick GaAs-on-Si samples the line E is the dominant luminescence feature. With increasing layer thickness the intensity of this defect-related line is reduced as compared to the other luminescence lines. Inspection of Fig. 6 further reveals that the structural as well as the optical properties of GaAs-on-Si considerably improve with increasing layer thickness. It is important to note, however, that the observed XRD linewidth of 1 mrad (≅ 200 s of arc) and consequently the defect density is still about one order of magnitude larger for a 4 μm thick layer than that of GaAs-on-GaAs.

The integrated intrinsic luminescence intensity of GaAs-on-Si provides information about the minority-carrier properties which are important for the quality of photonic devices made from the heteroepitaxial layers. In Fig. 7 we compare the integrated PL intensity of GaAs-on-Si with that of GaAs-on-GaAs grown under the same growth conditions for the temperature range 4.5 to 300 K. At 4.5 K the intrinsic luminescence intensity of the homoepitaxial GaAs is by two orders of magnitude superior to that of the heteroepitaxial GaAs. In the temperature range where we observe the exciton-related recombination, both types of GaAs layers show a similar decrease in

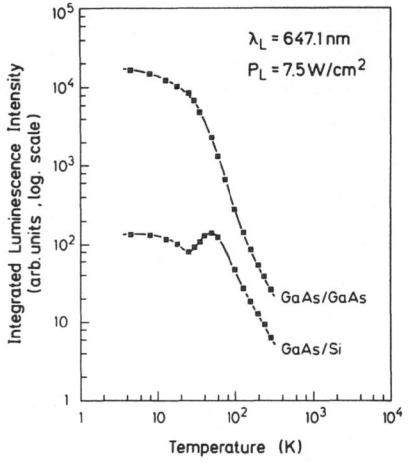

FIGURE 7. Comparison of integrated PL intensities of the intrinsic recombination process in GaAs-on-Si and in GaAs-on-GaAs as a function of sample temperature. The layer thickness is 4 μm in both cases.

intensity. However, in the temperature range where the change to band-to-band recombination occurs, we observe a distinct increase in intensity for the GaAs-on-Si only. For temperatures above 70 K both layer types again exhibit the same decrease in luminescence intensity with increasing temperature. At room temperature the PL intensity ratio between GaAs-on-GaAs and GaAs-on-Si is reduced from 100 : 1 at 4.5 K to about 4.5 : 1 at 300 K because of the peculiar intensity increase for GaAs-on-Si between 30 and 50 K. This change in luminescence intensity around 40 K is not well understood and still under investigation. The result that the PL intensity is inferior by only a factor of 4.5 at 300 K is very encouraging for further improvements of the crystal quality of GaAs-on-Si.

The GaAs-on-Si layers of thickness > 0.1 μm are tetragonally distorted at room temperature. The observed biaxial tensile strain in the GaAs layers is caused by the different thermal expansion coefficients of GaAs and Si. Therefore, although the GaAs layer is relaxed at the growth temperature of about 600 °C, it becomes strained upon cooling to room temperature after growth. In order to reduce this cooling induced strain, we have to lower the substrate temperature during MBE growth by several hundred degrees C. A substantial reduction of the substrate temperature to T_s = 300 °C is accomplished by depositing the constituent Ga and As layers of the (100) GaAs crystal in a monolayer by monolayer sequence (5). The shutters of the Ga and As_4 effusion cells are alternately opened for a time interval (typically between one and two seconds) which corresponds to the deposition of one monolayer of the respective material. For the growth of GaAs-onSi this modulated growth mode is applied after the deposition and the heat cycling of the 250 Å thick uniform GaAs layer. An excellent smooth surface morphology is obtained, and for the case of AlAs/GaAs multilayer structures sharp interfaces are produced. Although we have obtained high-quality AlAs/GaAs-heterostructures- and GaAs-on-GaAs at T_s=300 °C by the application of the modulated MBE growth mode, the structural and the optical properties of GaAs-on-Si prepared under the same conditions at T_s = 300 °C are still inferior by about a factor of 3 to 5 (FWHM X-ray and luminescence intensity) as compared to the material prepared by the conventional MBE process at T_s = 600 °C. Therefore, further improvements of the modulated growth mode with emphasis on the particularly important GaAs prelayer formation on Si and its heat cycling are underway in our laboratory.

ACKNOWLEDGEMENTS

The work was sponsored by the Bundesministerium für Forschung und Technologie of the Federal Republic of Germany.

REFERENCES

1. For an extensive survey see: "Heteroepitaxy on Silicon I", Mater. Res. Soc. Symp. Proc. 67, 1 - 273 (1986) and "Heteroepitaxy on Silicon", Mater. Res. Soc. Symp. Proc. 91, 1 - 508 (1987).
2. A.C. Gossard, P.M. Petroff, W. Wiegmann, R. Dingle, and A. Savage, Appl. Phys. Lett. 29, 323 (1976); T. Isu, D.S. Jiang, and K. Ploog, Appl. Phys. A 43, 75 (1988).
3. P.N. Uppal and H. Krömer, J. Appl. Phys. 58, 2195 (1985).
4. T. Sakamoto and G. Hashiguchi, Jpn. J. Appl. Phys. 25, L 78 (1986); D.E. Aspnes and J. Ihm, Phys. Rev. Lett. 57, 3054 (1987).
5. F. Briones, D. Golmayo, L. Gonzales, M. Recio, A. Ruiz, and J.P. Silveira, Inst. Phys. Conf. Ser. 91, 165 (1988); L. Gonzales, A. Ruiz, Y. Gonzales, A. Mazuelas, and F. Briones, this volume.
6. L. Tapfer and K. Ploog, Phys. Rev. B 33, 5565 (1986)
7. W. Stolz, F.E. G. Guimaraes, and K. Ploog, J. Appl. Phys. 63, 492 (1988).

GaAs ON Si:POTENTIAL APPLICATIONS

DON W. SHAW

Texas Instruments Incorporated
P.O. Box 655936, MS-147
Dallas, Texas 75265 USA

1. INTRODUCTION

Epitaxial gallium arsenide deposited on silicon substrates is a prime example of a mismatched epitaxial system. The crystal lattice constants of the two materials differ by about 4%, and the thermal expansion coefficients differ by more than a factor of two. Although such mismatched systems were considered infeasible only a few years ago, today they exemplify the revolution in epitaxial thin film capabilities that has spawned the promising new field called "Artificially Structured Materials".

Early attempts to deposit GaAs on Si were unsuccessful because of the inability to establish an oxide-free Si surface, chemical incompatibilities with the deposition processes, excessive deposition temperatures, and failure to use the proper buffer or intermediate layers. The first success was obtained by use of Ge buffer layers, but soon it was determined that the Ge could be effectively replaced with special GaAs intermediate layers deposited under conditions different from those employed for growth of the bulk of the GaAs layer thickness.

Molecular beam epitaxy with its sophisticated instrumentation and ultrahigh vacuum environment permitted preparation of oxide-free silicon surfaces. It also offered low temperature depositions, free of detrimental chemical interactions. This resulted in successful surface preparation procedures, but the lattice mismatch and associated misfit dislocations remained. Soon it was found that by appropriate choice of substrate crystallographic orientation and intermediate layer deposition regime many of the misfit dislocations could be confined to a thin region near the interface. As a result the thicker portions of the epitaxial structure could have a crystal perfection that, though inferior to homoepitaxial GaAs on GaAs substrates, was suitable for acceptable device performance. The technology has since appeared increasingly promising with demonstrations of several discrete devices (Table 1), LSI-level integrated circuits, and lasers in GaAs layers deposited on Si.

Two distinct areas are being considered for application of GaAs on Si; one where the Si serves simply as a passive substrate with advantages such as lower cost, larger area, and greater strength, and the other where active devices are fabricated in both GaAs and Si. Both potential applications will be considered in the following with particular attention devoted to the advantages of the materials combination and the limitations that must be overcome to achieve success.

Y. I. Nissim and E. Rosencher (eds.), Heterostructures on Silicon: One Step Further with Silicon, 61–74.
© *1989 by Kluwer Academic Publishers.*

TABLE 1. Devices demonstrated in GaAs on Si

Metal semiconductor field effect transistor
Modulation-doped field effect transistor
Heterojunction bipolar transistor (emitter up)
Heterojunction bipolar transistor (emitter down)
Resonant tunneling diode
Avalanche photodiode
Single quantum well laser
Multiple quantum well laser
Light emitting diode
Solar cell
Heteroface solar cell

2. SILICON AS PASSIVE SUBSTRATE FOR GaAs COMPONENTS

2.1. Advantages of Silicon Substrates

Endowed with years of preparation history and favorable characteristics such as a relatively high thermal conductivity, large wafer size and low cost, silicon is a very attractive substitute for GaAs. Consider wafer size. Only recently have 100 mm GaAs wafers become available. Because of the thermal stresses associated with GaAs's poor thermal conductivity, preparation of 125 mm and greater diameter GaAs wafers with acceptable crystal perfection will be difficult. On the other hand, silicon substrates are readily available now with diameters up to 200 mm. Also, in terms of fracture strength, Si is about 2.5 times less likely to break than GaAs.(1) This is an important consideration for large-area integrated circuit fabrication, where the number of bars available on a wafer is limited.

Currently gallium arsenide substrates are about 25 times more expensive than silicon. This would appear to represent a major cost advantage for using GaAs on Si; however, GaAs epitaxial structures are much more expensive than GaAs substrates, so the savings resulting from use of a silicon substrate rather than GaAs is only a fraction of the total cost of the overall epitaxial structure. The real cost advantage for GaAs on Si emerges with wafer sizes larger than 75 mm where GaAs wafers, if they were available, would be very expensive. Some GaAs applications, such as MESFET-based digital integrated circuits, can be fabricated with ion implanted active regions, and no epitaxial layers are required. In this case the cost comparison is between a GaAs substrate and a GaAs-on-Si epitaxial structure, and the latter is the more expensive approach because of the relatively high cost of epitaxial deposition. It may be argued that a better cost assessment would be obtained by comparing GaAs and Si epitaxial processes, with the difference assumed to be only in the costs of the starting materials. This could be valid, but only for manufacturing volumes substantially greater than exist for GaAs today.

Because of the increased levels of crystal imperfections and stresses, the advantages of Si as a passive substrate for GaAs must, in general, be obtained at the expense of reduced performance of the resulting devices. The significance of these limitations and the substantial progress that has been made toward their circumvention are discussed next.

2.2. Limitations of Silicon Substrates

Stresses, caused by thermal expansion mismatch, and dislocations resulting from differences in lattice dimensions are presently considered to be the main limitations to application of GaAs on Si. In the early stages of the technology, antiphase domain boundaries, associated with growth of a polar crystal on a nonpolar substrate, were considered to a major barrier. Some GaAs crystal orientations, including the most desirable {100}, consist of alternating monoatomic layers of gallium and arsenic atoms. Unless the silicon substrate surface is atomically flat (or all steps are an even number of atomic layers in height) and growth proceeds without two-dimensional nucleation, multiple nucleation will occur on the surface; and when these regions grow together to form a continuous film, the layers of a given species may or may not be in alignment with the same species from one separately nucleated region to the other. If two regions are misaligned, then the resultant boundary is called an antiphase boundary. These defects were commonly encountered in the early GaAs-on-Si structures and were once considered to be a major limitation to the technology. However, by 1985 convincing evidence had emerged that the antiphase boundaries could be suppressed.(2) This major development was achieved by initiating the growth with a thin GaAs prelayer deposited at low temperatures prior to subsequent growth of the bulk of the layers at higher temperatures (the two-step method (3)) and by use of an intentionally misoriented or "tilted" substrate with a surface orientation of about 4° from the (001) toward the [011].(4) Silicon surfaces with this orientation, when subjected to high temperatures (such as during oxide removal), rearrange to form steps with double-atomic-layer heights.(5) Thus with proper choice of substrate preparation and surface orientation, antiphase domains are not considered a serious limitation.

The relatively large (4%) difference in lattice dimensions of GaAs and Si can be accommodated by formation of misfit dislocations within the GaAs layer. If all of these dislocations propagate throughout the epitaxial film, the dislocation density would be so large (around 10^{12} cm^{-2}) that most applications would be impractical. Fortunately, through use of the growth procedures described above the dislocation density can be reduced to below 10^7 cm^{-2}. In the very early stages of growth the lattice mismatch is accommodated at least partially by compressive strain, and the GaAs lattice spacing contracts in an attempt to match the underlying Si spacing. After a few nanometers the strain energy exceeds that necessary to form dislocations, and misfit dislocations are generated. The two types of misfit dislocations, illustrated in Fig. 1, are observed to occur.(6) One of these (Type I) lies in the interface plane and does not generate threading dislocations that propagate into the epitaxial layer.

The same tilted surface orientation that inhibits formation of antiphase domains also serves to significantly reduce threading dislocations by favoring generation of the Type I dislocations that are localized at the interface region. Additional dislocation reduction can be obtained by use of strained intermediate layers such as superlattices. The strain serves to bend over dislocations that would otherwise propagate into the subsequently deposited epitaxial layers. These superlattices may be thin, alternating layers of different composition or alternating layers of the same composition but deposited at different temperatures.(8) Finally, some dislocation reduction can be obtained by

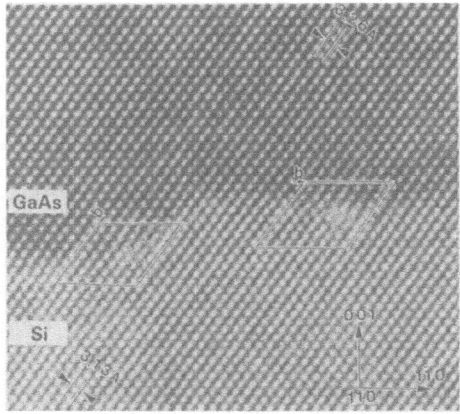

FIGURE 1. Transmission electron micrograph of GaAs/Si interface showing two basic dislocation configurations, Type I (right) and Type II (left).(7)

annealing in situ during growth of the thicker portions of the epitaxial layer or by a separate anneal after growth.(9)

Excessive dislocations in GaAs films on Si substrates may interfere with device processing and reliability by enhancing the diffusion rates of impurities and point defects. They also degrade device performance by serving as centers for carrier trapping or recombination. Minority carrier devices, such as lasers, solar cells, and bipolar transistors, are particularly sensitive to these effects. On the other hand some defects appear to be benign. For example, with avalanche photodiodes in GaAs on Si it was found that of the 10^7-10^8 cm^{-2} dislocations present in the epitaxial layers, only $\sim 10^4$ cm^{-2} were electrically active in causing increased leakage currents and smaller gains.(10)

While dislocations and antiphase domains were the first GaAs-on-Si limitations to be considered, it soon became apparent that stresses caused by differences in thermal expansion were at least as important as the crystal defects. When the structures are cooled to room temperature from the epitaxial deposition temperature (500-700°C), the GaAs, if not constrained, would contract about twice as much as the Si substrate. Consequently, in the composite structure, after cooling to room temperature, the GaAs layer is under considerable tensile stress at levels ranging from 5 X 10^8 to over 3 X 10^9 dyne cm^{-2}, depending on the deposition temperature. Since the GaAs lattice constant is greater than that of Si, any residual strain not relieved by formation of misfit dislocations at or near the deposition temperature will be compressive and will tend to reduce the net thermal-expansion-induced tensile stress observed after cooling to room temperature.(11)

Although the layer stress does not vary greatly with layer thickness, the elastic energy does increase. At a critical thickness, dependent on the growth temperature, the elastic energy increases to the point where cracking occurs. For typical growth conditions cracking is usually observed with layers thicker than about 5-7 microns. Some devices, such as solar cells, require relatively thick layers. Since the stresses, and hence the critical thickness for the onset of cracking, are determined mostly by the differences in the thermal contraction properties of GaAs and Si, low deposition temperatures are desirable. Unfortunately, reduced deposition temperature are generally accompanied by reduced deposition rates, compounding the problems of applying GaAs on Si for devices requiring thick layers.

The most obvious manifestation of the GaAs tensile stress is wafer warpage or bowing, where the epitaxial surface becomes concave. Warpage is reduced by decreasing the deposition temperature, increasing the substrate thickness, or patterning the deposition into an array of small islands. However, it should be noted that some techniques, such as increasing the substrate thickness, reduce the warpage but not the layer stress to any appreciable extent. The principal problem with warpage is the difficulty it imposes on fine-line lithography during device fabrication. Generally, LSI-level lithography requires that the wafer bow be maintained at less than 10 microns over a 2-inch wafer. The residual layer stress, is a serious reliability limitation for some devices such as lasers, as will be discussed later.

Difficulties are also encountered in processing GaAs-on-Si structures during device fabrication. Gallium arsenide preferentially cleaves along {110} planes, but silicon cleaves along {111}. Adjustments in wafer dicing and mask alignment may be necessary to conform to these different cleavage properties. The two materials have different sensitivities to the etchants employed in processing. Depending on the process conditions and temperatures, cross-doping or cross-contamination can impose restrictions, particularly when functional devices are present in both materials.

2.3. Discrete GaAs Devices on Si Substrates.

2.3.1. Solar cells. Solar cells were among the first devices fabricated and evaluated in GaAs on Si. Replacement of GaAs substrates with stronger, less-expensive Si was the primary motivation. However, the lower density of Si also makes it an attractive substrate for GaAs solar cells destined for space applications. Unfortunately, solar cells are minority carrier devices and highly dependent on the crystal perfection of the host material. The best GaAs-on-Si cells, fabricated in material with a dislocation density of 2×10^6 cm^{-2}, yielded 18% efficiency.(12) Since solar cells require relatively thick layers, they are prone to cracking. If the defect density and cracking can be controlled, cascade structures with a GaAs top solar cell and a silicon bottom cell offer a theoretical efficiency as high as 36% by more efficiently utilizing the available solar spectrum.

2.3.2. Lasers. Solid state lasers represent the most severe test of the quality of GaAs-on-Si structures. These minority carrier devices require low-stress, low-dislocation-density material with relatively complex multilayer epitaxial config- urations. Laser performance in GaAs on Si has evolved rapidly and steadily.(13) First to be demonstrated were pulsed devices operating at cryogenic temperatures. These were followed in succession by cw operation at 77 K, pulsed operation at room temperature, photopumped cw operation at room temperature, and finally by cw, injection devices emitting at room temperature (Fig. 2). Pulsed operation of quaternary lasers from a InP/GaInAsP/InP/Si structure has also been reported.(14) In spite of this progress, the demonstrated performance and reliability are still not adequate for real applications.

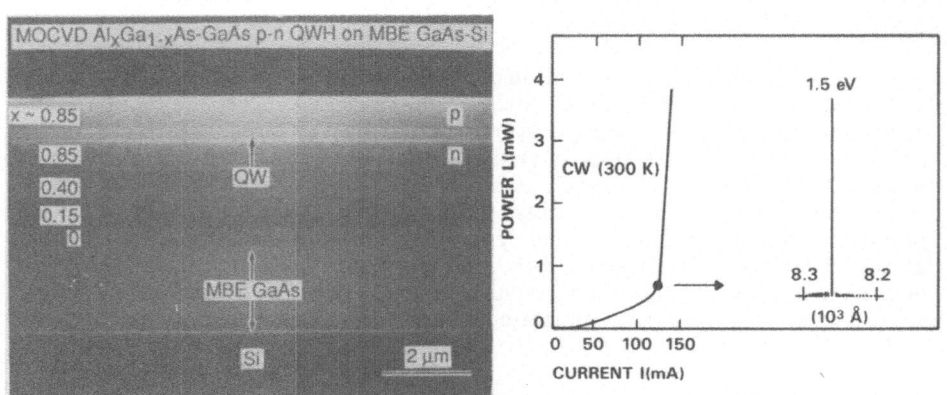

FIGURE 2. MOCVD $Al_xGa_{1-x}As$-GaAs p-n quantum well double heterostructure laser on MBE GaAs on Si.(17)

Lasers are subject to degradation if only a single dislocation penetrates the active region. For typical dimensions this places an upper limit on the dislocation density at about 3×10^4 cm^{-2}, about 100 times lower than the best values presently being achieved with GaAs on Si. The residual stress also contributes to degradation. For reliable laser performance the stress should not exceed 1×10^8 dyne cm^{-2},(15) which is considerably below the values currently being obtained. Stresses are reduced with patterned layers. Recent results (16) show that when the surface dimensions are restricted to values close to the layer thickness, the stresses are markedly reduced. For a stripe laser the stress becomes essentially uniaxial because the stress perpendicular to the laser stripe edge is negligible in comparison with the parallel stress, which remains at high levels. If the laser dimensions can be reduced in both directions, as for example with surface emitting lasers, then the expected reduction in stress should have a significant impact on stability and reliability.

One motivation for replacement of GaAs substrates with Si for discrete laser applications is silicon's greater thermal conductivity, which may permit upright rather than flip chip mounting. However, as will be discussed later, the greatest merit for lasers in GaAs on Si is associated with combined-function monolithic circuitry using lasers for intrachip and interchip communications.

2.3.3. Light emitting diodes.
Light emitting diodes have been fabricated in GaAs on Si.(18) These minority carrier devices operate at high current densities and require relatively thick layers. There has been little interest in using GaAs on Si for commercial discrete LED's. Large area arrays of GaAs LEDs for printing applications have been suggested as a potential application of GaAs on Si because of the availability of large Si wafers. However, most GaAs-on-Si LED research is aimed at combined function circuitry such as interchip communications, which will be discussed later.

2.3.4. Field effect transistors.
Field effect transistors (FETs) are the most widely evaluated devices in GaAs on Si. The majority carrier devices are relatively insensitive to crystal defects, and with the best devices there is little degradation in performance from that expected from FETs fabricated on GaAs substrates. However, GaAs FET operation depends on a high resistivity underlying substrate or buffer layer. In comparison with semi-insulating GaAs, Si substrates are much more conductive and require epitaxial growth of an effective high resistivity buffer for successful GaAs FET operation. Marginal buffer layers result in greater leakage currents and backgating effects.

Use of GaAs on Si has been proposed for GaAs power FETs to take advantage of the superior thermal conductivity of Si. Also it has been suggested, but not demonstrated, that GaAs-on-Si FETs might possess a measure of radiation (single event upset) resistance because carriers generated within the substrate by an ionizing particle may be reflected by the band edge discontinuity or trapped at dislocations at the interface. Generally, FETs in GaAs on Si exhibit transconductances that are approximately 80% of the values from all-GaAs control devices. However, in some cases the performances of GaAs-on-Si FETs are virtually indistinguishable with conventional GaAs transistors. For example, cutoff frequencies up to 55 GHz have been measured with GaAs-on-Si MESFETs.(19)

Modulation-doped FETs (MODFETs or HEMTs) have also been fabricated in GaAs on Si. Good transconductances (up to 275 mS/mm at 77 K) (20) have been measured; however, the values were always inferior to companion devices processed simultaneously on GaAs substrates. It is interesting that the drain I-V collapse problem that has plagued these devices was not observed with MODFETs in GaAs on Si.(21) Because the collapse phenomenon is generally associated with a particular defect (the D-X center) of GaAlAs, the results show that the high concentration of dislocations in GaAs/GaAlAs layers on Si substrates does not produce a large increase in such centers or their associated donor impurities. The MODFET results together with supporting Hall data also indicate that threading dislocations do not lead to substantial Si outdiffusion or autodoping.

2.4 GaAs Integrated Circuits on Si Substrates

2.4.1. Digital integrated circuits - MESFET.
Integrated circuits with large chip areas can benefit from the larger wafers available with GaAs on Si. As previously discussed there is a potential cost benefit associated with use of wafer sizes greater than those available with bulk GaAs. Economic advantages also result from reduced breakage, superior compatibility of GaAs-on-Si wafers with most semiconductor process equipment, and the potentially lower cost of the starting wafer. The latter, however, must be a qualified advantage because some digital circuits can be fabricated entirely in ion implanted bulk GaAs wafers,

thus avoiding a costly epitaxial growth step. Although GaAs-on-Si wafers with sizes up to 125 mm have been demonstrated, there has been little evaluation of the behavior of such large wafers during processing.

Impressive early results have been obtained with MESFET-based LSI-level circuits. Shichijo, et al. (22) demonstrated a functional 1K static random access memory (SRAM) containing over 7,500 transistors (Fig. 3). To achieve functionality the voltage thresholds had to be maintained within narrow limits for both enhancement- and depletion-mode MESFETs that comprise this circuit. The SRAM was fabricated entirely by ion implantation into a GaAs layer deposited on a two-inch Si substrate. Wafer bowing either before or after post-implantation annealing was not sufficiently serious to impede fabrication of such a dense circuit. The MESFET parameters were inferior to those on GaAs substrates, and the circuit yield was lower. Typically, the transconductances were about 80% of the values expected, and the measured backgating was much worse. Nevertheless, the overall circuit performance was quite respectable, with access times only slightly slower than those of circuits fabricated simultaneously in all-GaAs control wafers.

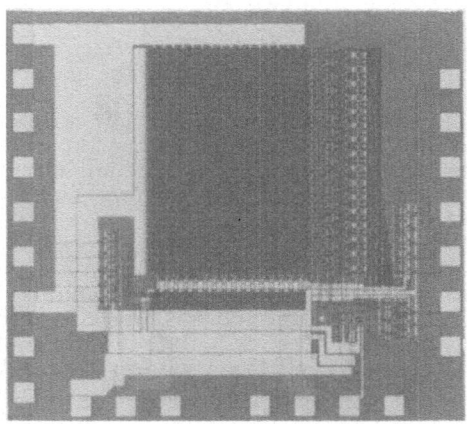

FIGURE 3. 1K static random access memory in GaAs on Si.(22)

2.4.2. Digital integrated circuits - bipolar. Successful fabrication of heterojunction bipolar integrated circuits in GaAs on Si would be expected to be more difficult than MESFET-based circuits because the bipolars are minority carrier devices. Nevertheless, functional circuitry has been demonstrated, at least at the SSI level. Fifteen-stage ring oscillators and read-only memories have been fabricated in GaAs/AlGaAs on Si using the difficult emitter-down (HI^{2}L) configuration that is most suitable for high-density integration.(23) All layers for these circuits were grown by MBE, and ion implantation was used to form the p$^+$ base regions. In the HI^{2}L configuration, the emitter contact is made to the Si substrate, and the current path is through the imperfect Si/GaAs heterojunction. However, because of the high doping levels on both sides of the heterojunction, the conduction properties were normal. The observation of identical emitter/base junction characteristics for the GaAs-on-Si and the all-GaAs control devices indicates that the dislocations threading from the Si substrate through the GaAs/AlGaAs layers do not appreciably degrade the performance of these heterojunction bipolar transistors (HBTs).

The GaAs/Si interface becomes an active junction in the novel concept of a GaAs-Si heterojunction bipolar transistor that combines a Si collector with a GaAs base and GaAlAs emitter.(24) Although this device would require formation of the critical GaAs base in the most defective portion of the epitaxial layer just at the interface, experimental measurements (25) of p-GaAs/n-Si junction properties provide encouragement for further investigation. The heterojunction device would merge the intrinsic high-performance potential of a GaAs/GaAlAs heterojunction bipolar transistor with the planar processing advantages of Si technology. Calculations indicate an f_{max} of 108 GHz for the GaAs/Si HBT as compared with 76 GHz for the conventional GaAs/GaAlAs HBT.

2.4.3 <u>Monolithic microwave integrated circuits</u>. Monolithic microwave integrated circuits (MMICs), consisting of GaAs components connected with metal microstrip transmission lines on a semi-insulating substrate, promise to be an important, large scale application for GaAs. Although these circuits are not densely integrated, they require large chips because of on-chip matching components such as inductors. The promise of using large diameter GaAs-on-Si wafers, which are resistant to breakage and have superior heat dissipation characteristics, is very attractive; but can the silicon substrate's relatively high electrical conductivity be tolerated? Indeed, the availability of GaAs in a high resistivity (10^8 ohm cm), semi-insulating form was a major factor in developing MMIC technology. Early work (26) on the use of Si as a transmission medium for hybrid circuits showed that of the two transmission line loss mechanisms - substrate loss and conductor loss - the conductor losses dominate at higher frequencies (> 30 GHz) as long as the substrate resistivity exceeds 2000 ohm cm. At mm-wave frequencies (90 GHz) the attenuation of microstrip lines on Si with a resistivity of 10,000 ohm cm can be quite low (0.6 dB/cm). (27)

The active elements in MMICs are usually majority-carrier MESFETs which are easily fabricated with acceptable performance in GaAs on Si. Application of GaAs on Si to MMICs then depends on the availability of high resistivity silicon. Such material is grown by the float zone process and traditionally has been available only in sizes less than 125 mm. Although float zone growth of 150 mm Si crystals is believed to be feasible, the cost may be excessive in comparison with the expected advantages of the larger wafers. Wafers of intermediate size (100 mm) would still be attractive for potential MMIC applications, but there remains an additional potential problem. High resistivity silicon is subject to resistivity reductions during high temperature processing. Fortunately, recent results (27) indicate that Si with a resistivity as high as 10,000 ohm cm can be stable at diffusion and thermal oxidation temperatures.

A single-stage MMIC feedback amplifier in GaAs on Si has been tested.(28) Operating between 8.5 and 11.5 GHz, the GaAs-on-Si MMIC showed a lower gain and smaller bandwidth than the GaAs-substrate control. However, computer simulations showed that the lower performance was due mostly to a leaky Schottky gate on the GaAs-on-Si test circuit and that the lower resistivity of the Si substrate (7000 ohm cm) had only a negligible effect on the microwave performance. No resistivity degradation occurred during growth of the GaAs layer or during subsequent processing. The superior thermal conductivity of the Si substrate was evident in thermal resistance measurements which indicated that the junction temperature of the GaAs-on-Si MMIC was about 15°C lower than the all-GaAs control.

The advantages and limitations associated with use of Si substrates for several of the above-discussed GaAs applications are summarized in Table 2.

TABLE 2. Use of Si substrates with GaAs device applications.

APPLICATION	ADVANTAGES	LIMITATIONS
MESFETs (Power)	Heat Dissipation	Performance degradation
Solar Cells	Larger wafer size Lower cost Reduced breakage Lighter weight Cascade cells	Cracking Reduced efficiency
Lasers	Heat Dissipation Monolithic integration with Si circuitry Optoelectronic integrated circuits	Reduced performance and reliability due to high dislocation density and stress
Digital integrated circuits	Larger wafer size Lower cost Reduced breakage Monolithic integration	Reduced yield Wafer bowing
Monolithic microwave integrated circuits	Larger wafer size Reduced breakage	Transmission losses

3. SILICON AS AN ACTIVE SUBSTRATE (MONOLITHIC INTEGRATION OF GaAs AND Si)

While simple replacement of GaAs substrates with stronger, less expensive Si is attractive for some applications, the more exciting potential occurs in applications that, by using the Si as an active rather than passive substrate, combine functional components of both GaAs and Si in a single monolithic chip. This can be an enabling technology that couples the high component density and process sophistication of Si with the high speed, high frequency, and optoelectronic capabilities of GaAs. Examples of such integration include intra- and interchip GaAs-based on-chip optical communications for complex Si ICs, "smart GaAs MMICs" with on-chip Si signal processing, and GaAs high speed processors joined with Si high density CMOS memory.

Success depends on the development of special processing regimes, e.g., growth of the GaAs on selected regions of a preprocessed silicon wafer containing fabricated devices or circuits. Because silicon processing generally requires higher temperatures than GaAs, the silicon circuits must be formed (at least up to metallization) before GaAs epitaxial deposition, during which the preprocessed Si circuitry must be protected by a inert mask. Ultimately, interconnects must be established between the Si and GaAs components. This requires that the upper surface of the GaAs layer be nearly coplanar with the Si surface (or the dielectric layer covering it). The approach being investigated to achieve such coplanarity employs deposition of the GaAs selectively into depressions or "trenches" etched to the optimum depth within the Si substrate.(23) When the GaAs is deposited by MBE, a polycrystalline film is simultaneously formed over the mask that protects the Si and defines the desired single-crystal GaAs area. The polycrystalline material must be removed during subsequent processing.

Deposition of GaAs into trenches etched into Si is fraught with potential problems. As mentioned previously, tilted substrate orientations are desired to minimize threading dislocations and eliminate antiphase domains; however, multiple orientations, exposed within the trench, are subject to simultaneous deposition. Depending on the trench etching process, the original tilted surface orientation may not be preserved at the bottom of the trench. The severity of these potential difficulties is being experimentally assessed.

3.1 High Speed GaAs with High Density Si

Assuming successful circumvention of the process limitations, it should be possible to develop monolithic integrated circuits containing regions with high density Si CMOS memory integrated with high speed GaAs components such as arithmetic logic processors, cache memory, or A/D-D/A converters. The GaAs logic and memory elements could be built with either MESFET or bipolar devices. These multimaterial ICs may be especially attractive for use at liquid nitrogen temperature.(29) Successful operation of a preliminary integrated circuit has been demonstrated with a composite ring oscillator containing 35 Si CMOS inverter stages and 12 GaAs MESFET buffered logic inverter stages all operating in concert.(30)

3.2. Optoelectronic GaAs with Si

The opportunity to monolithically combine GaAs optoelectronic components with Si circuitry is frequently mentioned in connection with GaAs on Si. Multiple silicon chips could communicate by high speed optical links either coupled through fibers or by free-space propagation. Single-chip repeaters for optical communications could be fabricated. These applications require use of minority carrier GaAs components, such as lasers, that are the most demanding on material quality. A device-level demonstration vehicle integrating a AlGaAs/GaAs light emitting diode with a Si MOSFET fabricated in the underlying substrate has already been tested.(31) The LED was modulated at rates up to 27 Mbit/s (limited by the large area of the particular Si test device used). But serious applications will require integration of lasers to obtain the requisite output powers and bandwidth, and the barriers to laser performance and reliability imposed by the high dislocation density and stress of GaAs on Si must be circumvented. Presently the stresses measured in GaAs on Si are around 10^9 dyne cm^{-2}, about an order of magnitude greater than considered acceptable for lasers. (15)

4. CONCLUSIONS

In spite of the large mismatch in lattice dimensions and thermal expansion properties, remarkable progress has been made in epitaxially depositing GaAs on Si substrates. Presently the epitaxial layer quality, although clearly inferior to that of GaAs deposited on GaAs, is adequate for laboratory-scale fabrication of a variety of electronic and optoelectronic devices that exhibit reasonable performances. Two broad areas for applications of GaAs on Si are being considered. In one the Si serves to replace conventional GaAs substrate wafers with larger, less expensive and stronger Si substrates. The other, a longer range and more difficult application, involves fabrication of active devices both in the GaAs epitaxial layer and in the underlying Si substrate. The rapid progress already demonstrated with GaAs-on-Si devices and circuits provides encouragement and stimulus for investigations of other promising mismatched epitaxial systems.

ACKNOWLEDGEMENTS

The author wishes to express his appreciation to H. Shichijo, R.J. Matyi, H-L. Tsai, and H.Q. Tserng for helpful discussions and comments.

REFERENCES

1 C.P. Chen and M.H. Leipold, Proc. 18th IEEE Photovoltaic Specialists Conf., 1985, pp 310-316.

2. H. Kroemer, Mat. Res. Soc. Symp. Proc. 67, 1, 1986.

3. M. Akiyama, Y. Kawarada, and K. Kaminishi, J. Crystal Growth 68, 21 (1984).

4. R. Fischer, W.T. Masselink, J. Klem, T. Henderson, T.C. McGlinn, M.V. Klein, H. Morkoc, J.H. Mazur, and J. Washburn, J. Appl. Phys. 58, 374 (1985). R. Fischer, H. Morkoc, D.A. Neumann, H. Zabel, C. Choi, N. Otsuka, M. Longerbone, and L.P. Erickson, J. Appl. Phys. 60, 1640 (1986).

5. T. Sakamoto and G. Hashiguchi, Japan. J. Appl. Phys. 25, L78 (1986).

6. N. Otsuka, C. Choi, L.A. Kolodziejski, R.L. Gunshor, R. Fischer, C.K. Peng, H. Morkoc, Y. Nakamura, and S Nagakura, J. Vac. Sci. Technol. B 4, 896 (1986).

7. H.L. Tsai and J.W. Lee, Appl. Phys. Lett. 51, 130 (1987).

8. J.W. Lee, Proceed. 1986 Int. Symp. on GaAs and Related Compounds, Inst. Phys. Conf. Ser. 83, 111 (1987).

9. J.W. Lee, H. Shichijo, H.L. Tsai, and R.J. Matyi, Appl. Phys. Lett. 50, 31 (1987).

10. N. Chand, J. Allam, J.M. Gibson, F. Capasso, F. Beltram, A.T. Macrander, A.L. Hutchinson, L.C. Hopkins, C.G. Bethea, B.F. Levine, and A.Y. Cho, J. Vac. Sci. Technol. B 5, 822 (1987).

11. D.W. Shaw, Mat. Res. Soc. Symp. Proc. 91, 15 (1987).

12. A. Yamamoto and M. Yamaguchi, to be published, Mat. Res. Soc. Symp. Proc., 1988.

13. G.W. Turner, H.K. Choi, J.P. Mattia, C.L. Chen, S.J. Eglash, and B-Y. Tsaur, to be published, Mat. Res. Soc. Symp. Proc. 1988.

14. M. Razeghi, to be published, Mat. Res. Soc. Sym. Proc. Soc., 1988.

15. A.R. Goodwin, P.A. Kirkby, I.G.A. Davies, and R.S. Baulcomb, Appl. Phys. Lett. 34, 647 (1979).

16. H.P. Lee, S. Wang, Y-H. Huang, and P. Yu, Appl. Phys. Lett. 52, 215 (1988). B.G. Yacobi, C. Jagannath, S. Zemon, and P. Sheldon, Appl. Phys. Lett. 52, 555 (1988). S. Sakai, S.S. Chang, R.J. Matyi, and H. Shichijo, to be published, Mat. Res. Soc. Symp. Proc. (1988).

17. D.G. Deppe, N. Holonyak, Jr., D.W. Nam, K.C. Hsieh, R.J. Matyi, H. Shichijo, J.E. Epler, and H.F. Chung, Appl. Phys. Lett. 51, 637 (1987).

18. R.M. Fletcher, D.K. Wagner, and J.M. Ballantyne, Mater. Res. Soc. Symp. Proc. 25, 417 (1984). A. Hashimoto, Y. Kawarada, T. Kamijoh, M. Akiyama, N. Watanabe, M. Sakuta, IEDM Tech. Digest, 1985, pp. 658-661.

19. M.I. Aksun, H. Morkoc, L.F. Lester, K.H.G. Duh, P.M. Smith, P.C. Chao, M. Longerbone, and L.P. Erickson, Appl. Phys. Lett. 49, 1654 (1986).

20. R.J. Fischer, T. Henderson, J. Klem, W.T. Masselink, W.F. Kopp, H. Morkoc, and C.W. Litton, Electron. Lett. 20, 945 (1984).

21. R.J. Fischer, W.F. Kopp, J.S. Gedymin, and H. Morkoc, IEEE Trans. Electron Devices ED-33, 1407 (1986).

22. H. Shichijo, J.W. Lee, W.V. McLevige, and A.H. Taddiken, Proceed. 1986 Int. Symp. GaAs and Related Compounds, Inst. Phys. Conf. Ser. 83, 489, 1987.

23. H. Shichijo, L.T. Tran, R.J. Matyi, and J.W. Lee, Mat. Res. Soc. Symp. Proc. 91, 201 (1987).

24. J. Chen, M.S. Unlu, H. Morkoc, and D. Verret, Appl. Phys. Lett. 52, 822 (1988).

25. M.S. Unlu, G. Munns, J. Chen, T. Won, H. Unlu, H. Morkoc, G. Radhakrishan, J. Katz, and D. Verret, Appl. Phys. Lett. 51, 1995 (1987).

26. A. Rosen, M. Caulton, P. Stabile, A.M. Gombar, W.M. Janton, C.P. Wu, J.F. Corboy, and C.W. Magee, RCA Review 42, 633 (1981).

27. J. Buechler, E. Kasper, P. Russer, and K.M. Strohm, IEEE Trans. Elec. Dev. ED-33, 2047 (1986).

28. M. Eron, G. Taylor, R. Menna, S.Y. Narayan, and J. Klatskin, IEEE Electron Device Lett. EDL-8, 350 (1987).

29. H.K. Choi, G.W. Turner, and B-Y. Tsaur, Mat. Res. Soc. Symp. Proc. 91, 213 (1987).

30. H. Shichijo, R.J. Matyi, and A/J/ Taddiken, to be published, IEEE Electron. Device Lett., 1988.

31. H.K. Choi, G.W. Turner, T.H. Windhorn, and B.-Y. Tsaur, IEEE Electron Device Lett. EDL-7, 500 (1986).

Ge, GaAs AND InSb   HETEROEPITAXY ON (100) Si

D.C. HOUGHTON, J.-M. BARIBEAU, T.E. JACKMAN, J. MCCAFFREY, T. SUDERSENA
RAO ,J.B. WEBB, D. PEROVIC[x], G.C. WEATHERLY [x] AND J.P. NOAD[*]

Division of Physics, National Research Council of Canada, Ottawa, Canada, K1A 0R6
[x]Department of Metallurgy and Materials Science, University of Toronto,
Totonto, Canada, M5S 1H4
[*]Communications Research Centre, Ottawa, Canada, K2H 8S2

1. INTRODUCTION

Recently the growth of III-V semiconductors on Si substrates has received considerable attention. In particular heteroepitaxy of systems with larger lattice mismatch than GaAs on Si has been reported (1). The possibility of integrating both high speed and optoelectronic, III-V devices with Si technology is extremely attractive. Furthermore large diameter Si wafers provide low cost robust and virtually defect free substrates. However, many materials problems arise due to the large lattice parameter and thermal expansion coefficient differences and difficulties in the nucleation of polar semiconductors on non-polar surfaces (2).

The introduction of Ge intermediate layers in the growth of GaAs on Si to has been suggested as a possibility to minimize the lattice and thermal mismatches at the III-V/Si heterointerface (2,3). In the first section we discuss the MBE growth techniques necessary for the synthesis of low defect density, relaxed Ge buffer layers on (100) Si (3-6). It is widely recognized that the presently available quality of GaAs and Ge films on Si is not acceptable for many optoelectronic device applications. Defect reduction by 2-3 orders of magnitude is required. With this goal in view, the design of strained layer superlattices (SLS's) as dislocation barriers (2, 7-9) is discussed in the second section. We then briefly describe the use of GaAs and GaAs/Ge coated (100) Si wafers with SLS dislocation barriers as substrates for InSb epitaxial growth by Metallorganic Magnetron Sputtering (MOMS).

2. EXPERIMENT

The Si and Ge epitaxial layers were deposited by MBE on (100) Si substrates in a Vacuum Generators V80/V80H system following procedures described previously (4,6). Briefly, the Si wafers were treated in a UV ozone reactor for $\approx$ 45 minutes to minimize surface hydrocarbon contamination prior to loading in the UHV chamber. Oxide removal, under a Si flux of ≈ 0.01 nm s^{-1} at $\approx 900°C$, was achieved in 15 min, followed by cooling to a growth temperature in the range 450-600°C. Typically Si and Ge fluxes provided a growth rate of 0.5 nm s^{-1}, which was controlled using Sentinel III optical sensors.

GaAs was deposited on Si and Ge epilayers 75 mm Si diameter (100) wafers using Indium free holders in a V80H MBE system. The III-V growth chamber was connected by a UHV transfer system to the Si-Ge V80 deposition system described above and transfers were carried out at 200°C, typically within 5 minutes. The substrates were immediately heated to 600°C in the GaAs growth chamber before cooling to 500°C for the deposition of 200 nm of GaAs at 0.2 nm s^{-1}. The substrate temperature was then raised to 580°C for the growth of 1.5-3.0 μm of GaAs at 0.35 nm s^{-1}. In$_{0.15}$Ga$_{0.85}$As/GaAs SLS's dislocation filters (3 periods of 4/15 nm thickness, respectively) were used in selected samples.

75

Y. I. Nissim and E. Rosencher (eds.), Heterostructures on Silicon: One Step Further with Silicon, 75–83.
© 1989 by Kluwer Academic Publishers.

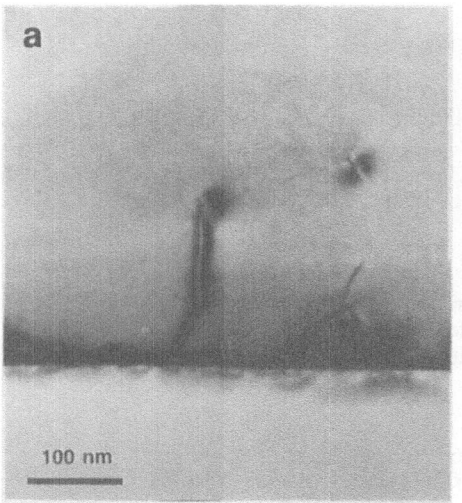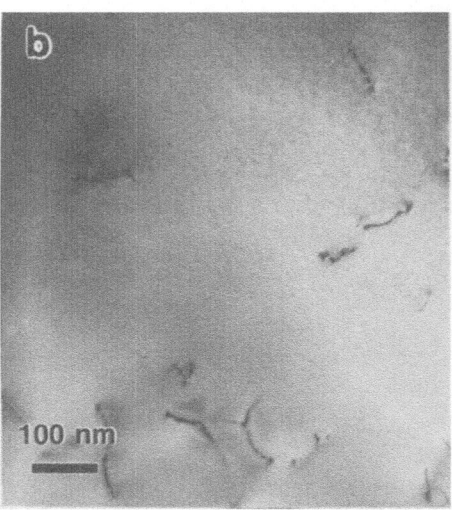

FIGURE 1. a) Cross-sectional and b) plan view TEM micrographs of a Ge film on (100) Si.

InSb was deposited on GaAs/Si and GaAs/Ge/Si substrates using the technique of Metalorganic Magnetron Sputtering (MOMS) which is described elsewhere (1). In this technique an antimony target is reactively D.C. sputtered using trimethyl indium as the reactive gas at a pressure of 3×10^{-3} T and substrate temperature of 425°C. InSb epilayers 2 to 3 μm thick were deposited at a growth rate of typically 4 A s^{-1}.

The multilayer structures were examined in cross-section and plan-view TEM in a Philips EM430 operated at 300 kV. Double crystal and conventional X-ray diffraction were used to assess crystalline quality, measure the composition and dimensions of SLS's and determine residual strains. RBS/ion channeling using 2.0 MeV ^{4}He$^+$ was also used on selected epitaxial layers to determine defect density as a function of depth.

3. RESULTS AND DISCUSSION
3.1. Ge and Ge$_x$Si$_{1-x}$ on (100) Si

The MBE conditions for the epitaxial growth of high quality Ge and Ge-Si alloys on (100) Si have been recently reported by Baribeau et al. (4-6) and by Sheldon et al. (3). We have investigated techniques such as composition grading by either varying the Ge fraction x smoothly from 0 to 1 or by step grading in 20% increments (4-6). These methods generally produced poor epitaxial layers due to the excessive dislocation generation in the interval $0.3 < x < 0.70$. The highest quality Ge on (100) Si and surprisingly the most reproducible method was found to be the deposition of pure Ge directly on Si in the temperature range 450 °C$<$T$<$600°C. The TEM micrographs of Figure 1 show typical defect distribution routinely observed in pure Ge epilayers on (100) Si. The threading dislocation density at the upper Ge surface is 10^7 cm^{-2} and the surface morphology was specular under Nomarski illumination. Defect density could be further reduced by post-growth annealing treatments. Considerable dislocation interaction takes place upon annealing, which results in a cell like network (6). Dislocation densities in Ge/Si heterostructures as low as 5×10^6 cm^{-2} have been observed. The mechanism for defect reduction can be understood in terms of the difference in thermal expansion coefficients (4).

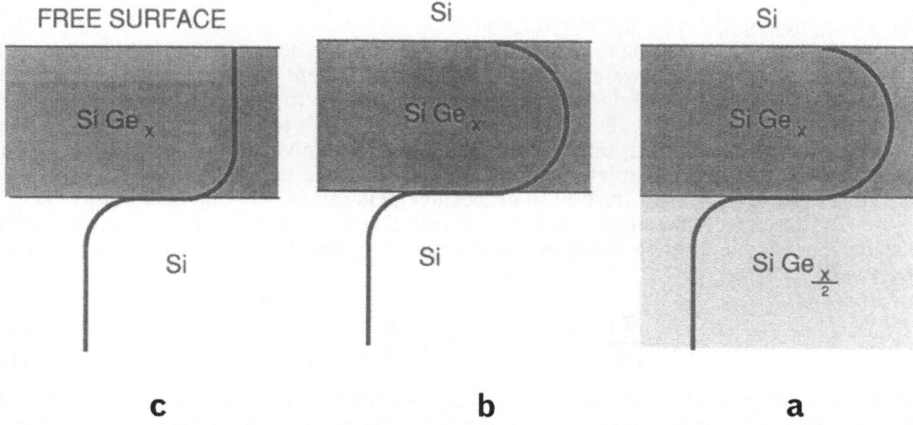

FIGURE 2. Schematic diagrams for the force balance model applied a capped strained layer on a relaxed buffer (a), a capped strained layer (b), and an uncapped strainedlayer(c).

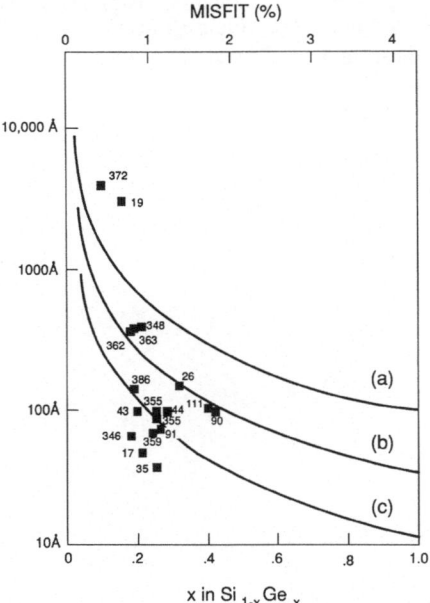

FIGURE 3. Plot of strain and strained layer thicknesses for the superlattices and epilayers studied in this work . The solid curves represent the three critical strain configurations drawn in Figure 2 given by Eq. (1).

3.2. The design of SLS's for defect reduction

The use of misfit strain to reduce threading dislocation densities in epitaxial films was first considered by Matthews et al. (10-13). Several recent experimental studies have evaluated strained-layer superlattices as dislocation barriers to reduce the defect density in mismatched epitaxial layers (8-10). The efficiency of SLS's to block dislocations clearly depends on their dimensions, intrinsic misfit strain and number of mismatched interfaces. Other growth and material related parameters are substrate temperature, initial substrate threading dislocation density, the ease of dislocation glide and the stacking fault energy (12).

The force on a threading dislocation penetrating a strained interface is a function of both the intrinsic misfit strain, f, and the thickness of the strained layer (11). The critical strain is given by

$$\varepsilon_c = \frac{b(1 - v \cos^2 \alpha)}{4\pi h_c (1 + v) \cos \lambda} \left\{ \ln\left(\frac{h_c}{b}\right) + 1 \right\}$$

(1)

where h_c is the critical thickness, b is the Burgers vector of the threading dislocation, v is Poisson's ratio, α is the angle between b and the dislocation line and λ is the angle between b and the direction in the interface perpendicular to the intersection of the slip plane and the interface. For typical 60° misfit dislocations of the type a/2<011>, $\cos \lambda = \cos \alpha = 1/2$, $b \cong$ 4Å and in Ge/Si, v is $\cong$ 0.3. Equation (1) describes the force balance criterion for the mechanical stability of a strained layer entrained in unstrained epilayers and is shown schematically in Figure 2 (b). This value of critical strain is exactly one half of that predicted by Matthews and Blakeslee (10) who assumed that the misfit strain was partitioned between the layers in a superlattice i.e. $\varepsilon_c = f/2$. This condition is met by a SLS on a buffer layer with a misfit half of that in the SLS, Figure 2 (c). Figure 2 (a) shows the critical strain configuration for an uncapped strained layer, ie. $\varepsilon_c = 2f$ in eq. (1), (11).

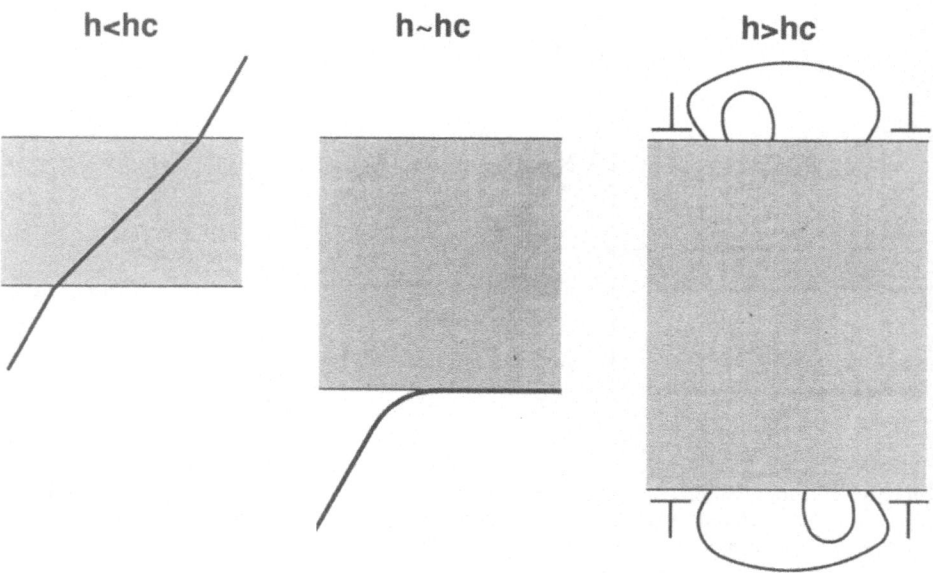

FIGURE 4. Schematic illustration of a threading dislocation interacting with the strain field of a strained layer as it exceeds its critical thickness.

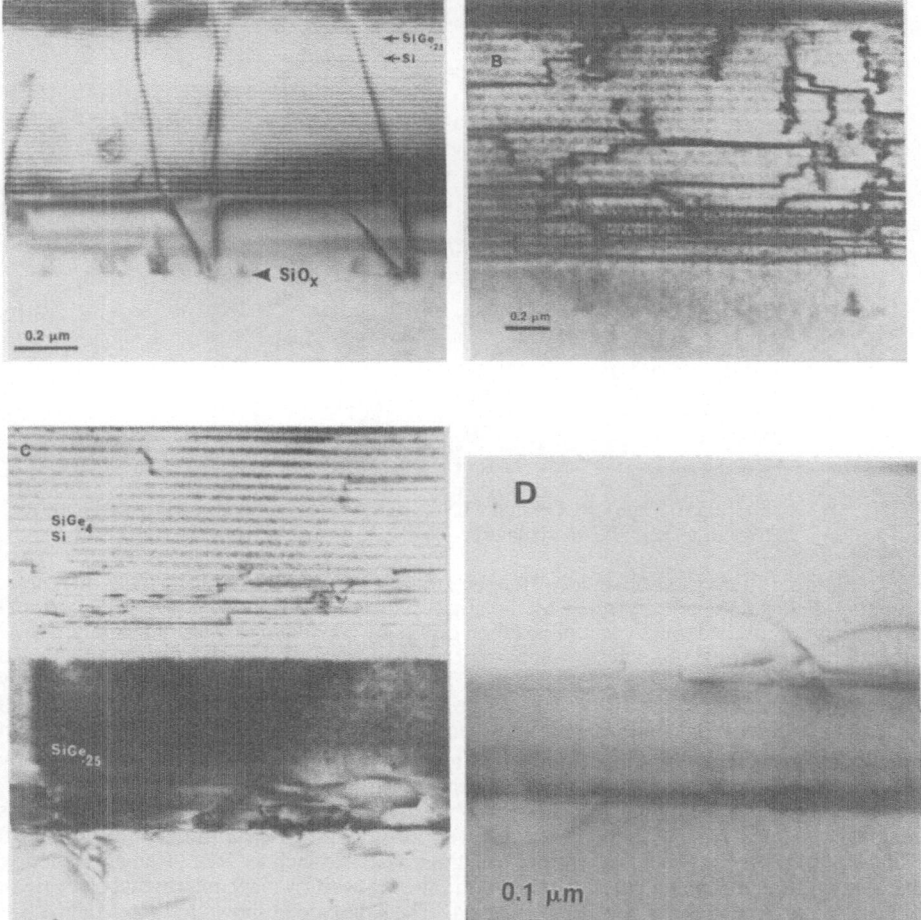

FIGURE 5. Cross sectional TEM micrographs of threading dislocations with SiGe SLS. a) a
SLS (sample 35), h<h$_c$, b) a SLS (sample 26) h=h$_c$, c) a SLS grown on a
relaxed buffer (sample 111), h=h$_c$, (d) a single epilayer (sample 19), h>h$_c$.

For a SiGe/Si SLS the critical thickness is defined by Figure 2 (b) and by using ε_c = f in eq.
(1). This curve is plotted in Figure 3 (b) and represents the critical strain to elongate a
threading dislocation along a pair of strained interfaces in a SLS and is the upper bound to
mechanical stability of the SLS.

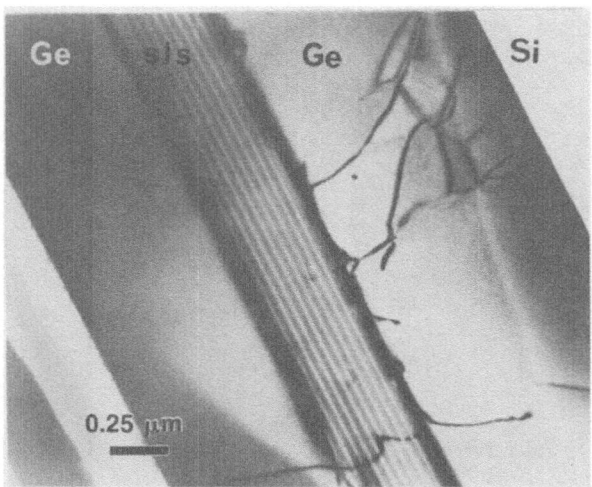

FIGURE 6. XTEM micrograph of a Ge rich SLS introduced to reduce threading dislocation density in a Ge/Si heterostructure.

Figure 4 illustrates schematically three strain regimes. Below the critical thickness, threading dislocations are deflected by the strain fields at heterointerfaces. As the critical thickness is approached, misfit accommodation occurs by bending into the strained interface. Exceeding the critical thickness causes the nucleation of dislocation loops and misfit dislocations at the interfaces. Also shown in Figure 4 are the strain states of several new SLS's designed to investigate dislocation reduction by misfit strain. To avoid problems due to nucleation of misfit dislocations found in many previous studies of relaxation (14-23) an intentional high threading dislocation density (10^7-10^8 cm^{-2}) was introduced, either by incomplete removal of surface oxide or by using a partially relaxed $Si_{1-x}Ge_x$ buffer layer. Observations of threading/misfit dislocations in the SLS's was by imaging both in cross sectional and plan-view TEM. Direct observation by TEM of strain relaxation by misfit dislocations avoids the problems of other techniques (20,22) such as RBS, Raman scattering and X-ray diffraction which monitor strain with lower sensitivity. The micrographs of Figure 5 illustrate the three regimes defined in Figure 4. The superlattice shown in bright-field TEM in Figure 5(a) is below the misfit strain-thickness required by eq. (1) to significantly deflect the threading dislocations which propagate vertically through the SLS from the substrate-epilayer interface. The SLS's in Figure 5(b) and (c) however are close to their critical thicknesses as indicated by eq. (1) and clearly show threading dislocations bent through 90° to lie along the $Si_{1-x}Ge_x$ interface as misfit dislocations. The $Si_{0.85}Ge_{0.15}$ epilayer in Figure 5(d) is well above the critical thickness curve (12) and has partially relaxed producing interfacial misfit dislocations. SLS's with thicknesses above the relevant curve in Figure 3 may indeed increase the threading dislocation density.

The micrograph in Figure 6 is a Ge rich SLS inserted in a Ge epilayer on (100) Si. Here the misfit strain accommodated by the $Ge_{0.8}Si_{0.2}$ alloy layers is tensile not compressive as in the alloy layers in Figure 5 (a-e). A substantial fraction of the threading dislocations are blocked by this SLS which is not optimized as it is considerably below the critical thickness curve of Figure 4 (b). The geometry and both magnitude and sign of intrinsic strain in a SLS are evidently important considerations in the design of effective dislocations filters. The experimental data suggest that misfit-thickness combination close to that given by the curve in Figure 4 (b) is necessary for effective reduction of threading dislocation densities.

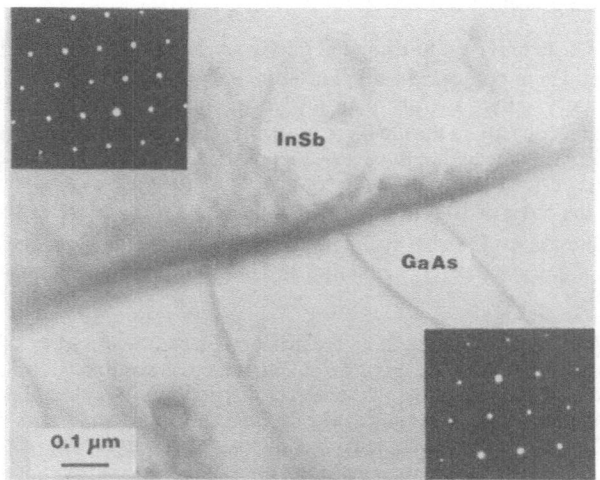

FIGURE 7. XTEM micrograph revealing the defects adjacent to the InSb/GaAs interface in a
InSb/GaAs/Si heterostructure. (Selected area diffraction patterns shown in inset.)

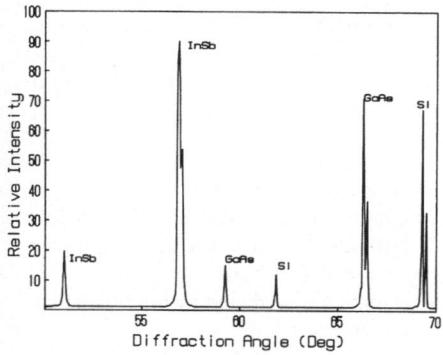

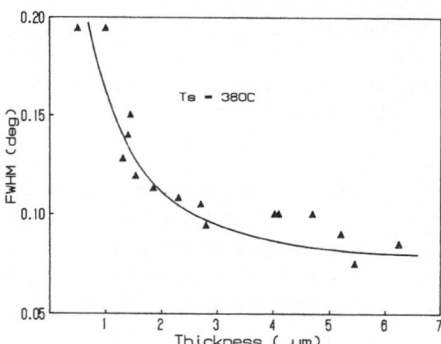

FIGURE 8. Single crystal X-ray diffraction scan
of the heterostructure in Figure 6.

FIGURE 9. Reduction in InSb K_β X-ray
linewidth with increasing
thickness for InSb on GaAs.

3.3. III-V epitaxy on (100) Si and (100) Ge/Si

The growth of GaAs on (100) Si using Ge buffer layers has been reported by several authors (2,3). GaAs grown on optimized Ge/Si buffer layers (defect density $\sim 10^7$ cm^{-2}) were consistently of higher crystalline quality than grown directly on (100) Si under similar conditions (2). The defect density in the Ge layer is $\sim 5 \times 10^7$ cm^{-2} and further extended defects originate at the Ge/GaAs interface. However InGaAs/GaAs SLS's were used in these structures and are clearly effective as barriers to some of these defects. The uppermost GaAs has a defect density in the order of 10^7 cm^{-2} and RBS/channeling measurements showed bulk-like χ_{min} and dechanelling rate. Thermal cycling of similar heteroepitaxial layers in the range 700-950° C lead to a significant defect reduction in GaAs epilayer and improved quality determined by P.L. (2).

Optimized GaAs/Si and GaAs/Ge/Si heterostructures were then used as substrates for the deposition of InSb using MOMS (1). Figure 7 displays a cross sectional TEM micrograph of an InSb/GaAs/Si heterostructure showing the defect distribution in both epilayers and the unresolved high dislocation density region at the InSb/GaAs interface. This micrograph shows that despite a 15% lattice mismatch, fairly good quality InSb films can be produced on GaAs coated (100) Si. The dislocation density in the GaAs layer is approximately 10^8 cm^{-2} and about 6×10^8 cm^{-2} at the surface of a 2μm thick InSb layer. The tangle of dislocation debris (with density as high as 10^{11} cm^{-2}) at the InSb/GaAs interface extends about 0.25 μm in the InSb films. The steady state defect density of 10^8-10^9 cm^{-2} a few microns from the interface is surprisingly low for this large mismatch.

Single crystal X-ray diffraction was also used to characterize these multilayers. Figure 8 shows the X-ray spectrum of the sample shown in Figure 7. The full width at half maximum (FWHM) of the Cu Kβ peaks increase in successive layer (0.065°, 0.101° and 0.125° in Si, GaAs and InSb respectively) indicating an increase in mosaic spread due to higher defect densities (4). Figure 9 shows the decrease in the FWHM in InSb on GaAs with thickness indicating a significant improvement at thickness above 2 μm. RBS/channeling measurements on both the InSb and GaAs and Ge intermediate layers were carried out and compared to bulk single crystals for reference. These results confirmed the TEM and XRD data indicating a steady state defect density after about 2 μm rising to a maximum at the GaAs/InSb interface.

4. SUMMARY

The main results reported in this work can be summarized as follows.

1- The direct deposition of Ge on (100) Si in the temperature range 450-600°C produces smooth epitaxial films with a defect density 10^7-10^8 cm^{-2}. Poast growth annealing at ~700°C can be further reduce the defect density to about 5×10^6 cm^{-2}.

2- The critical thickness for Si$_{1-x}$Ge$_x$ strained layers in Si are predicted by the Matthews and Blakeslee (11) force balance model using $\varepsilon_c = f$.

3- SLS's close to the critical thickness are effective in reducing threading dislocation densities.

4- InSb epitaxy on GaAs/Si and GaAs/Ge/Si substrates gives comparable crystalline to GaAs on Si.

5. REFERENCES

1. T. Sudersena Rao, J.B. Webb, D.C. Houghton, J.M. Baribeau, W.T. Moore and J.P. Noad, Appl. Phys. Lett. *in press*, and J.B. Webb and C. Halpin Appl. Phys. Lett. **47** (1985) 831.
2. J.M. Baribeau, D.C. Houghton, P. Maigné, W.T. Moore, R.L.S. Devine, M.W.

Denhoff, R.J. Stoner, J. McCaffrey and T.E. Jackman. in MRS Symposium Heteroepitaxy on Si
II, edited by J.C.C. Fan, J.M. Phillips and B.-Y. Tsaur (MRS Press 1987, Vol 91) p.175.

3. P. Sheldon, B.G. Yacobi, S.E. Asher, K.M. Jones, M.J. Hafich and G.Y. Robinson. J. Vac. Sci. Technol. A4 (1986) 889.
4. J.M. Baribeau, T.E. Jackman, D.C. Houghton, P. Maigné and M.W. Denhoff, J. Appl. Phys. in press.
5. J.M. Baribeau, D.C. Houghton, T.E. Jackman and J. McCaffrey. J. Electrochem. Soc. in press.
6. J.M. Baribeau, T.E. Jackman, P. Maigné, D.C. Houghton and M.W. Denhoff. J. Vac. Sci. Technol. A5 (1987) 1898.
7. P.L. Gourley, T.J. Drummond and B.L. Doyle, Appl. Phys. Lett. 49 (1986) 1101.
8. Y.G. Chai, R. Chow, J. Appl Phys. 53 (1982) 1229.
9. N. Elmasry, J.C.L. Tam, T.P. Humphreys, N. Hamaguchi, N.H. Karam and S.M. Bedair, Appl. Phys. Lett. 51 (1987) 1608.
10. J.W. Matthews and A.E. Blakeslee, Journal of Cryst Growt h 27 (1974) 118.
11. J.W. Matthews and A.E. Blakeslee and S. Mader, Thin Solid Films 33 (1976) 253.
12. J.W. Matthews, Journal of Vac. Sci. and Technol. 12 (1975) 126.
13. W.A.H. Jesser and J.W. Matthews, Philos. Mag. 15 (1967) 1097.
14. E. Kasper, H.J. Herzog and H. Kibble, Appl. Phys. 8 (1975) 199.
15. H.-J. Herzog, H. Jorke, E. Kasper and S. Mantl, J. Electrochem. Soc. (to be published).
16. R.H. Miles, T.C. McGill, P.P. Show, D.C. Johnson, R.J. Hauenstein, S.W. Nieh and M.D. Strathman, Appl. Phys. Lett. 52 (1988) 916.
17. R. People and J.C. Bean, Appl. Phys. Lett. 47 (1985) 322.
18. A.T. Fiory, J.C. Bean, R. Hulland, S. Nakahara, Phys. Rev. B. 31 (1985) 4063.
19. I.J. Fritz, S.I. Picraux, L.R. Dawson, T.J. Drummond, W.P. Laidig and N.G. Anderson, Appl. Phys. Lett. 46 (1986) 967.
20. B.W. Dodson and J.Y. Tsao, Appl. Phys. Lett 51 (1987) 1325.
21. P.L. Gourley, I.J. Fritz and L.R. Dawson, Appl. Phys. Lett. 52 (1988) 377.
22. I.J. Fritz, Appl. Phys. Lett. 51 (1987) 1080.
23. P.M.J. Marée, J.C. Barbour, J.F. Van der Veen, K.L. Kavanagh, S.W.T. Bulle-Lieuwma and M.P.A. Viegers, J. Appl. Phys. 62 (1987) 4413.

HETEROEPITAXY OF CdTe ON GaAs-ON-Si

G. RADHAKRISHNAN, A. NOUHI, and J. KATZ

Jet Propulsion Labortory, California Institute of Technology
4800 Oak Grove Dr., Pasadena, CA 91109

1. INTRODUCTION

CdTe is a semiconductor with numerous applications, the most prominent of which is its use as a buffer layer for the growth of epitaxial films of HdCdTe for infrared detector arrays. However, bulk CdTe is extremely difficult to grow and the non-availability of high quality, large area, inexpensive, single crystal CdTe has slowed the development of HgCdTe detectors. Infrared (IR) device processing requires large areas of single crystal material, and the problems associated with bulk CdTe are now being circumvented by utilizing alternative substrates on which CdTe can first be grown epitaxially. This in turn serves as a buffer layer for the growth of HgCdTe.

InSb[1-3], GaAs[4-13], Si[14, 15], and sapphire[16-18] have all in fact been employed as substrates and are all available in large area and high crystalline quality. InSb has the smallest lattice mismatch to CdTe (0.05%), while the lattice mismatches of GaAs and Si with CdTe are quite high (14% and 19% respectively). However, for applications that might involve back illumination of the substrate, InSb is unsuitable due to its low band gap of 0.22 eV which causes it to absorb IR radiation at wavelengths shorter than 5 μm. Sapphire, undoped GaAs, and Si, offer the benefit of being transparent over a wide range in the infrared, starting at their individual bandgaps and extending upto 6 μm, 18 μm and 25 μm respectively. This makes them suitable substrates for both short and long wavelength IR detectors. In addition, a comparison of the thermal expansion coefficients of CdTe, GaAs, Si, and sapphire indicates the close match between the thermal expansions of CdTe and GaAs, which is an added advantage in favor of GaAs. The recent interest in integrating GaAs and Si technologies leads to yet another possible substrate for growing CdTe, namely the GaAs/Si substrate. The HgCdTe/CdTe/GaAs/Si heteroepitaxy offers the possibility of silicon compatible infrared focal plane arrays. Additionally, fast GaAs optoelectronic circuitry can be fabricated on the same wafer.

Until now there has been only one report on the growth of CdTe on the GaAs/Si substrate. Bean et al.[19] grew epitaxial films of CdTe and HgCdTe on (100) GaAs/Si using a combination of three techniques, namely, molecular beam epitaxy (MBE) for the growth of GaAs on Si, and congruent evaporation in ultrahigh vacuum and vapor transport respectively for the subsequent growth of CdTe and HgCdTe.

Here we report the growth of CdTe on both (100) and (111) GaAs/Si. The growth of CdTe on (111) GaAs/Si is the first such attempt to be reported. In addition, the growth technique is novel in that it involves a combination of MBE and metalorganic chemical vapor deposition (MOCVD)[20], thus offering the combined advantages of these two well established growth methods. MOCVD is employed to grow CdTe layers, thus offering the advantages of very high throuput and low cost for the production of large area substrates for IR detectors. In addition, the initial growth of GaAs on Si is achieved by MBE which offers the ability

Y. I. Nissim and E. Rosencher (eds.), Heterostructures on Silicon: One Step Further with Silicon, 85–91.
© *1989 by Kluwer Academic Publishers.*

of highly controlled nucleation, and is therefore recognized as one of the most successful methods of producing epitaxial films of GaAs on Si.

2. PROCEDURE

2.1. Growth method

The first step in this heteroepitaxy is the growth of a 2 μm thick layer of GaAs on n-type, 2 inch diameter Si(100) and Si(111) substrates using MBE. The growth was conducted in a Riber-2300 MBE system. The Si (100) substrates were tilted 4° off-axis toward the (011) direction and had a resistivity of < 0.1 Ω.cm, while the Si(111) substrates were directly on-axis and had a higher resistivity of 7 Ω.cm. Briefly, an initial degreasing of the Si wafers (trichloroethylene, acetone, and methanol) was followed by a series of oxidation and oxide removal steps. Oxide coated Si wafers were loaded into a N_2 load chamber. The wafers were then spin-etched with dilute hydrofluoric acid-ethanol to remove the oxide and immediately transferred into the growth chamber. The growth procedure involved a two-step process, slow growth at 0.1 monolayer/s, at about 300°C, followed by a faster growth at the rate of 1 μm/h, at about 500°C. Alternative growth schemes have also been successfully attempted for the MBE growth of GaAs on Si. In brief, these methods involve temperature cycling during growth, in-situ annealing, and an initial growth of a 300 Å buffer layer with an alternate shuttering of the Ga and As sources. In all of the above growth methods that were employed, the layers of GaAs were always doped with Si, to doping levels of 1×10^{18} cm^3.

CdTe epilayers were grown by the MOCVD technique on MBE grown (100) and (111) GaAs/Si substrates. The metal-organic sources for Cd and Te were dimethyl cadmium and diethyl tellurium respectively. The CdTe layers were grown at atmospheric pressure in an MOCVD system having a horizontal quartz reactor with an RF heated graphite susceptor. These growth runs were conducted at 400°C and under growth conditions that were optimized for single crystal GaAs substrates. Further improvement in material quality is expected as the process parameters are optimized for GaAs/Si substrates. The overall flow rate of the hydrogen carrier gas was 3 1/min and the partial pressures of both the Cd and Te sources were 10 torr. The (100) GaAs/Si and (111) GaAs/Si substrates were degreased (trichloroethylene, acetone, and methanol) and rinsed with deionized water prior to loading into the MOCVD reactor for growth.

2.2. Material Characterization

Figures 1 a) and 1 b) show the results of scanning electron microscopy (SEM), and x-ray diffraction scans of typical films of CdTe grown on (100) GaAs/(100) Si substrates. X-ray diffraction measurements indicate that both the (100) and (111) orientations of CdTe are observed on the (100) GaAs/(100) Si substrates. In addition, the CdTe epilayers are observed to grow exclusively in the (100) or (111) orientations. Each individual film is epitaxial and single crystalline as determined by rocking curve data. In view of the fact that these growths were performed under identical substrate preparation and growth conditions, the exact reasons for the growth of CdTe in both the (100) and (111) orientations are not clear at this time and this subject is under further investigation. Similar observations have been made during the growth of CdTe on (100) GaAs substrates where both the (100)[5,6] and (111)[7,8] orientations of the CdTe have been observed. However, in this case the orientation of the CdTe is believed to be related to the presence or absence of an oxide layer on the GaAs substrate[11]. The SEM micrographs indicate a smooth and featureless surface in the case of the (111) CdTe on (100) GaAs/Si, however the surface morphology of (100) CdTe on (100) GaAs/Si features small mounds.

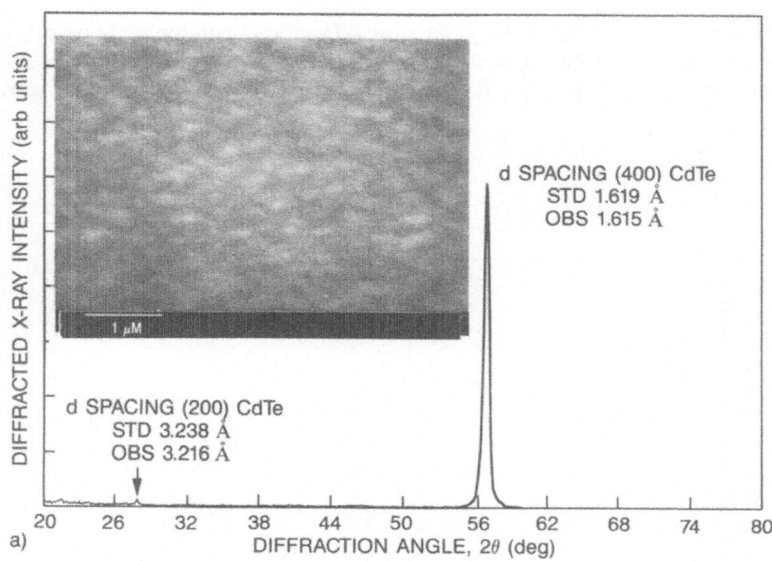

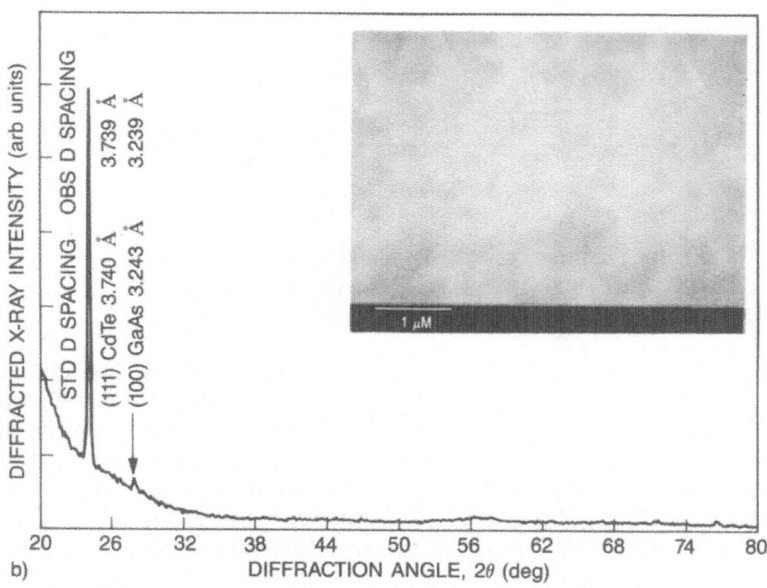

Figure 1. Surface morphologies and x-ray diffraction scans of
a) (100) CdTe and b) (111) CdTe grown on a (100) GaAs/(100) Si substrate.

Figure 2 shows the surface morphology and x-ray diffraction scan, representative of CdTe grown on a (111) GaAs/ (111) Si substrate. The surface morphology looks featureless. The x-ray diffraction measurements indicate the orientation of CdTe on the (111) GaAs/Si substrate is (111).

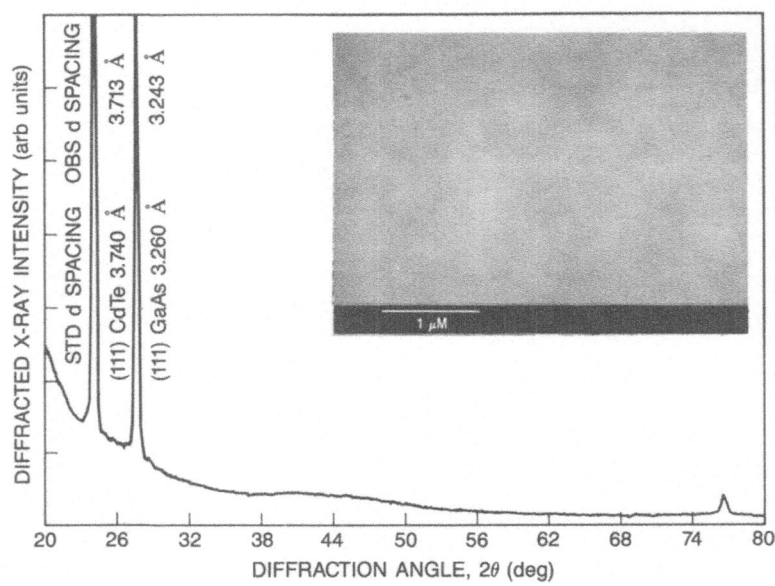

Figure 2. Surface morphology and x-ray diffraction of (111) CdTe grown on a (111) GaAs/(111) Si substrate.

Double crystal x-ray rocking curve measurements were performed for CdTe on both the (100) and (111) GaAs/Si substrates. The Fe $K_{\alpha 1}$ line was used, with its (400) reflection for the (100) CdTe and the (333) reflection for the (111) CdTe samples. Incident beam dimensions were 1 x 0.5 mm^2 and the intrinsic rocking curve full-width at half maximum (FWHM) linewidth of the diffractometer was 12 arc seconds. The measurements indicate that the films of CdTe on both the (100) and (111) GaAs/Si substrates are single crystalline. However, the FWHM of peaks of both (100) and (111) CdTe are relatively large, nearly 1 degree, which is indicative of a large amount of strain still present in these films. Much smaller FWHMs have been observed for CdTe on (100) GaAs/Si, 150 arc seconds for a 3.5 μm layer[19], and for CdTe on (100) and (111) GaAs, 100 arc seconds for a 5 μm thick (100) CdTe on (100) GaAs[9], and 78 arc seconds for a 10 μm thick (111) CdTe on (111) GaAs[10]. This leads us to believe that a combination of growth parameters, and film thicknesses of less than 3 μm may be responsible for the strain still present in our CdTe films.

Shown in figure 3 are the photoluminescence (PL) spectra of both the (100) and (111) orientations of CdTe on (100) GaAs/(100) Si.

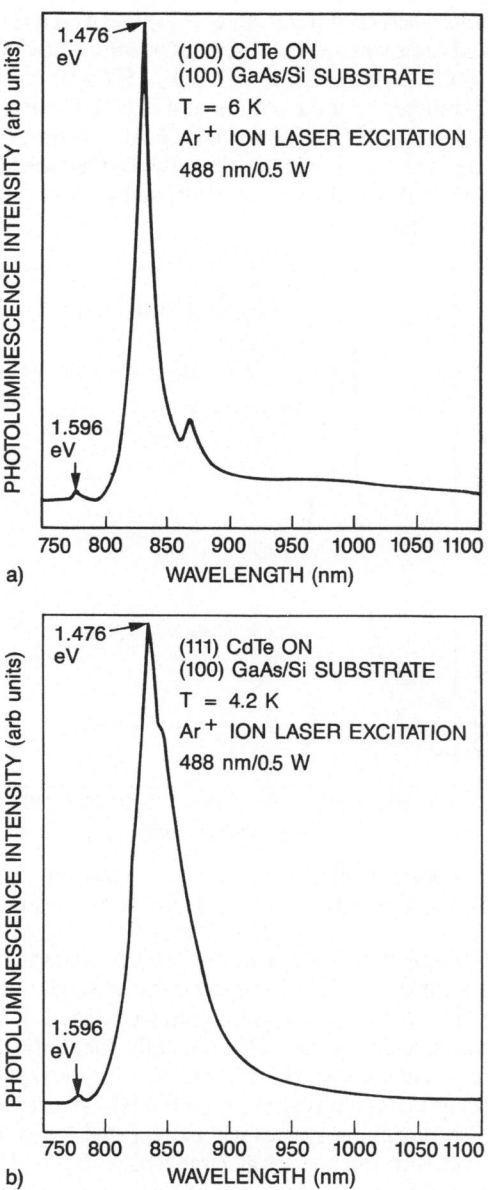

Figure 3. Photoluminescence spectra of CdTe, a) (100) CdTe and b) (111) CdTe grown on a (100) GaAs/(100) Si substrate.

Figure 4 shows the photoluminescence spectrum of (111) CdTe on a (111) GaAs/Si substrate. The excitation source in all cases was an Ar[+] ion laser operating at 488 nm. All three spectra are characterized by a low intensity peak at 777 nm (1.596 eV) and a broader peak with a much higher intensity centered around 840 nm (1.476 eV). Comparing these PL spectra to the PL spectra of (100) and (111) CdTe on (100) GaAs substrates[12,13], the small peak at 777 nm can be attributed to an excitonic transition, while the broad peak centered around 840 nm is a defect related peak which may be associated with recombination through an impurity-cadmium vacancy complex.

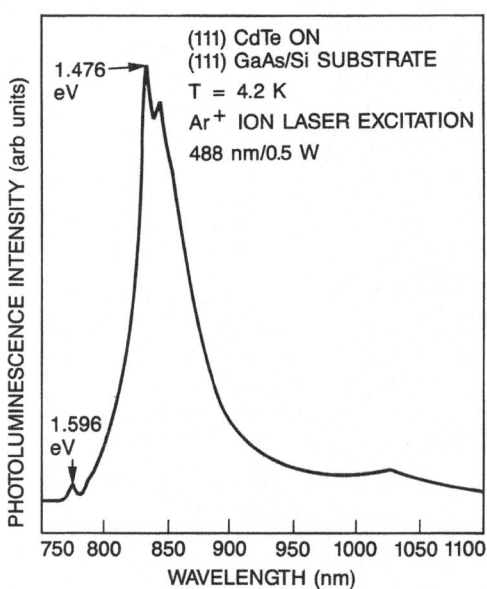

Figure 4. Photoluminescence spectrum
of (111) CdTe grown on a (111) GaAs/Si substrate.

We have investigated the effect of CdTe layer thickness on material quality by measuring the PL spectra for four samples of CdTe, all grown under similar conditions, on a (100) GaAs/(100) Si substrate. Thicknesses of these samples were 0.5, 1.0, 1.5, and 2.0 μm. The PL spectra show no measurable differences. This is mainly due to the fact that the intensity of the excitonic peak observed in our spectra is rather low, which makes it difficult to measure any change that might occur in its energy position or FWHM as a result of strain relaxation with increasing film thickness. Similar studies will be repeated for even thicker CdTe films of improved quality, in fact with thicknesses as large as 3.5-10 μm. This is the view of the fact that the narrowest FWHMs on rocking curves of CdTe on GaAs and GaAs/Si substrates have in fact been reported for films with such thicknesses[9,10,19], indicative of a release of strain at these thicknesses. We have also performed control runs by growing CdTe on (100) bulk GaAs substrates. In this case the orientation of CdTe was always found to be (100). The PL spectra show the same features as CdTe on GaAs/Si substrates, and preliminary results of a two-hour proximity anneal at 450°C of a 1.5 μm thick CdTe sample grown on a (100) bulk GaAs substrate indicate a slight enhancement in the intensity of only the defect related peak, thus supporting the identification of the defect peak with a cadmium vacancy.

We are currently pursuing postgrowth annealing studies, as well as a study of the effects of various growth parameters such as Cd:Te source ratios in order to improve the material quality and increase the ratios of excitonic to defect PL intensities.

In conclusion, we have successfully demonstrated the combined implementation of MBE and MOCVD for growing CdTe on GaAs/Si substrates. This technique offers the potential for monolithically integrating infrared detector technology with GaAs and Si technologies. This is the first report of the growth of CdTe on a (111)GaAs/(111)Si substrate. Further work is in progress in our laboratory on varying the growth conditions and attempting postgrowth annealing studies.

The research described in this paper was carried out by the Jet Propulsion Laboratory (JPL), California Institute of Technology, and was sponsored by the Strategic Defense Initiative Organization, Innovative Science and Technology Office, and the National Aeronautics and Space Administration. The work was performed as part of JPL's Center for Space Microelectronics Technology.

The authors wish to thank Dr. N.M. Haegel of the University of California, Los Angeles, for the photoluminescence measurements.

REFERENCES

1. R.F.C. Farrow, G.R. Jones, G.M. Williams, and I.M. Young, Appl. Phys. Lett. 39, 954 (1981).
2. W.E. Hoke, P.J. Lemonias, and R. Traczewski, Appl. Phys. Lett, 44, 1046 (1984).
3. C.H. Wang, K.Y. Cheng, S.J. Yang, and F.C. Hwang, J. Appl. Phys. 58, 757 (1985).
4. I.B. Bhat, N.R. Taskar, and S.K. Ghandhi, J. Vac. Sci. Technol. A4, 2230 (1986).
5. R.N. Bicknell, R.W. Yanka, N.C. Giles, J.F. Schetzina, T.J. Magee, C. Leung, and H. Kawayoshi, Appl. Phys. Lett. 44, 313 (1984).
6. K. Nishitani, R. Okhata, and T. Murotani, J. Electron. Mater. 12, 619 (1983).
7. H.A. Mar, K.T. Chee, and N. Salansky, Appl. Phys. Lett. 44, 237 (1984).
8. J.T. Cheung, M. Khoshnevisan, and T. Magee, Appl. Phys. Lett. 43, 462 (1983).
9. J.M. Ballingall, W.J. Takei, and B.J. Feldman, Appl. Phys. Lett. 47, 599 (1985).
10. R.N. Bicknell, K.A. Harris, J.W. Cook Jr., J.F. Schetzina, and W.S. Takei, Bull. Am. Phys. Soc. 30, 210 (1985).
11. J.M. Ballingall, M.L. Wroge, and D.J. Leopold, Appl. Phys. Lett. 48, 1273 (1986).
12. D.J. Leopold, J.M. Ballingall, and M.L. Wroge, Appl. Phys. Lett. 49, 1473 (1986).
13. R. Srinivasa, M.B. Panish, and H. Temkin, Appl. Phys. Lett. 50, 1441 (1987).
14. H. Zogg and S. Blunier, Appl. Phys. Lett. 49, 1531 (1986).
15. R.L. Chou, M.S. Lin, and K.S. Chou, Appl. Phys. Lett. 48, 523 (1986).
16. W.E. Hoke, R. Traczewski, V.G. Kreismanis, R. Korenstein, and P.J. Lemonias, Appl. Phys. Lett. 47, 276 (1985).
17. T.H. Myers, Y. Lo, R.N. Bicknell, and J.F. Schetzina, Appl. Phys. Lett. 42, 247 (1983).
18. H.S. Cole, H.H. Woodbury, and J.F. Schetzina, J. Appl. Phys. 55, 3166 (1984).
19. R.C. Bean, K.R. Zanio, K.A. Hay, J.M. Wright, E.J. Saller, R. Fischer, and H. Morkoc, J. Vac Sci. Technol. A4, 2153 (1986).
20. G. Radhakrishnan, A. Nouhi, and J. Liu, "Growth and Characterization of CdTe on GaAs/Si Substrates", SPIE Meeting, January 1988, Los Angeles, California.

HETEROEPITAXIAL GROWTH OF (Al)GaAs ON InP BY MOVPE

A. ACKAERT, P. DEMEESTER, I. MOERMAN and R. BAETS

University of Gent, (L.E.A.-IMEC), St. Pietersnieuwstraat 41,
B-9000 Gent, Belgium

1. INTRODUCTION

There is currently a great interest in the heteroepitaxial growth of GaAs on Si, although other lattice mismatched material combinations are also investigated (e.g. InP on GaAs or Si, GaAs on InP). In this paper, results will be presented on the heteroepitaxial growth of (Al)GaAs on InP. This is a very promising combination for the realisation of high efficiency tandem solar cells, where a AlGaAs cell (1.9 eV) is grown on a InGaAsP cell (1.1 eV) lattice matched to InP. A second important application is the fabrication of optoelectronic integrated circuits, namely the integration of GaAs electronic circuits with InP long wavelength optoelectronic devices.

The first report on the heteroepitaxial growth of GaAs on InP was published in 1984 by Razeghi and group [1]. They showed, by means of X-ray diffraction, that it was possible to grow crystalline GaAs on InP. The real breakthrough of this technology came in 1987 when it was shown by the researchgroup of NEC that high quality optoelectronic integrated circuits could be obtained [2]. They demonstrated a long wavelength optical receiver/transmitter where the optical devices were made in lattice matched InP material and the electronic circuits in GaAs lattice mismatched material.

One of the major problems in the GaAs on InP technology is the large difference in lattice constant which results in a high density of misfit dislocations in the GaAs. Those dislocations seriously degrade the quality of minority carrier devices (e.g. laser diodes). A second problem is the difference in linear thermal expansion coefficient (LTE) between GaAs and InP (factor 1.45). This gives rise to a tensile stress in the GaAs layer grown on the InP substrate and may result in wafer bowing and crack formation. This problem however is much less severe than in the GaAs on Si technology since in that case the LTE coefficients ratio is 2.5. The other problems encountered during GaAs on Si heteroepitaxial growth are absent in the GaAs on InP growth : there is no stable oxide on InP and both semiconductors are polar. Some major properties of Si, GaAs and InP are given in table 1 for comparison.

	Si	GaAs	InP
Lattice parameter a (nm)	0.543	0.565	0.587
Crystal structure	cubic non-polar	zincblende polar	zincblende polar
Lin. therm. exp. α (10^{-6}K)	2.60	6.86	4.75
Density d (g.cm^{-3})	2.328	5.317	4.810

Table 1 : Properties of Si, GaAs and InP.

Y. I. Nissim and E. Rosencher (eds.), Heterostructures on Silicon: One Step Further with Silicon, 93–99.
© 1989 by Kluwer Academic Publishers.

2. GROWTH PROCEDURE

Metal Organic Vapour Phase Epitaxy (MOVPE) is used for the heteroepitaxial growth of GaAs on InP. The commercial system uses a small horizontal research reactor working at atmospheric pressure and with an infra-red heated silicon susceptor. For fast switching of the gases, a vent/run system is used with nearly zero dead-volume. The metalorganic sources are TriMethylGallium (TMG), TriMethylAluminum (TMA) and DiEthylZinc (DEZ, p-doping source). For the hydrides we use arsine (AsH_3) and silane (SiH_4, n-doping source).

Typical parameters which are used for the growth of (Al)GaAs device structures are summarized in table 2. Figure 1 gives the growth procedure we developed for the growth of GaAs on InP. The samples are heated up to 660°C (or 720°C) after a long purge under H_2 at room temperature. When the temperature reaches 420°C, the AsH_3 is switched to the reactor. Growth is started from 450°C on where TMG is also switched to the reactor. After one minute the TMG is switched off again and this results in a thin GaAs nucleation layer (approx. 10 nm). This layer is important to protect the InP substrate against P outdiffusion (there is no PH_3 available in the system). When the temperature reaches 650°C, the growth of a 3 μm GaAs bufferlayer is started. The bufferlayer should be thick enough in order to reduce the dislocation density in the final device structure which is grown on top of this bufferlayer. This growth procedure resulted in good layer quality, as will be shown in the next paragraph.

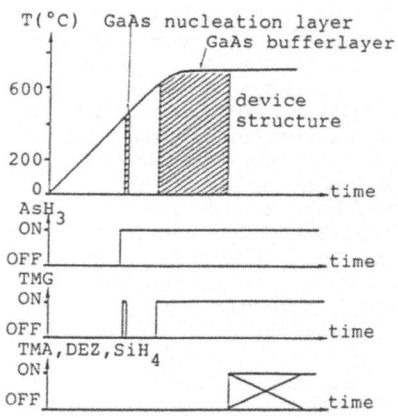

Figure 1 : Procedure for the growth of GaAs on InP.

Reactor pressure	1 atm
Growth temperature	660–720°C
Growth velocity	100 nm/min
Mole fraction	AsH_3 : $2\cdot10^{-3}$
	TMG : $2\cdot10^{-4}$
V/III ratio	10

Table 2 : Growth parameters.

3. LAYER CHARACTERIZATION

An extensive characterization of simple layer structures of GaAs on InP has been done by using : optical microscopy, X-ray diffraction, Transmission Electron Microscopy (TEM), photoluminescence, Hall- and C/V measurements.

A microscopic picture of the morphology of a 3 µm thick GaAs layer on InP is shown in figure 2. Although the morphology is mirrorlike, there is some background roughness which is not observed for GaAs on GaAs but which is less pronounced than for GaAs on Si growth. It should be mentioned here that best morphology not necessary means best electrical and optical quality of the material. When the layers were grown thicker (6µm), no cracks appeared. The morphology became bad when no nucleation layer was used and when growth was started at a high temperature. This is due to P outdiffusion from the substrate. X-rays simple diffraction showed the crystallinity of the material.

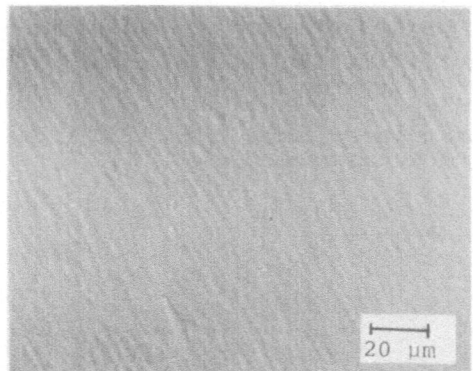

Figure 2 : Morphology of a 3 µm GaAs
layer on InP.

There are different methods to measure the dislocation density in crystalline material, namely : wet chemical etching (KOH, DSL, AB, etc.), cross-sectional TEM and plan view TEM. There is some indication that plan view TEM gives the best quantitative measurement of dislocation density and in some cases there may be a discrepancy of orders of magnitude between the different techniques. Plan view TEM was done at Plessey Research (Caswell,UK) on a thin (<1 µm) GaAs epitaxial layer on InP (figure 3). We observed a high dislocation density : $1.5 \cdot 10^8$ stacking faults and $7 \cdot 10^8$ dislocations per cm^2. This clearly shows that there is a high dislocation density near the heterointerface. Further results will show that it is possible to obtain good quality material when using a thicker bufferlayer.

We have used both Hall and C/V measurements to investigate the electrical quality of the material. The layer structure consisted of a 3 µm undoped GaAs bufferlayer and a 200 nm n-doped GaAs active layer. Hall measurements resulted in a doping concentration of $1.5 \cdot 10^{17}$ cm^3 and a room temperature mobility of 3700 $cm^2/V.s$. These results are in agreement with calibration measurements done on thick GaAs layers on GaAs substrates. The doping profile obtained from C/V measurements is shown in figure 4. The background doping level in the undoped bufferlayer is in the range of low 10^{14} cm^3. Hall measurements of 6 µm thick GaAs undoped layers were impossible due to the depletion of the layer at those low background concentrations.

The photoluminescence spectra of a GaAs on InP and GaAs on GaAs layer are compared in figure 5. We observe a broadening of the peak and a reduction

Figure 3 : Plan view TEM of a 1μm GaAs
layer on InP.

of the intensity for the heteroepitaxial layer. This is due to a reduction
of the optical quality of the material.

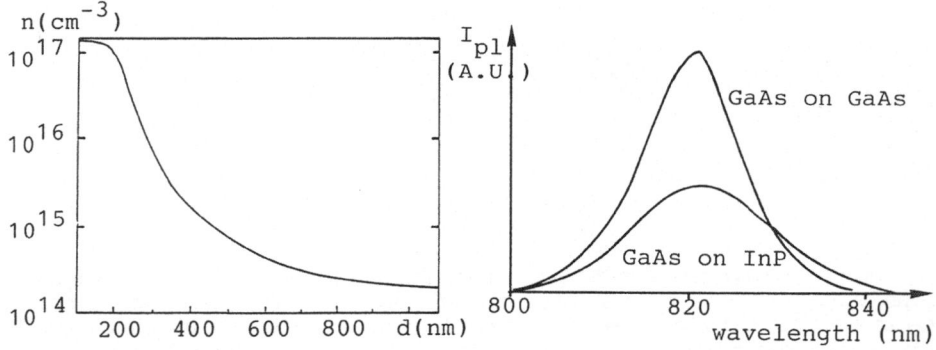

Figure 4 : Doping profile of an
active layer structure.

Figure 5 : Photoluminescence
spectra at 77K.

In this paragraph we have shown that it is possible to obtain GaAs on
InP with reasonable morphology and crystalline quality. The high dislocation
density near the substrate-layer interface does not strongly degrade the
electrical quality of the material but the optical quality is reduced.

4. GaAs MESFETs ON InP SUBSTRATES

The GaAs MESFET technology on GaAs substrates is now well developed and
already transferred to production. This is not true for InP FETs on InP, a
technology which is still in a research environment. One of the main problems
is the formation of a Schottky barrier on InP or on InGaAs material. Because
the intrinsic barrier height between the metal and the semiconductor is too
low, it is impossible to use a reverse biased schottky contact to modulate
the current through the device. To overcome this problem one can introduce
an artificial barrier by using a thin (native) oxide layer. Other possibilities
are to use a p-n junction to obtain a JFET (Junction FET) or growing another

material on top of InP or InGaAs with a higher barrier height (e.g. lattice matched InAlAs, lattice mismatched GaAs, etc.). Here we come to the first application of the GaAs on InP technology, where the schottky contact is formed on a GaAs on InP layer. One step further is to make the whole transistor in GaAs by growing a GaAs active layer (+ bufferlayer) on the InP. In this way it is possible to combine the well-developed GaAs electronic circuit technology with high quality long wavelength optoelectronic devices based on InP technology. It is this aspect of the GaAs on InP technology we will describe now in detail.

The layerstructure of the processed GaAs on InP MESFET consisted of a thin nucleation layer, a 3 μm undoped bufferlayer and on top of it a 200 nm GaAs active layer. Ohmic (AuGe/Ni) and schottky (TiW/Au) contacts are deposited, after mesa etching for device isolation. The use of an InP substrate did not cause any problems during the standard processing of the GaAs MESFETs.

Measurements were done on both GaAs on GaAs and GaAs on InP MESFETs to compare both technologies. An input-output characteristic for a GaAs on InP MESFET is shown in figure 6. We can deduce a transconductance (g_m) of 90 mS/mm and a threshold voltage (V_m) of -3 V for a 1.5 μm gatelength. This compares very well with our results obtained on GaAs. We also measured the high frequency response of those devices. The H_{21}-parameter (output current/input current) versus frequency is shown in figure 7. The cut-off frequency (H_{21} = 1) is equal to 7.1 GHz, and seems to be independent of the substrate material.

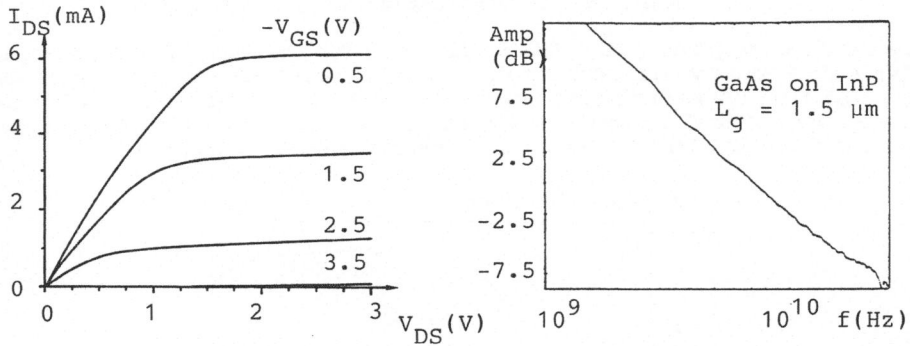

Figure 6 : Input-output curve.   Figure 7 : High frequency response.

5. GaAs SOLAR CELLS ON InP SUBSTRATES

A second important application of the GaAs on InP technology is the fabrication of a GaAs solar cell on an InP substrate. Those solar cells, important from a device point of view, are also a very useful tool in the evaluation of the material quality. If it is possible to grow good quality GaAs solar cells on InP, no major problem is expected to grow AlGaAs cells on InGaAsP cells. It is thus possible to combine materials with a bandgap varying between approximately 1 and 2 eV. This is very important for the fabrication of tandem solar cells with matched bandgaps, which may be used in concentrator systems and for space applications.

A GaAs on InP or GaAs on GaAs heterointerface solar cell consists of a 3 μm n⁺-doped GaAs bufferlayer, a 3 μm n-doped GaAs base layer, a 0.8 μm p-doped GaAs emitter layer, a 80 nm p-doped Al.₈Ga.₂As window layer and a 100 nm p⁺-doped GaAs top layer. AuGe/Ni is used as a broad area bottom contact and Zn/Au as top contact. An anti-reflection coating is used after removal of the absorbing top layer.

The current-voltage characteristics under AM 1.5 and 1 sun illumination show a reduction of short-circuit current and open-circuit voltage for a GaAs on InP cell, compared to a GaAs on GaAs cell (figure 8). This results in a reduction of the efficiency from 16 % to 11.5 %. Note that both cells were grown and processed during the same run.

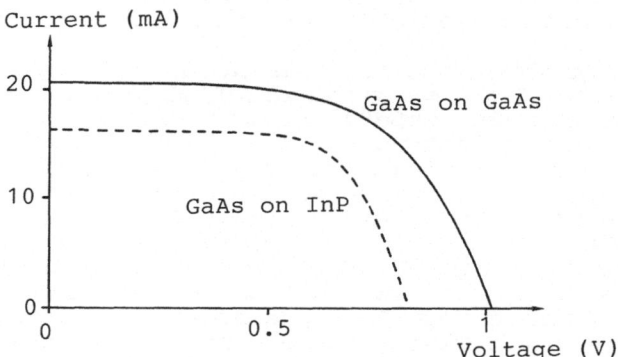

Figure 8 : Current-voltage characteristic of a GaAs
solar cell on GaAs and on InP (AM 1.5, 1 sun)

The spectral response (under AM1.5 and 1 sun) for the same cells is shown in figure 9. We observe an overall reduction of the spectral response, which was expected from I-V measurements. There is however a stronger reduction for the low energy spectral response, which can be due to a reduction in minority carrier diffusion length. The results are summarized in table 3.

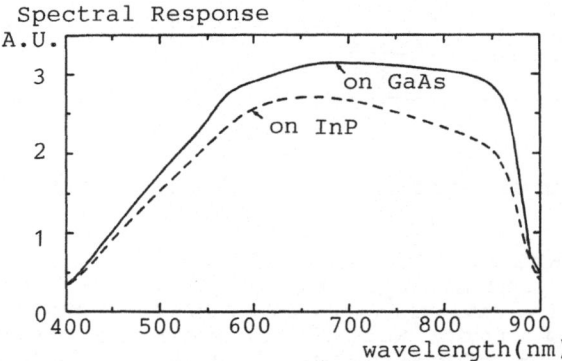

Figure 9 : Spectral response of a GaAs solar cell
on GaAs and on InP.

(AM 1.5)	GaAs on GaAs	GaAs on InP
efficiency	16%	11.5%
fill factor	61%	68%
short circuit current	28.3 mA/cm^2	20.6 mA/cm^2
open circuit voltage	1.005 V	0.815 V

table 3 : Comparision between GaAs on GaAs and GaAs on InP solar cells.

6. CONCLUSION

In this paper we have reported on the heteroepitaxial growth of high quality GaAs on InP by MOVPE. By using a thin GaAs protection layer grown at 450°C, it was possible to avoid the use of PH$_3$ to stabilise the InP surface. Good morphology and crystalline material was obtained. In spite of a high dislocation density near the GaAs/InP interface, it was possible to grow MESFETs on InP with comparable performances as GaAs on GaAs transistors. The high dislocation density resulted however in a reduction of the optical quality (photoluminescence intensity). This was confirmed by the reduction of the efficiency of GaAs on InP solar cells.

In conclusion we can say that with the described technology for the growth of GaAs on InP it will be possible to obtain high quality long wavelength optoelectronic integrated circuits and high efficiency tandem solar cells.

Acknowledgement : The authors thank P.D. Hodson (Plessey) for TEM measurements, M. Mauk for solar cell measurements and F. De Pestel, C. Eeckhout, Y. Gigase, D. Lootens and P. Van Daele for device processing.

A. Ackaert and I. Moerman wish to thank the IWONL (Instituut ter aanmoediging van het Wetenschappelijk Onderzoek in Nijverheid en Landbouw) for financial support. Part off this work is supported by the European Community RACE project number 1033.

7. REFERENCES

1. M. Razeghi et al; Low pressure metallo-organic chemical vapor deposition of GaInAsP alloys, Semicon. and Semimetals, 22 (1985), pp 299-378.
2. A. Suzuki et al; Long wavelength PINFET receiver OEIC on a GaAs on InP heterostructure, Electronic Letters 23 (18), 1987, pp 954-955.
3. K. Kasahara et al; MBE growth and charaterization of n-GaAs on InP substrates and its device application, Inst. Phys. Conf. Ser. No. 91, 1987, pp 195-198.

SiGe/Si SUPERLATTICES: STRAIN INFLUENCE AND DEVICES

E. KASPER

AEG RESEARCH CENTER ULM, SEDANSTR. 10, D-7900 ULM, FRG

ABSTRACT

SiGe/Si superlattices are a promising material for a silicon based integration of conventional integrated circuits with heterojunction devices. Si and Ge are lattice mismatched by 4.2 %. This paper describes conditions for stable superlattice growth and explains methods for strain adjustment in the superlattice. Devices are described and essential results given. Ultrathin superlattices and their potential for observation of zone folding effects are mentioned.

1. INTRODUCTION

Conventional silicon integrated circuits (Si-IC) are based on homojunction transistors. On the other hand III/V-devices successfully exploit the advantages of heterojunctions. For future electronics the marriage of mature conventional IC technology with heterojunction devices will continue the rapid increase in performance. SiGe/Si superlattices belong to the most promising heterostructure couples for silicon based electronics.

2. STRAINED LAYER EPITAXY

The cubic lattice cell of both materials, Si and Ge, and of the completely miscible alloy is of the diamond type. The room temperature lattice constant a_o of Si amounts to 0.5431 nm. Ge is lattice mismatched to Si by 4.17 % (a_{Ge} = 0.56575 nm, T = 300 K). The lattice mismatch of the $Si_{1-x}Ge_x$ alloy can be expressed in a first approximation by Vegard's law. The lattice mismatch is only slightly temperature dependent because of the small thermal expansion coefficients of diamond type materials. The higher thermal expansion coefficient of Ge leads to a slight increase of the lattice mismatch of Ge to 4.36 % (at T = 800 K) compared to the room temperature mismatch of 4.17 %.

Y. I. Nissim and E. Rosencher (eds.), Heterostructures on Silicon: One Step Further with Silicon, 101–119.
© 1989 by Kluwer Academic Publishers.

Nature gives two answers to uncracked accommodation of lattice mismatch. For layers thinner than a critical thickness t_c the accomodation is performed by strain ϵ distorting a cube with side length a into a tetragon with base length $a_{\shortparallel}$ and height $a_{\perp}$. In an isotropic material the height to base ratio is given by

$$a_{\perp}/a_{\shortparallel} = 1 - \frac{1 + \nu}{1 - \nu} \; \epsilon \tag{1}$$

with Poisson's ratio ν. Strained Layer Superlattices (SLS) are created by stacking such thin strained layers.

For layers thicker than the critical thickness t_c accommodation of mismatch takes partly place by generation of a misfit dislocation network lying in the interface. In the SiGe material system metastable strained layer growth /1-3/ shifts the critical thickness especially at growth temperatures below 900 K to higher values as compared with equilibrium.

2.1 Virtual Substrate

Let us consider a Si/SiGe superlattice on a Si substrate. The straightforward way would be to place the superlattice directly on the substrate /4-6/. Although this direct placing of the superlattice on the Si substrate has some advantages as high crystal perfection and ease of processing, it cannot be used for a general approach /2/ because of limited thickness range of the superlattice (e.g. 10 nm for a Si/Ge superlattice /6/) and inadjustable strain distribution (SiGe layers compressed, Si layers unstrained /2/).

The strain can be adjusted by a virtual substrate /7/ consisting of the Si substrate and a SiGe buffer layer between substrate and superlattice (Fig. 1). Often a rather thick and graded buffer layer is used which is inconvenient for MBE-growth methods and for device technology. We instead proposed a thin, homogeneous buffer and gave design rules for that buffer layer /8/.

2.2 Stability of a Strained Layer Superlattice (SLS)

Growing a SLS on different virtual substrates will result in different strain energies stored in the superlattice. Only the superlattice structure with the lowest energy content will be stable up to infinite thickness. All other structures will be unstable above a critical thickness t_{cs} of the whole superlattice, although the individual layers of the SLS are

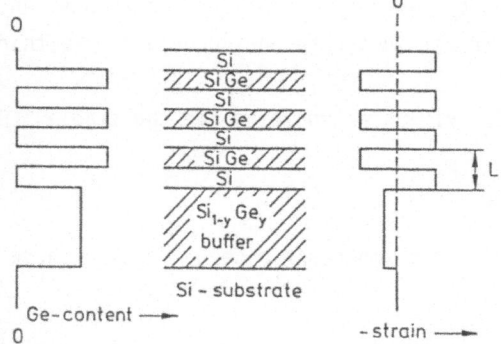

Fig. 1: SiGe/Si superlattice
on a Si substrate. An
incommensurate buffer
between superlattice
and Si substrate allows
for strain adjustment.

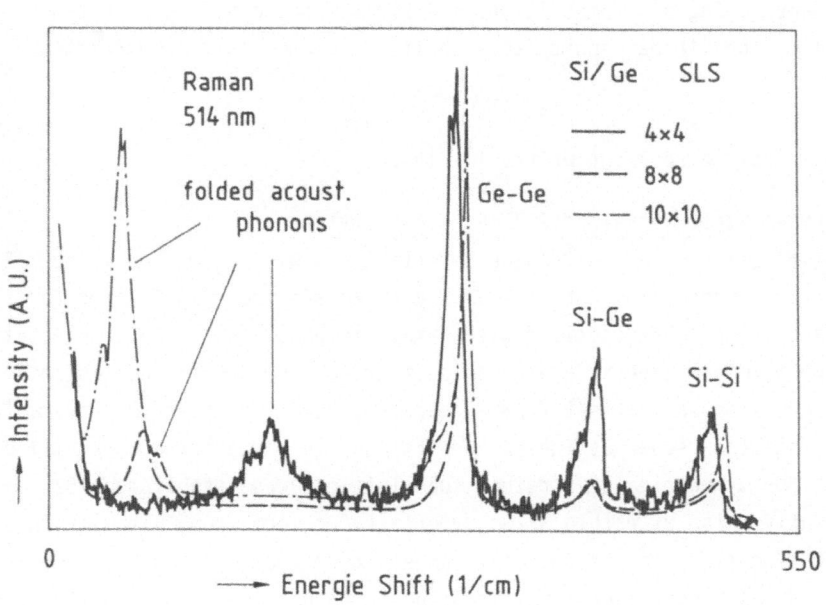

Fig. 2:   Raman spectra /10/ of ultrathin Si/Ge superlattices with
periods ranging from 8 ML to 20 ML (1 ML = a/4). Folded
acoustic phonons and three optical phonons (Ge-Ge, Si-Ge,
Si-Si) modes are visible.

thinner than the critical thickness t_c for layer growth. Dislocations tend to move to the interface between virtual substrate and superlattice within these unstable SLSs.

Minimizing the elastic energy, E_h, yields a condition for the stable SLS

$$\varepsilon_1 = (t_2/L)$$
$$\varepsilon_2 = (t_1/L) \qquad (2)$$

Strain symmetrization is required for the important case of equally thick $(t_1 = t_2 = L/2)$ layers

$$\varepsilon_1 = -\varepsilon_2 = \eta/2 \qquad (3)$$

If we discuss strain symmetrization in context with SLS stability we always refer to equ. (2) if unequally thick layers are used. Unsymmetrically strained SLSs are unstable above a critical superlattice thickness t_{cs}. This critical thickness is roughly the same as that for an alloy layer of the same integral composition. The strain in the SLS is adjusted by the choice of the virtual substrate with its in-plane lattice constant $a_{||}$

$$\varepsilon_i = \frac{a_{||} - a_i}{a_i} \qquad (4)$$

with a_i lattice constant of the ith layer.

2.3 Symmetrically strained Si/SiGe superlattices

Symmetrically strained Si/SiGe superlattices were grown on virtual SiGe substrates consisting of the Si substrate and the thin SiGe buffer /5/. The layers were characterized with X-ray methods, TEM, RBS, electrical transport measurements and Raman scattering. The strain induced shift of the conduction band offset from a near flat band condition for unsymmetrically strained superlattice (Si unstrained) to a staggered (type II) band offset for symmetrically strained superlattices was established /9/.

Recently, also ultrathin Si/Ge superlattices were grown with strain symmetrization /10/. The periodicity ranged from four monolayers (4 ML = a_0) to 20 ML as confirmed by folded acoustic phonon modes in Raman spectroscopy (Fig. 2). The thickness of the superlattices was 0.2 µm, twenty times higher than obtainable with unsymmetrically strained Si/Ge superlattices.

3. HETEROSTRUCTURE DEVICES

First heterostructure devices from MBE-grown SiGe/Si material were fabricated since 1984 (table 1). The first described pin detector /11/

Table 1 : Essential steps in silicon based heterostructure
device work from SiGe—MBE material

Year	Author	Device	Remarks
1984	S.Luryi et al. /11/	Ge—PIN Detector	incommen-surate Ge
1985	T.P.Pearsall et al./12/	p—MODFET	commensurate SiGe
	H.Daembkes et al./13/	n—MODFET	symmetrical strain
1986	H.Temkin et al. /14/	Waveguide PIN—Detector	Si/SiGe superlattice
	T.P.Pearsall et al. /15/	Waveguide APD Detector	Si avalanche region+super-lattice det.
1987	S.S.Iyer et al. /16/ *	Heterojunction Bipolar Transistor (HBT)	SiGe base
	H.Daembkes /17/	SiGe/Si— MISFET	low temper-ature growth ($<450\ ^{o}C$)
1988	T.Tatsumi et al. /18/	Collector Top—HBT	two step diff-erential epitaxy
	H.C.Liu et al. /19/	Resonant Tunneling Diode	hole tunnel-ing
	J.F.Luy et al. /20/	Hetero—MITATT	SiGe avalanche region
		QWITT	quantum well injection

* At the same meeting (2nd Int. Symp. Si—MBE) HBT results were also reported in the review papers of H. Temkin and H. Daembkes.

should be classified as an example of incommensurate epitaxy, since no use
has been made of the properties of a Ge/Si interface. But from 1985 on /12/
the properties of SiGe/Si interfaces grown by strained layer epitaxy were
exploited. Although silicon based strained layer devices are only fabricated since these few years, infrared (IR) detectors and modulation doped
field effect transistors (MODFET) have obtained properties competing with
or superior to existing Si devices. These both types of devices are described in the following two sections (3.1, 3.2). Since 1987 other heterojunction device principles were demonstrated with Si/SiGe material. These
demonstrators are described in sections 3.3, 3.4.

3.1 Ge IR Photodetectors on a Si chip

As is well known, the celebrated silicon technology has not been able to
produce an on-chip IR photodetector for long-wave length fiber-optic communication /21/. The obvious difficulty lies in the fact that silicon bandgap
is wider than the photon energy in the range of silica-fiber transparency
(λ = 1.3 μm - 1.55 μm). So far, the only practical way of employing silicon technology for fiber-optic communication has been to combine silicon
integrated circuits (IC) with germanium or InGaAsP detectors on a separate
chip.

S. Luryi et al. /11/ reported about a single-crystal Ge/Si structure on
a Si chip which worked as an efficient photodetector in the wave length
region of up to 1.5 μm (Fig. 3). For an incident light wave length of
1.45 μm they have measured a quantum efficiency of 41 % at T = 300 K. The
top three layers (Fig. 3, left) form a germanium pin diode which is separated from the Ge/Si interface by a buffer layer of high conductivity. A
disadvantage of this structure for MBE growth was the rather thick diode
structure ($\approx$ 4 μm). This was avoided by the waveguide geometry of the later narrow diode structures realized by Temkin et al. /14/ and Pearsall et
al. /15/. Light guiding to achieve strong optical absorption is obtained
by the higher refractive index of SiGe alloys or Si/SiGe superlattices compared to Si (Fig. 4). In a simple cleaved pin configuration uniform break-
downs of up to 38 V were measured with reverse leakages of $< 10^{-3}$A/cm^2 at
-10 V. At 1.3 μm wave length internal quantum efficiencies of over 50 %
were obtained. With the addition of an avalanche multiplication layer (Fig.
5) very spectacular results have been obtained. APD gains of up to 50 have
been produced, gain band with products of over 48 GHz have been achieved

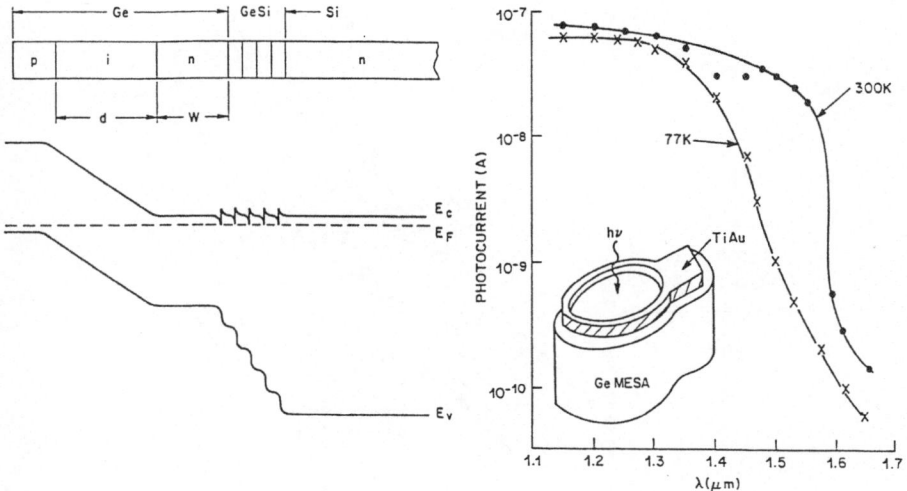

Fig. 3:   Ge pin photodetector on a Si chip. Left side: Structure and
           schematic band diagram. Right side: Photoresponse spectra at
           300 K and 77 K /11/

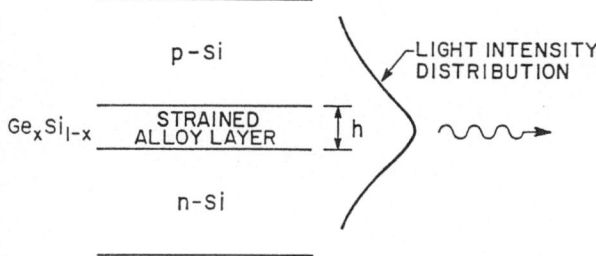

Fig. 4: Waveguide pin
photodetector: Upper
part: Light guiding by
an SiGe alloy layer
embedded in Si. Lower
part: Structure of the
Si/SiGe superlattice
photodetector on an n-Si
chip /14,21/.

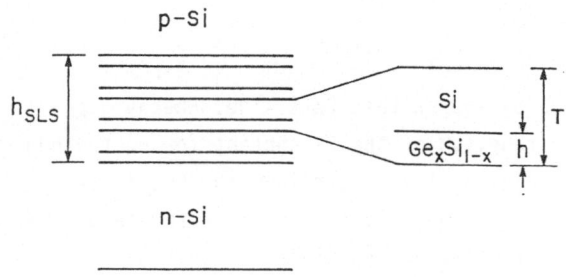

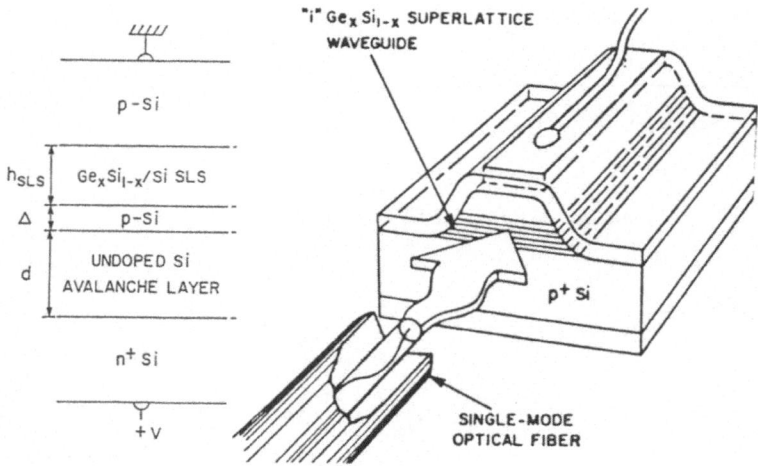

Fig. 5: Waveguide avalanche photodiode (APD). Left side: Layer structure. Right side: Coupling to the fiber /15,21/

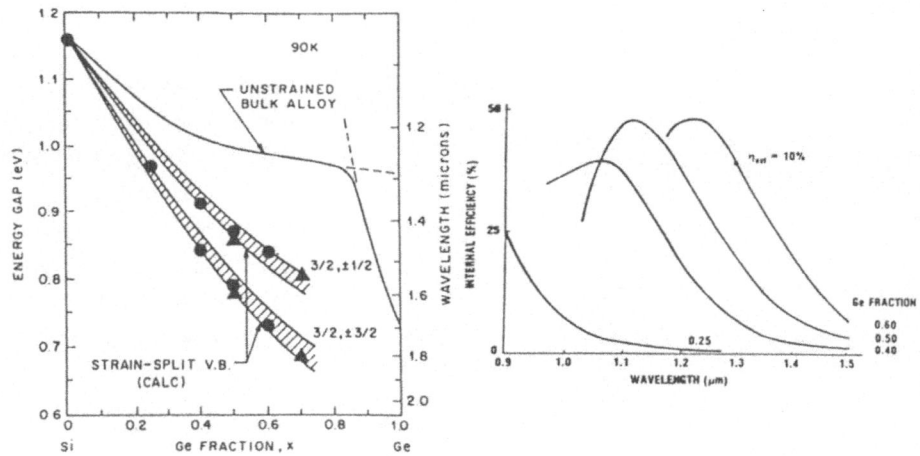

Fig. 6: $Si_{1-x}Ge_x$/Si SLS on Si substrates. Left side: Energy gap of
the SiGe alloy as function of the Ge content for unstrained
bulk alloys and for alloy layers strained to the silicon
lattice constant /22/. Right side: Spectral response of SLS
photodiodes with different Ge fractions in the SiGe layers.

and detectors have sensed 1.275 μm optical signals over a 45 km fiber link at 800 MHz with bit error rates of $> 10^{-9}$.

This series of photodetectors set several landmarks for the technological development of silicon based superlattice devices. Whereas the first pin photodetectors were based on incommensurate growth, the following waveguide detectors utilized fully the potential of strained layer epitaxy. The strained layer superlattice (SLS) of the APD structure /15/ was 600 nm thick with individual thicknesses of the Si and SiGe layers (mismatch η = 2.5 %) of 29 nm and 3.3 nm, respectively. Coupling light from the fiber was facilitated by the large numerical aperture ($\cong$ 0.5) of the SLS absorption waveguide. Strain effects decrease the energy gap of the SiGe alloy (Fig. 6) and shift the photoresponse more to the infrared as expected only from the Ge content x /22/. The band structure remains Si-like (conduction band minimum in < 100> direction, Δ point) also for strained pure Ge. In the unstrained alloy the band structure changes at x = 0.85 from Si-like (x < 0.85) to Ge-like (conduction band minimum in <111> direction).

3.2 Modulation doped field effect transistors (MODFET)

In III/V-technology the MODFET (also called high electron mobility transistor HEMT or twodimensional electron gas field effect transistor TEGFET) has very successfully driven microelectronics to higher speed. Fig. 7 shows a typical modulation doped structure as used by the Bell group for the investigation of the p-channel MODFET /12/.

The cladding layers on both sides of the SiGe channel are acceptor (B) doped where the doping is set back from the interfaces by a spacer region. Holes from the acceptor jump into the channel and create there a twodimensional hole gas (DHG) which exhibits high mobility because of separation from the Coulomb scattering centers (ionized acceptors). The p-channel MODFET of the Bell group /12/ was directly grown on a Si substrate. The good properties for a p-type device could be rather well explained by modelling of this MODFET type. Especially stressed was the fabrication of the device with IC compatible techniques. An n-channel MODFET in the same configuration could not been realized because of the near flat conduction band of SiGe/Si heterostructures directly grown on Si substrates. An AEG-group /13/ designed and realized an n-channel MODFET which utilized the strong strain influence on the band offsets. Fig. 8 shows the principal structure of this MODFET type with symmetrical strain distribution between

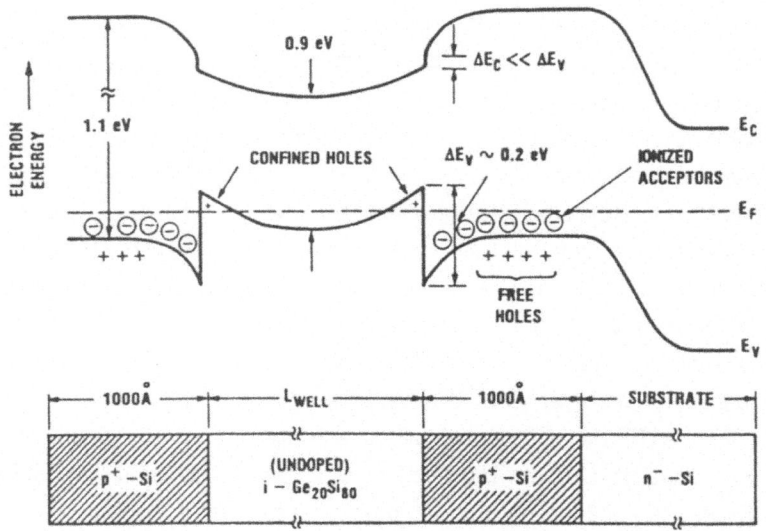

Fig. 7:   Silicon based modulation doped structure with SiGe p-channel.
           On both sides of the channel acceptor   doped Si-cladding
           layers are arranged. Doping is set back from the interfaces
           by a spacer layer /12/.

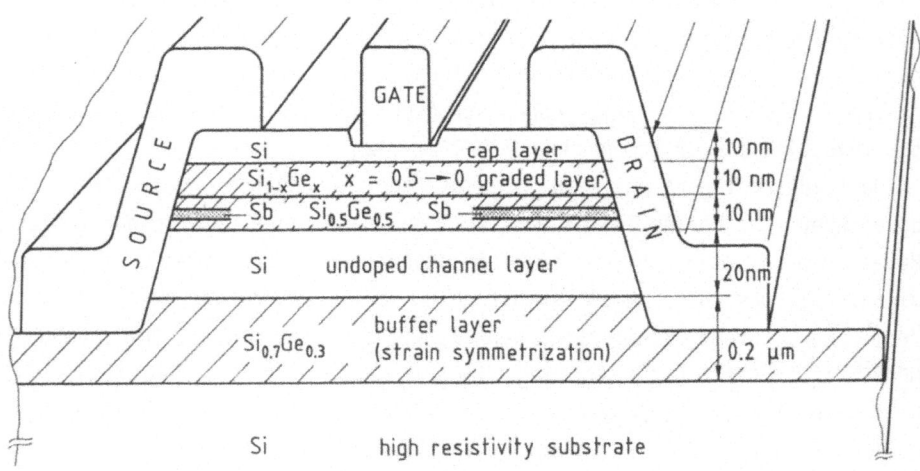

Fig. 8:   Silicon based n-channel MODFET. The basic structure consists of
           the undoped Si channel and the -spike doped (Sb) SiGe layer.
           Strain symmetrization was obtained by the SiGe buffer layer on
           the Si substrate /23/.

Si (tensile strain) and $Si_{0.5}Ge_{0.5}$ (compressive strain) layers. Strain adjustment was performed by an incommensurate buffer layer on the si substrate. This virtual substrate consisting of the Si substrate and the $Si_{0.7}Ge_{0.3}$ buffer layer offers to the following device layers an in plane lattice constant of a $Si_{0.75}Ge_{0.25}$ bulk alloy substrate. Notice the effective Ge content (25 %) of the virtual substrate to be smaller than the Ge content of the buffer. This is because of the residual compressive strain in the buffer layer. The buffer layer composition depends on layer thickness, growth temperature and eventual annealing steps. One cladding layer was employed with n-type doping by an Sb-spike. A graded layer and a Si cap protected the device and made it compatible with standard lithography steps. Clear characteristics, good pinch off behaviour and superior external transconductances of up to 50 mS/mm for a 1.6 μm gate length design /23/ were achieved.

The novel application of strain dependent band offsets and the excellent device characteristics of this first strain adjusted SiGe device encourage further work in several directions. We will give some examples of such developments although only preliminary results are available at the moment /17/.

(i) Complementary MODFET's: In Si technology complementary (CMOS) field effect transistors are widely used for advantageous digital applications. The foregoing descriptions have demonstrated the technical realizations of both, p-channel and n-channel MODFET's. But these are based on different structures. A proposal of a complementary MODFET structure is given /17/ consisting of a Si channel for electrons and SiGe channel for holes. The structure employs strain symmetrization and needs adaption of IC techniques such as ion implantation, rapid thermal annealing and lateral insulation to the SiGe material system (Fig. 9).

(ii) Very low growth temperatures: Generally, there is a tendency of semiconductor technology toward low temperature processing. The necessity of low temperature processing is enhanced with heterostructures or superlattice devices. We used the n-channel MODFET structure as test vehicle for examination of the device quality of Si-MBE material grown at temperatures far below the usual 550 °C - 600 °C. Growth and doping of the SiGe layers were performed below 350 °C in these experiments /24/. All devices showed good characteristics with a complete pinch-off behaviour, a distinct ohmic and saturated region. No looping was detected. Only in some cases a certain

bias dependent shift of the characteristics toward higher currents was observed, which might indicate the presence of some traps.

(iii) MODFET with insulating gate (Si/SiGe MISFET): Conventional Si
MOSFETs are processed at temperatures well above 900 - 1000 °C for rather
long times. SiGe/Si modulation doped strained heterostructures are assumed
to completely degrade at this temperature due to the diffusion of the dopant and the Ge atoms. Therefore low temperature oxides are needed to enable the fabrication of a MOS gate contact. Besides photo enhanced CVD processes direct oxidation of the silicon surface might be a way to produce a
thin SiO_2 surface of sufficient quality. In a feasibility study a single
heterointerface SiGe/Si MODFET structure, similar to that shown in Fig. 8,
was fabricated in the way described before. The layer sequence was grown
at extremely low temperatures well below 450 °C and doped by spontaneous
incorporation of the Sb dopant. The gate was slightly recessed and the free
surface then treated in a pure oxygen plasma. The usual gate metal was then
deposited and patterned. The device is not fully pinched off and shows a
certain amount of drift. This is probably due to the inferior quality of
the insulating layer and may be improved. These results are to demonstrate
the feasibility of an modulation doped hetero FET with insulating gate.

(iv) Multi-quantum-well FETs: To multiply exploit the advantages of the
2-DEG, multi-quantum-well FET structures were grown. In these structures
several quantum-well channels are operated in parallel. Undoped silicon
layers are sandwiched between spike doped $Si_{0.5}Ge_{0.5}$ layers. The thickness
of the undoped Si layers was chosen thin enough so that only one 2-DEG is
formed in each well. Doping and thickness of the $Si_{0.5}Ge_{0.5}$ layers are adjusted so that all free carriers are transferred into the quantum-well
channels to avoid any disadvantageous bypass in the doped layers. From
this design an increased saturation current, higher transconductance, reduced parasitic resistances and output conductance are expected /17/.

3.3 Two terminal devices

For mm-wave operation (> 30 GHz) two terminal devices have some advantages. At 100 GHz the homo junction Si-IMPATT diode /25/ is known as the far-
most powerful semiconductor device for power generation. Monolithic integration of twoterminal devices with microstrips has recently led to the
first 100 GHz Si mm-wave IC's (SIMMWIC) for receiving and transmitting
/26/. Since one year attempts were made to introduce SiGe/Si heterojunc-

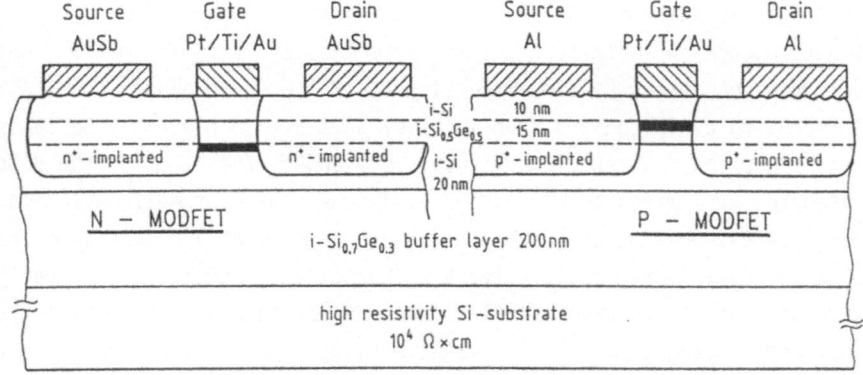

Fig. 9: Proposed structure of a complementary n- and p-channel SiGe/Si MODFET. /17/.

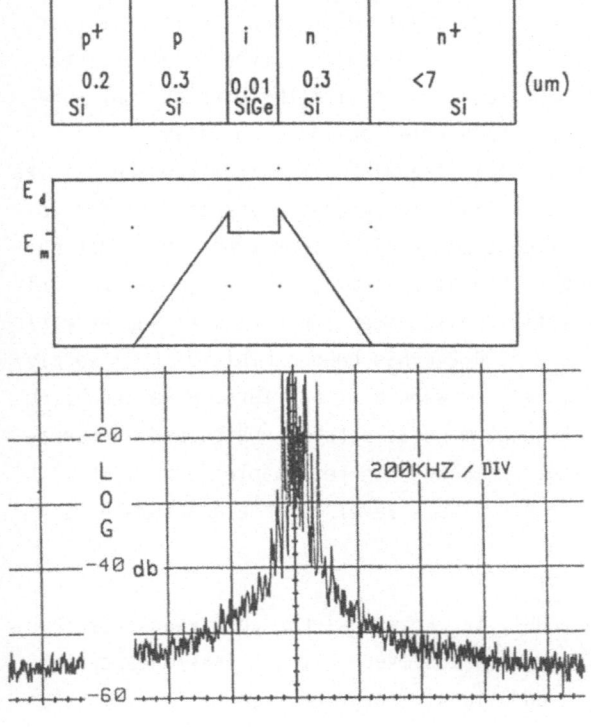

p^+	p	i	n	n^+	
0.2	0.3	0.01	0.3	<7	(um)
Si	Si	SiGe	Si	Si	

Fig. 10: Heterojunction MITATT. Structure, electric field and spectrum of output power at 74 GHz for a device from wafer B1298 /20/.

tion concepts in this new field of Si based mm-wave circuits.

(i) Heterojunction MITATT: Starting from the layout of an IMPATT diode (impact avalanche transit time) the layout of the heterojunction MITATT mixed tunneling avalanche and transit time) adds a thin (10 nm) SiGe layer between the p and n drift region (Fig. 10). Under reverse bias the SiGe well operates as a mixed tunneling and avalanche region of well defined dimensions. Output powers of 20 mW at 75 GHz and 1 mW at 100 GHz were obtained with preliminary device types. The noise properties considerably better than that of comparable IMPATT's let expect heterojunction MITATT's as promising source of low noise mm-wave power.

(ii) QWITT (quantum well injection and transit time) diode: Transit time devices require an injection mechanism which is an avalanche multiplication process in IMPATT's or a tunneling process in TUNETT's or a mixed process in MITATT's. In 1987 it was proposed /27/ to utilize the resonant tunneling of a double barrier quantum well as injection mechanism. Fig. 11 shows the principal structure and the conduction band diagram of the first SiGe-QWITT diode /20/. In this unique design the SiGe drift region is simultaneously used as buffer for strain adjustment. Only then the SiGe barriers of the Si quantum well are effective for electrons (type II heterointerface). Current-voltage measurements at 77 K confirm the barrier configuration. At room temperature only slight barrier action is observed.

(iii) Resonant tunneling diode (RTD): Transistor structures which utilize resonant tunneling in their emitter-base region are very promising for logic, multifunctional and high frequency applications /28/. The first observation of resonant tunneling in the SiGe/Si material system was recently reported /19/. Negative differential resistance with a peak-to-valley ratio in current of 1.8 at 77 K and 2.2 at 4.2 K has been exhibited by a sample with a 3.3 nm wide SiGe quantum-well between 6 nm barriers. Hole tunneling was obtained with strained SiGe quantum-wells between unstrained Si barriers. The positions of the current peaks are in reasonable agreement with calculations of the positions of heavy hole levels in the quantum-well.

3.4 Heterobipolar Transistor (HBT)

The heterobipolar transistor (HBT) is known since many years /29/ and extremely impressive results have been achieved in the AlGaAs/GaAs system. The basic idea is to provide an additional energy barrier ΔE_v to holes injected from the p-base into the n-emitter. The SiGe material is ideal to

n$^+$	i	i	i	i	i	n$^+$
Si	Si	SiGe$_{0.75}$	Ge	Si	Ge	Si
sub	driftregion	buffer	bar.2	well	bar.1	contact
	180nm	10nm	1nm	2nm	1nm	8nm

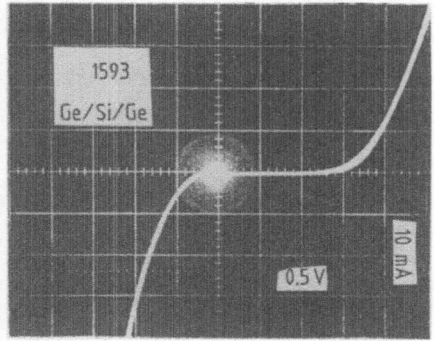

Fig. 11:   SiGe-QWITT.

Structure, and current-voltage characteristics at 77 K /20/.

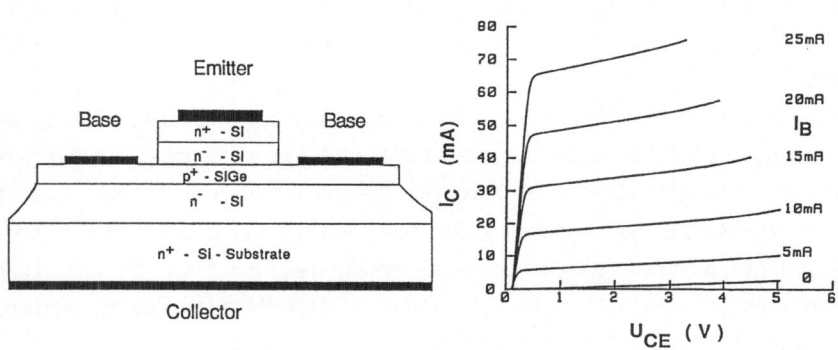

Fig. 12:   Heterobipolar transistor (HBT). Scheme of the layer sequence and typical collector current-voltage characteristics /32/.

be used as the basis layer of silicon based HBTs. The whole difference in band gap is available as difference in the valence band ΔE_V, if unsymmetrical strain distribution (Si unstrained, SiGe fully compressed) is chosen. Rather conservative estimations of the expected high frequency behaviour indicate that for devices with a more conservative base width of about 0.15 µm and a Ge content of 15 % a maximum frequency of oscillation of about 40 GHz /30/ is achieveable. By using todays spacer technology and a thin base even higher performances will be possible. So the SiGe/Si HBT is the natural candidate to further extend the existing high performance of todays Si bipolar technology and will be of high importance for future industrial applications. At least five companies (AEG, AT&T, British Telecom, IBM, NEC) started work in this direction. An IBM-group /16/ reported first results at the 2nd Int. Symp. on Si-MBE (Oct., 1987). Fully functional heterojunction bipolar transistors have been fabricated and with a 100 nm $Si_{0.88}Ge_{0.12}$ base current gains of 15 at room temperature and 25 at at 90 K were achieved.

An NEC group /18/ published a collector top design and a novel realization using a two step differential epitaxy. With differential epitaxy /31/ the lateral dimensions of the transistor are defined by a prepatterned substrate. Room temperature current gains of 15 and 250 were achieved with $Si_{0.7}Ge_{0.3}$ base layers of 5 x 10^{19}/cm^3 and 8 x 10^{17}/cm^3 base doping, respectively. Investigations of our group /32/ concentrate on SiGe-HBT's with selectively etched base contacts and -doped base regions (Fig. 12).

4. OUTLOOK

Existing device work with Si/SiGe is principally integratable with integrated circuits (IC) but this integration is not yet performed. Todays silicon based conventional microelectronics does not explicitly make use of heterostructures. It was proposed /33/ that future silicon based electronics will contain three semiconductor material regions (Fig. 13): Silicon substrate, conventional IC, heterostructure or superlattice device region. The function of the substrate itself is confined to give
- mechanical and chemical stability
- growth ordering information for the epitaxial layer, which needs perfect crystal lattice
- electrical insulation via p/n-junctions or selective oxide structures
- thermal conductivity for effective heat removal.

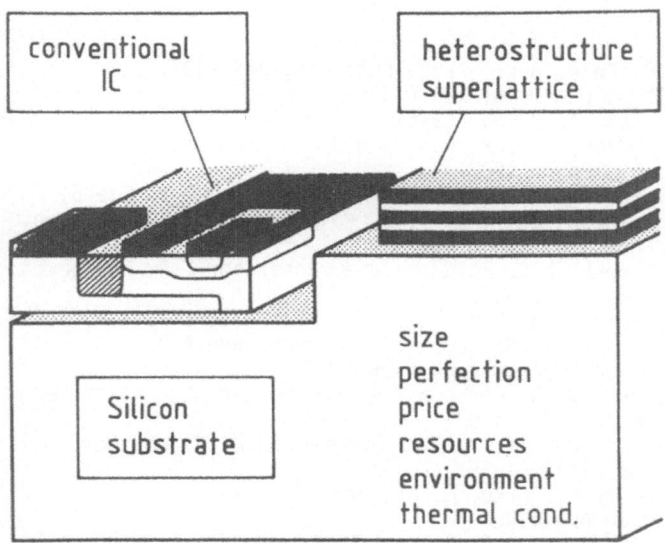

Fig. 13: Silicon based superlattice IC material concept /33/.

The conventional IC part is necessary, because of availability of
- large complexity circuits.

The conventional IC will cover the majority of the chip area.
The heterostructure or superlattice region which covers only a minor part of the chip will be used for
- high performance cores
- rapid intrachip connections
- novel device functions.

In a simplified manner one can say that the heterostructure region determines the ultimate performance, the conventional IC part allows for extremely high complexity and the silicon substrate combines good physical properties with low price and environmental harmlessness.

Speculations about a silicon based integration of microelectronics and optoelectronics grow since predictions of a quasi-direct bandgap /34-36/ led to first experiments with ultrathin Si/Ge superlattices which exhibited zone folding effects /10,37,38/. Further work is ongoing which should clarify the potential of zone folding effects for device applications.

Acknowledgement: Cooperation with my collegues from the AEG Research Center Ulm is gratefully acknowledged. Work was partly sponsored by the Federal Ministry of Technology (SiGe material) and by the ESPRIT programme of the European Commission (No. 305).

REFERENCES

/1/ R. People and J.C. Bean: Appl. Phys. Lett. 47, 322 (1985),
    Erratum: Appl. Phys. Lett. 49, 229 (1986)
/2/ E. Kasper: Surf. Sci. 174, 630 (1986)
/3/ J.Y. Tsao, B.W. Dodson, S.T. Picraux and D.M. Cornelison:
    Phys. Rev. Lett. 59, 2455 (1987)
/4/ E. Kasper, H.-J. Herzog and H. Kibbel: Appl. Phys. 8, 199 (1975)
/5/ J.C. Bean, L.C. Feldman, A.T. Fiory, S. Nakahara and I.K. Robinson:
    J. Vac. Sci. Technol. A2, 436 (1984)
/6/ T.P. Pearsall, J. Bevk, L.C. Feldman, J.M. Bonar and J.P. Mannaerts:
    Phys. Rev. Lett. 58, 729 (1987)
/7/ E. Kasper, H.-J. Herzog, H. Dämbkes and G. Abstreiter: Mat. Res. Soc.
    Proc. Vol. 56, ed. by J.M. Gibson, G.C. Osbourne and R.M. Tromp:
    p. 347, Mat. Res. Soc., Pittsburgh (1986)
/8/ E. Kasper, H.-J. Herzog, H. Jorke and G. Abstreiter: Superlattices
    and Microstructures 3, 141 (1987)
/9/ G. Abstreiter, H. Brugger, T. Wolf, H. Jorke and H.-J. Herzog:
    Phys. Rev. Lett. 54, 2441 (1985)
/10/ E. Kasper, H. Kibbel, H. Jorke, H. Brugger, E. Frieß and
     G. Abstreiter: to be published
/11/ S. Luryi, A. Kastalsky and J.C. Bean: IEEE Trans. ED-31, 1135 (1984)
/12/ T.P. Pearsall, J.C. Bean, R. People and A.T. Fiory: Proc. 1st Int.
     Symp. Si-MBE, ed. by J.C. Bean: Electrochem. Soc. Proc. Vol. 85-7,
     Pennington, N.J. (1985), p. 400
     and T.P. Pearsall and J.C. Bean: Appl. Phys. Lett. 48, 538 (1986)
/13/ H. Dämbkes, H.-J. Herzog, H. Jorke, H. Kibbel and E. Kasper: IEEE
     Trans. ED-33, 633 (1986)
/14/ H. Temkin, T.P. Pearsall, J.C. Bean, R.A. Logan and S. Luryi:
     Appl. Phys. Lett. 48, 963 (1986)
     and S. Luryi, T.P. Pearsall, H. Temkin and J.C. Bean: IEEE EDL-7,
     104 (1986)
/15/ T.P. Pearsall, H. Temkin, J.C. Bean and S. Luryi: IEEE EDL-7, 330
     (1986)
/16/ S.S. Iyer, G.L. Patton, S.L. Delage, S. Tiwari and J.M.C. Stork:
     Proc. 2nd Int. Symp. Si-MBE, ed. by J.C. Bean and L. Schowalter:
     Electrochem. Soc., Pennington, N.J. (1988), p. 114
     and S.S. Iyer: IEEE Trans. MTT (1988)
/17/ H. Dämbkes: same Proc. as /16/, p. 15
/18/ T. Tatsumi, H. Hirayama and N. Aizaki: Appl. Phys. Lett. 52, 895
     (1988)
/19/ H.C. Liu, D. Landheer, M. Buchanan and D.C. Houghton, to be published
/20/ J.-F. Luy, H. Jorke, A. Casel and E. Kasper: results presented at ITG
     Diskussionssitzung "Heterostruktur-Bauelemente", Bad Soden, April 88,
     will be published
/21/ S. Luryi and S.M. Sze, Possible Device Application of Si-MBE, in Si-
     MBE, ed by E. Kasper and J.C. Bean: CRC Press, Boca Raton (USA), 1988
/22/ R. People: Phys. Rev. B32, 1405 (1985)
/23/ E. Kasper and H. Dämbkes: Inst. Phys. Conf. Ser. No. 82, 93 (1987)
/24/ H. Jorke: to be published
/25/ J.-F. Luy: IEEE Trans. ED-34, 1084 (1987)
/26/ J. Büchler, E. Kasper, J.-F. Luy, P. Russer and K. Strohm:
     submitted IEEE Trans. MTT - Dec. 88
/27/ V.P. Kesan, D.P. Neikirk, B.G. Streetman and P.A. Blakey: IEEE EDL-8,
     129 (1987)

/28/ M. Kelly: 4th Workshop MTT-Chapter, IEEE (West German Section),
     Sept. 29th, 1987
/29/ H. Krömer: Proc. IEEE 70, 12 (1982)
/30/ C. Smith and A.D. Welbourne: Proc. IEEE Bipolar Circ. and Techn.
     Meeting, Minneapolis (1987)
/31/ E. Kasper, H.-J. Herzog and K. Wörner: J. Crystal Growth 81, 458
     (1987)
/32/ P. Narozny, M. Hamacher, H. Dämbkes, H. Kibbel and E. Kasper:
     will be published
/33/ E. Kasper: Silicon Germanium-Heterostructures on Silicon Substrates:
     in Advances in Solid State Physics, Vol. 27, p. 265, ed. by P. Grosse,
     Vieweg, Braunschweig, 1987
/34/ U. Gnutzmann and K. Clausecker: Appl. Phys. 3, 9 (1974)
/35/ S. Froyen, D.M. Wood and A. Zunger: Phys. Rev. B36, 4547 (1987)
/36/ I. Morrison and M. Jaros: to be published
/37/ T.P. Pearsall, J. Berk, L.C. Feldman, J.M. Bonar, J.P. Mannaerts and
     A. Durmazd: Phys. Rev. Lett. 58, 729 (1987)
/38/ R. Zachai, E. Frieß, G. Abstreiter, E. Kasper and H. Kibbel:
     to be published

RELAXATION OF Si/Si$_{1-x}$Ge$_x$ STRAINED LAYER STRUCTURES

C.J.GIBBINGS, C.G.TUPPEN, M.A.G.HALLIWELL,
M.HOCKLY, S.T.DAVEY and M.H.LYONS

British Telecom Research Laboratories, Martlesham Heath,
Ipswich, IP5 7RE, U.K.

1. INTRODUCTION

Silicon-germanium strained layer structures have recently aroused much interest. High efficiency injection in heterojunction bipolar transistors and high electron mobilities in modulation doped FETs were first demonstrated in III-V systems. Similar devices in the Si-Ge materials system will have the added advantages of silicon processing.

The relaxation of strained layers is important in determining the maximum temperatures used in device processing. Relaxed layers may also play a part in device design, as they enable the strain in device layers to be tailored to the particular application. Although the conduction band offset is close to zero in a Si/Si$_{.5}$Ge$_{.5}$ heterojunction on silicon (1), Daembkes *et al* (2) were able to achieve an offset by growing a modulation doped FET structure on a relaxed Si$_{.75}$Ge$_{.25}$ buffer layer. The growth of buffer layers may therefore play a crucial role in Si/Ge technology, and it is important to develop simple techniques to measure the germanium content, the degree of relaxation and the density of dislocations threading through the buffer layer.

The strained layer growth of superlattices was considered by Hull *et al* (3). It was shown that the critical thickness of a superlattice stack is equal to that for a single alloy layer of the same average composition, so long as the layers making up the superlattice do not exceed the critical thickness with respect to each other. In a previous paper (4) a range of superlattices above and below the critical thickness was studied by x-ray diffractometry, Raman spectroscopy and transmission electron microscopy. The x-ray and Raman measurements have since been refined, and this paper reports higher precision data. The use of chemical etching, Raman depth profiling and a study of a superlattice grown on heavily defected silicon will also be presented.

2. GROWTH OF SUPERLATTICES

Silicon/germanium structures were grown in a VG Semicon V80 MBE system, equipped with Temescal electron beam evaporators and an Inficon Sentinel flux controller. Sb doped (001) Wacker substrates with resistivity less than 20mΩcm were used for this work.

Several series of Si/Si$_{1-x}$Ge$_x$ superlattices were grown, with 15, 30, 60 and 120 periods. Both the silicon and the alloy layers were nominally 10nm thick, and the alloy compositions studied were x=0.2 and x=0.4. The layer data are summarised in table 1.

3. X-RAY DIFFRACTOMETRY

Figure 1a shows a typical superlattice rocking curve obtained using a double crystal diffractometer. All the superlattices in this study

Y. I. Nissim and E. Rosencher (eds.), Heterostructures on Silicon: One Step Further with Silicon, 121–128.
© *1989 by Kluwer Academic Publishers.*

showed a strong substrate peak and a series of superlattice peaks, one of which could be identified as the zero order superlattice reflection (Bragg reflection from a unit cell equal to the mean value for the superlattice). The angle between the zero order superlattice peak and the substrate peak (Z) was used to calculate the dimensions of the mean superlattice unit cell. To obtain simultaneous equations in the 'a' and 'c' dimensions of the tetragonally distorted unit cell, Z must be measured for at least two reflections. Further reflections can be used to increase the precision of the measurement. The superlattice period (D) is obtained from the spacing between superlattice peaks (S). Full details of these calculations are given in Halliwell et al (5).

Symmetric (004) and asymmetric (224, 115 and 044) reflections were used, with 004 and 044 InP first crystal reflections respectively. High and low incidence angles were used for the asymmetric reflections. The values of Z and S so obtained were used (5) to obtain the relaxation (R%), alloy composition (x) and superlattice period values shown in table 1.

As the relaxation increases the superlattice peaks broaden, as shown in figure 1b. This places a limitation on the accuracy of the technique at large relaxations, as the peak positions are less well defined.

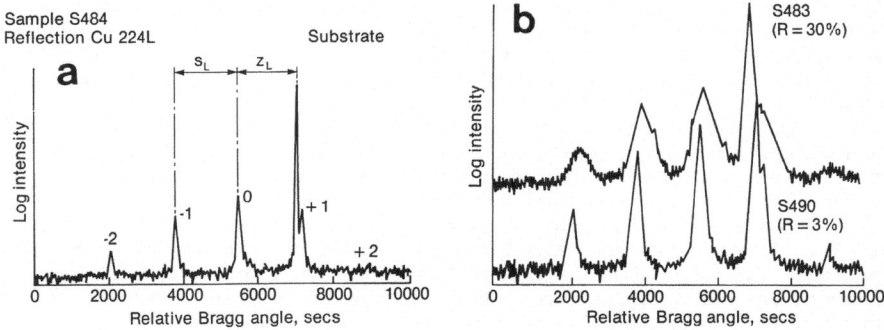

FIGURE 1. a) Typical x-ray rocking curve. b) Comparison of unrelaxed (s490) and relaxed (s483) superlattice rocking curves.

4. RAMAN SPECTROSCOPY

The Raman system and scattering geometry used for this work has been discussed elsewhere (6). A typical Raman spectrum is shown in figure 2a. Peaks above $200cm^{-1}$ are due to optical phonon scattering in the alloy layers (Ge-Ge, Si-Ge and Si-Si) and the silicon layers (Si) of the superlattice. As the superlattice structure relaxes the biaxial compression on an alloy layer decreases due to the generation of mismatch dislocations and the overlying silicon layers in the superlattice are placed under biaxial tension. We have used a combination of Raman spectroscopy and electrochemical etching (7) to study this effect as a function of depth into the sample.

Flat bottomed wells were electrochemically etched in superlattice s421 to a range of depths, and the Raman spectrum was measured for each well. As the sampling depth of the 457.9nm laser light used in this work is approximately $0.6\mu m$, i.e. much less than the thickness of the 120 period superlattice, this enables the relaxation to be plotted as a function of depth (figure 4). It can be seen that the mismatch dislocations must be largely confined to within 0.5μm of the silicon- superlattice interface. This agrees with Hull *et al* (3) who showed that the strained layer superlattice should relax as a complete unit with respect to the substrate.

We have thus shown that the biaxial tension in the uppermost silicon layers of the superlattice is equal and opposite to the relief of biaxial compression in the top alloy layers, to a first approximation. This enables the alloy composition of the superlattice to be deduced from the Si and Si-Si peak energies (5). The values of x and R% obtained in this way are summarised in Table 1. The superlattice periods listed in this table were deduced from the folded longitudinal acoustic modes (the 1b, 3a and 3b peaks in figure 2b - see Halliwell *et al* (5)).

5. TRANSMISSION ELECTRON MICROSCOPY

The procedures for obtaining TEM cross-sections such as figure 3 are detailed in Halliwell *et al* (5). It was found that approximately 70% of mismatch dislocations lay along <110> directions in the lowest interface. The remaining dislocations were mainly confined to the interfaces in the lowest third of the superlattice, and also lay along <110> directions. Burgers vector analysis showed that nearly all dislocations were of the glissile 60° type, which lie in {111} glide planes as they thread through the superlattice.

Relaxation values were obtained from TEM dislocation densities using the counting procedure outlined in (5). These results are listed in table 1. If the degree of relaxation is small the estimate is less accurate because of the large mean dislocation spacing and the limited area being sampled. Electron diffraction patterns containing super-lattice spots were used to measure the period (table 1).

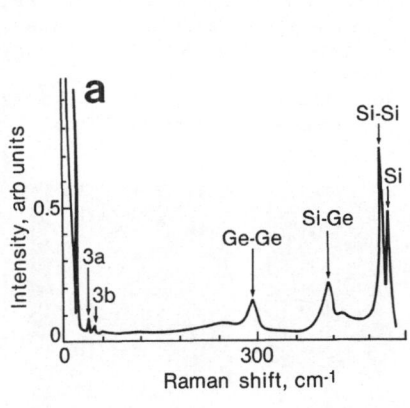
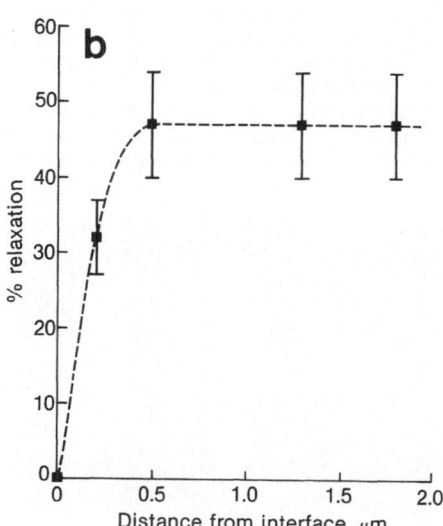

FIGURE 2. a) Raman spectrum for sample s421. b) Raman depth profile for sample s421.

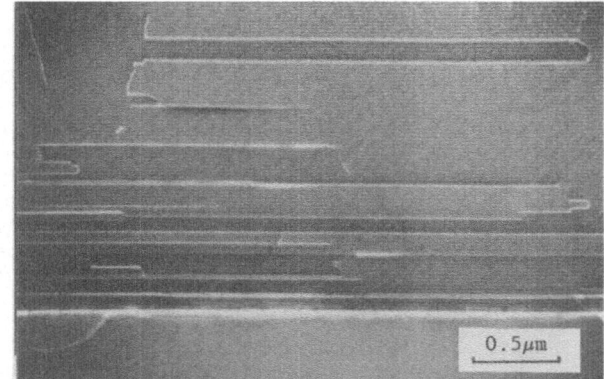

FIGURE 3. TEM cross-
-section of s421.

6. CHEMICAL ETCHING

The use of CrO_3/HF mixtures to reveal defects by preferential etching
is a well established procedure in silicon technology. The diluted
Schimmel etch (8) has proved most successful for thin epitaxial layers.
We have found that this etch can also be used to reveal networks of
mismatch dislocations in $Si/Si_{1-x}Ge_x$ structures over a range of
compositions. At high germanium concentrations (x > 0.4) the dissolution
rate of the etch appears to become much smaller, although measurement is
complicated by enhanced dissolution in heavily dislocated material. Pure
germanium shows no detectable dissolution in diluted Schimmel etch.

Figure 4 shows the results of etching a series of superlattices with
x=0.2 in dilute Schimmel etch for 2 minutes. The characteristic cross-
hatch pattern of mismatch lines is always seen when other techniques
indicate significant relaxation (>1% for x-ray measurements). Below this
relaxation level the density of lines decreases rapidly with thickness,
and for layers such as s485 only small cross-shaped features are seen.
The technique therefore appears to be more sensitive than x-ray
diffractometry in detecting partial relaxation.

Defect revealing of heavily relaxed superlattices with x=0.4 gives
rise to a very uneven surface pattern up to 1μm deep (figure 5). This

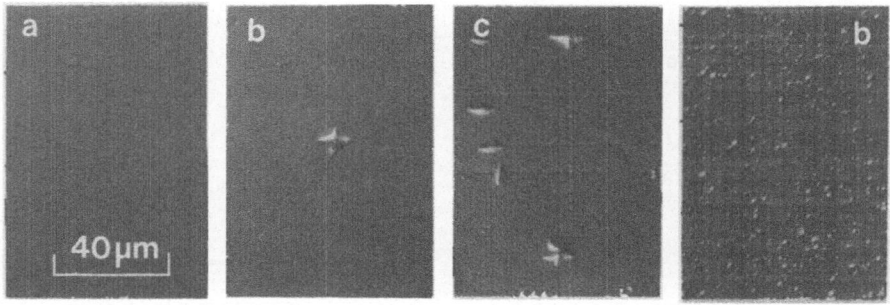

FIGURE 4. Nomarski micrographs of dilute Schimmel etched superlattices
with x = 0.2 : a) s484 b) s485 c) s486 d) s483

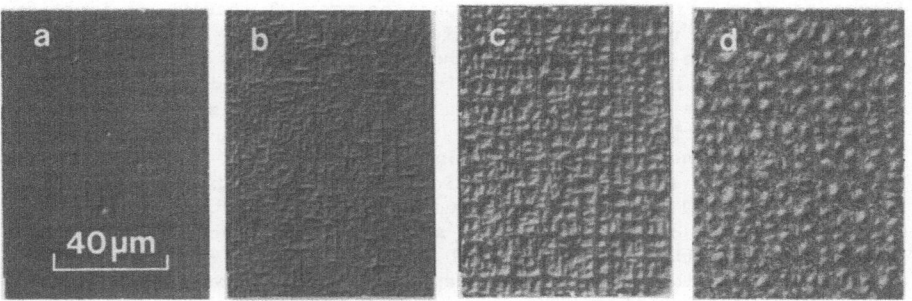

FIGURE 5. Nomarski micrographs of dilute Schimmel etched superlattices
with x = 0.4 : a) s436 b) s419 c) s420 d) s421

may be due to the large number of dislocations present. Alternatively
the complex etch patterns may be caused by a combination of fast
dissolution in strained areas followed by 'lift-off' of alloy layers as
the faster etching silicon layer is dissolved from beneath them.

Although the diluted Schimmel etch patterns still provide a good
indication of relaxation at x=0.4, better results might be obtained
using a defect revealing etch with less selectivity between silicon and
germanium. Another approach is to use the electrochemical defect etching
technique first developed for silicon and silicon-germanium MBE layers
by Tuppen *et al* (7). Precise depth control and clearly resolved etch
patterns can be obtained in this way.

7. DISCUSSION

The superlattice periods obtained by x-ray diffractometry, Raman
scattering and TEM agree well. Germanium content has been measured by x-
ray, Raman and RBS techniques, all giving good agreement. The relaxation
values agree less well. At small relaxations the x-ray technique is most
precise. Above 20% relaxation the precisions of Raman and x-ray methods
are similar because of the broadening of x-ray peaks. The agreement
between these two techniques implies that the relaxation in the top
0.5μm (Raman) differs only slightly from the mean relaxation of the
whole superlattice (x-ray); i.e. dislocations giving rise to relaxation
must be close to the substrate interface. This has been confirmed both
by Raman depth profiling and by TEM.

In single alloy layers relaxation proceeds as the <110> dislocation
segments at the alloy/substrate interface lengthen. Generally, at the
ends of such interfacial segments there are arms lying in the {111}
plane that thread to the surface. Tuppen *et al* (4) have shown that
superlattices appear to relax much less than single alloy layers of the
same mean alloy composition and thickness. It is apparent from the TEM
micrograph (figure 3) that the dislocation arms are much less free to
glide in a superlattice because alternating strain fields make them
double back along silicon/alloy interfaces. This factor contributes to
the low degree of relaxation, and suggests that, for a given effective
lattice constant, a superlattice will have a greater density of
threading dislocations.

Chemical etching provides a rapid means of assessing a $Si/Si_{1-x}Ge_x$
structure. There are three types of pattern on the etched surfaces of

figure 4: pits, lines and cross-shaped features. Threading dislocations give rise to etch pits in silicon defect reveal etching because strained material dissolves faster than the bulk. Figure 4d shows a high density (10^7cm^{-2}) of similar etch pits, which are interpreted as dislocations threading through the superlattice. TEM shows a similar density of threading dislocations.

Although the chemical etch may only dissolve part way through the superlattice, a mismatch dislocation at the substrate/superlattice interface is still observed to influence the etch pattern. This can be explained by the relief of strain above the interfacial dislocation, giving rise to a line in the chemical etch pattern.

The interpretation of the cross-shaped features is not clear, although their alignment with the (110) axes and their increase in density as the superlattice thickness increases imply that they are associated with mismatch dislocations. Indications of partial relaxation below the measured 'critical thickness' (such as on s485) agree with the recent work by Fritz (10) who shows that the disagreement between measured (11) and theoretical critical thicknesses can be explained in part by a lack of sensitivity in determining relaxation. A finite density of mismatch dislocations is therefore expected below the People and Bean (11) critical thickness.

The average spacing (L) of mismatch dislocations (60° type) at a $\text{Si}/\text{Si}_{1-x}\text{Ge}_x$ interface with a relaxation R% is given by:

$$L = (2.1 \times R\% \times x)^{-1} \ \mu m \qquad (1)$$

s486 is predicted to have L=1μm. The etch lines appear to be more widely spaced, presumably because mismatch dislocations tend occur in bunches, leading to an underestimate of relaxation from chemical etching. Better agreement is expected at low densities, but the other techniques used here are not sensitive enough to confirm this.

8. OXIDE REMOVAL AND RELAXATION

The data for table 1 were taken from samples near the centre of the 100mm diameter wafers used in this study. In most cases the superlattice properties were relatively constant across the wafer. However, Hirst topography of s486 showed that the structure was much more relaxed within 2cm of the edge. This was confirmed by dilute Schimmel etching (figure 6) and by recording X-ray rocking curves at 1cm intervals across

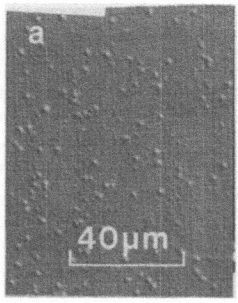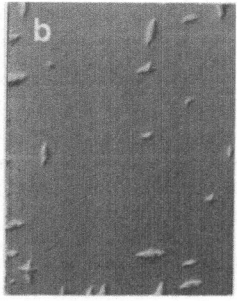

FIGURE 6. Nomarski micrographs of dilute Schimmel etched superlattice s486 at (a) edge and (b) centre of wafer.

the wafer. The FWHM of the peaks increased from <50sec at the centre to >500sec 3cm from the centre. A relaxation of 90±10% at the edge was measured using Raman spectroscopy.

A SIMS analysis of the centre and edge of s486 was undertaken to investigate the effectiveness of the initial oxide removal. The interfacial oxygen levels were found to be ~$1.4.10^{13}$cm^{-2} at the centre and ~$1.4.10^{14}$cm^{-2} at the edge. Incomplete removal of SiO_2 has been found to give rise to dislocation densities in the range 10^6-10^7cm^{-2} (12). It is therefore likely that the difference in relaxation between the edge and the centre is due to a variation in the threading dislocation density. These dislocations can serve as a source of misfit dislocations (13). As the multiplication of dislocations is believed to be an important mechanism in strain relief (14,15) a large initial density of threading dislocations would be expected to enhance relaxation. A more controlled set of experiments is required to confirm this.

9. CONCLUSIONS

High precision x-ray diffractometry and Raman spectroscopy have been used to measure superlattice period, alloy composition and relaxation for a series of superlattices. Good agreement was obtained, generally within experimental error. TEM was able to give precise measurements of superlattice period, and to measure the ratio of silicon to alloy layer width accurately. X-ray diffractometry gave precise results for relaxation and alloy composition, while Raman offered similar precision for relaxation > 25%. Raman spectroscopy and TEM have also been used to study strain relief as a function of depth in a superlattice.

Superlattices have been found to relax to a lesser extent than single alloy layers of the same average composition. TEM study indicates that this is due to alternating strain fields impeding dislocation movement.

$Si/Si_{1-x}Ge_x$ structures can be rapidly assessed by chemical etching, which may be especially valuable for relaxation below the detection limit of x-ray diffractometry.

The above techniques have been used to show that there is a correlation between poor oxide removal and enhanced relaxation. This is believed to be caused by a high density of threading dislocations.

10. ACKNOWLEDGEMENTS

We would like to thank Miss S.M.Casey of Trinity College, Dublin for her work on electrochemical etching and P.J.Skevington for SIMS analysis. We acknowledge the Director of Research and Technology, British Telecom plc for permission to publish.

REFERENCES

1. People R and Bean JC: Appl.Phys.Lett. 48, 538-540, 1986.
2. Daembkes H, Herzog H-J, Jorke H, Kibbel H and Kasper E: IEEE Trans. Electron Devices ED-33, 633, 1986.
3. Hull R, Bean JC, Cerdeira F, Fiory AT and Gibson JM: Appl.Phys.Lett. 48, 56-58, 1986.
4. Tuppen CG, Gibbings CJ, Davey ST, Lyons MH, Hockly M and Halliwell MAG: Proc. of 2nd Int. Conf. on Si MBE, Honolulu, Electrochemical Society, NJ, 1987.
5. Halliwell MAG, Lyons MH, Davey ST, Hockly M, Tuppen CG and Gibbings CJ: to be published.

6. Davey ST, Spurdens PC, Wakefield B and Nelson AW: Appl.Phys.Lett. 51
   758, 1987.
7. Tuppen CG, Gibbings CJ and Ayling CL: Proc. of 2nd Int. Conf. on Si
   MBE, Honolulu, Electrochemical Society, NJ, 1987.
8. ASTM Standard F47 - 82
9. Fiory AT, Bean JC, Feldman LC and Robinson IK: J.Appl.Phys. 56,
   1227, 1984.
10. Fritz IJ: Appl.Phys.Lett. 51, 1080, 1987.
11. People R and Bean JC: Appl.Phys.Lett. 47, 322, 1985 and erratum:
    Appl.Phys.Lett. 49, 229, 1986.
12. Gibbings CJ and Tuppen CG: to be published in MRS Symp.Proc. 102,
    1988.
13. Dodson BW and Tsao JY: Appl.Phys.Lett. 51, 1325, 1987.
14. Matthews JW and Blakeslee AE: J.Cryst.Growth, 27, 118-125, 1974.
15. Kvam EP, Eaglesham DJ, Maher DM, Humphreys CJ, Bean JC, Green GS
    and Tanner BK: to be published in MRS Symp.Proc. 104, 1988.

TABLE 1. Experimental results

Sample	No. of Periods	Period (Å) x-ray	Period (Å) Raman	Period (Å) TEM	%Ge in alloy x-ray	%Ge in alloy Raman	% Relaxation x-ray	% Relaxation Raman	% Relaxation TEM
s487	15	185	190		21.3	22	0.0	4.5	
s488	30	180	180	173	20.5	21	0.7	2.6	0.1
s493	60	182	182		19.5	21	2.2	2.6	
s490	120	180	180	177	20.5	21	4.5	4.7	1
s484	15	180	189		20.3	22	0.8	0	
s485	30	183	186		20.4	21	1.0	2.3	
s486	60	180	182		19.9	21	3.0	5.2	5
s483	120	180	194	179	20.0	22	30	53	38
s436*	15	214	240		(34)	34	46	23	
s419*	30	183	186		(40)	40	71	56	
s420	60	183	180	184	40.7	39	47	55	46
s421	120	185	188	181	40.8	40	49	45	49
error R<20%		2	3	5	0.5	2	0.5	15	(×2)
error R>25%		4	3	5	1	2	5	10	5

Growth temperature 600°C, except for s487, s488, s493 and s490 (all
550°C).
* only one x-ray reflection used, so accuracy is lower.

UNSTRAINED VS. STRAINED LAYER EPITAXY:
THICK Ge LAYERS AND Ge/Si SUPERLATTICES ON Si(100)

M. OSPELT, K.A. MÄDER, W. BACSA, J. HENZ, H. VON KÄNEL

Laboratorium für Festkörperphysik, ETH-Zürich, 8093 Zürich
Switzerland

ABSTRACT
    Smooth unstrained Ge layers (up to 1 μm) have been grown on
Si(100) by using low substrate temperatures T_s ($\leq$ 420 °C) for
the first few 100 Å of growth, thus suppressing the onset of
3-dimensional growth usually accompanying the generation of
misfit dislocations. By increasing T_s up to 550 °C during
growth the layers can be made to exhibit excellent crystallinity as evidenced by an RBS minimum channeling yield X_{min} of
3.5%. By contrast to these thick Ge layers high quality
strained-layer Si/Ge superlattices can only be fabricated by
keeping the thickness of the individual Ge layers below their
critical thickness ($\approx$ 6 monolayers). In addition the total
thickness has to be kept below the critical thickness of the
corresponding $Si_{1-x}Ge_x$ alloy. We have grown strained-layer
Si/Ge superlattices with periods exceeding 25 Å and with a Ge
thickness below 6 monolayers. Crystalline and interface quality have been confirmed by RBS and channeling ($X_{min} \approx$ 5%) and
by X-ray diffraction, respectively. The superlattices have
also been examined by means of Raman scattering. Up to three
orders of folded acoustic phonons have been observed. Furthermore, interface and Ge-modes are shifted due to the strain in
the Ge layers.

1. INTRODUCTION

    Thick relaxed Ge layers on Si are of considerable interest
from different points of view. First of all, germanium itself
has still important applications in device technology and the
growth of device quality Ge on Si would be a step further in
opto- and microelectronic integration. Furthermore, Ge can
serve as a possible link between GaAs and Si, rendering
possible the monolithic integration of GaAs devices on silicon. Since the lattice mismatch between Ge and GaAs is almost
zero, smooth Ge buffers on Si can be used as substrates for
the subsequent growth of GaAs. Whereas thick relaxed Ge layers
on Si exhibit their own lattice constant, thin Ge layers (up
to 6 monolayers) can be grown in a pseudomorphic way (1). In
Si/Ge superlattices (SL's) entailing such thin Ge layers the
combination of strain and the new periodicity (bandfolding
effects) leads to novel electronic and optic properties (2-4).
In the present work we show that using suitable growth conditions both thick relaxed Ge-layers and SL's with excellent
crystalline quality can be achieved. Moreover, in the former
case 3-dimensional growth can be completely suppppressed resulting in as smooth surfaces as for the SL's.

Y. I. Nissim and E. Rosencher (eds.), Heterostructures on Silicon: One Step Further with Silicon, 129–136.
© 1989 by Kluwer Academic Publishers.

2. EXPERIMENTAL

We use 3 inch diameter, (100)-oriented and phosphorus doped
(0.05-5.0 Ωcm) Si wafer as substrates. They are cleaned thermally by heating to 830 °C for ∼ ½ h and subjecting them to a
low intensity Si flux (total amount of deposited Si: ∼ 10 Å).
Subsequently a Si-buffer of 500 - 1000 Å thickness is grown.
The substrates prepared in this manner exhibit a smooth 2x1 +
1x2 reconstructed surface, verified in situ by RHEED. The
silicon is evaporated from an electron-gun with 270° electron
deflection, the germanium from a Knudsen cell. The growth
rate for Si is monitored by a quartz crystal balance, the temperature dependence of the Ge rate has been calibrated by RBS.
The base pressure in our MBE machine is 10^{-11} mbar, not exceeding 10^{-10} mbar during growth.

3. GROWTH AND CRYSTALLINE QUALITY

One of the greatest problems when growing thick relaxed Ge
layers on silicon is the suppression of 3-dimensional growth.
Due to the large mismatch between Ge and Si of 4%, the
critical thickness h_c for pseudomorphic growth is very small.
It has been found to lie in the range of 6 monolayers (1).
When the thickness of a Ge layer is extended beyond h_c, the
layers start to relax by forming dislocations which may act
as nucleation centers, depending on the exact growth conditions. The resulting 3-dim. growth mode causes a considerable
roughening of thicker films. Using a relatively high evaporation rate combined with a low growth temperature is essential
for obtaining smooth films. Starting growth conditions for our
thick Ge layers have been an evaporation rate of about 2 Å/s

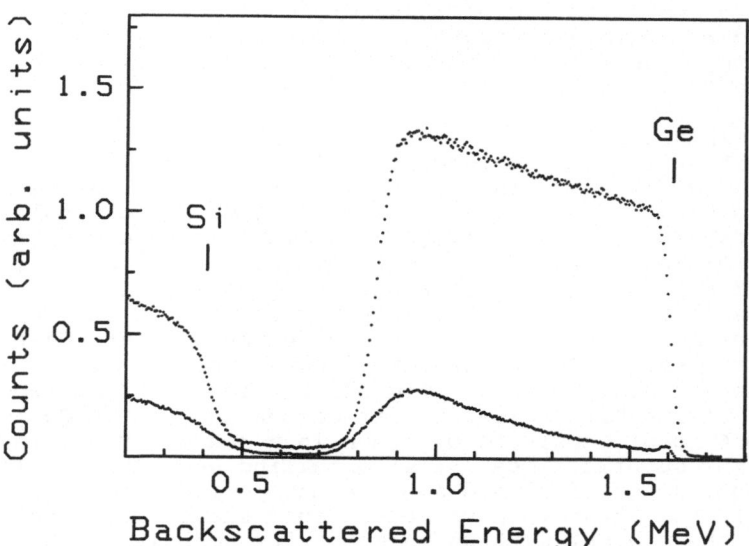

Fig. 1: <110> channeling and random RBS spectra (2 MeV He)
of a 1 μm relaxed Ge layer on Si (100). Minimum
channeling yield X_{min} = 3.5%.

at a substrate temperature T_S of 420 °C. After the deposition
of about 1000 Å, T_S can be increased in steps up to 550 °C.
Proceeding in this way, islanding can be suppressed and films
without detectable roughness (Nomarski interference contrast
microscopy) can be grown. After 1000 Å they already show a
sharp RHEED pattern of a 4x2 reconstructed surface, which is
typical for a relaxed Ge surface. Despite the use of low temperatures, the crystalline quality of our thick Ge layers is
excellent, as can be seen in fig. 1. It shows the channeling
spectrum of a 1 μm thick relaxed Ge layer grown by the
procedure described above. Crystallinity can be measured by
comparing the random yield and the channeling yield in a RBS
measurement. Channeling spectra are taken by aligning the beam
along a major crystallographic direction. The minimum yield
X_{min}, which is the channeling yield normalized to the yield
obtained with a random beam orientation, is a good means to
describe the crystalline quality of a material. The lower
X_{min}, the better is the crystallinity, because less beam
particles are scattered back by irregularities in the material. An ideal crystal has a X_{min} of ~3% (5). As can be seen
in fig. 1, the first few hundred Å of Ge near the interface,
i.e. the low energy part of the Ge-signal, are of minor
crystalline quality expressed by a X_{min} of ~20%. This is an
indication for a high density of dislocations. But as the Ge
layer gets thicker, its quality increases considerably and at
the surface the minimum yield becomes as low as 3.5% in the
<110> direction, a value which is comparable to those obtained
from perfect crystals. Furthermore the low energy slope of the
Ge-signal in the random spectrum (fig. 1) is only broadened
because of the loss in energy resolution due to energy
straggling in the layer. There is no indication for interdif-
fusion of Ge and Si.

In contrast to thick Ge layers, the growth of Si/Ge strained
layer superlattices (SL's) is only possible by keeping the
thickness of an individual Ge layer below the critical
thickness h_c. Going beyond h_c results in strain relaxation and
deterioration of the crystalline quality, as mentioned before.
We have therefore grown $Ge_m Si_n$ SL's with Ge thicknesses
ranging from 4 to 6 monolayers and Si thicknesses between 20
and 70 Å (4≤ m ≤ 6, 15 ≤ n ≤ 50, where m and n denote the
number of monolayers of Ge and Si, respectively). We have
found the optimal growth temperature to lie between 460 and
480 °C. The evaporation rates are 0.5 Å/s for Ge and Si. The
SL's grown under these conditions exhibit smooth surfaces,
equal to those of thick relaxed Ge layers. Besides h_c, there
is also a critical thickness H_C for the superlattices as a
whole, which we assume to be equal to that one of the corresponding $Ge_x Si_{1-x}$ alloy (6). The latter depends strongly on
the Ge content in the alloy. Values for H_C can be found in
refs. 7 and 8. By keeping the total thickness below H_C, we
have grown SL's with thicknesses up to 5000 Å. As long as the
conditions for pseudomorphic growth are fulfilled, the SL's do
not relax, which means, that the interface betweeen the
substrate and the SL should not be disturbed by dislocations.
This behavior is in contrast to the growth of thick relaxed Ge
layers. Fig. 2 shows the RBS and channeling spectra in differ-

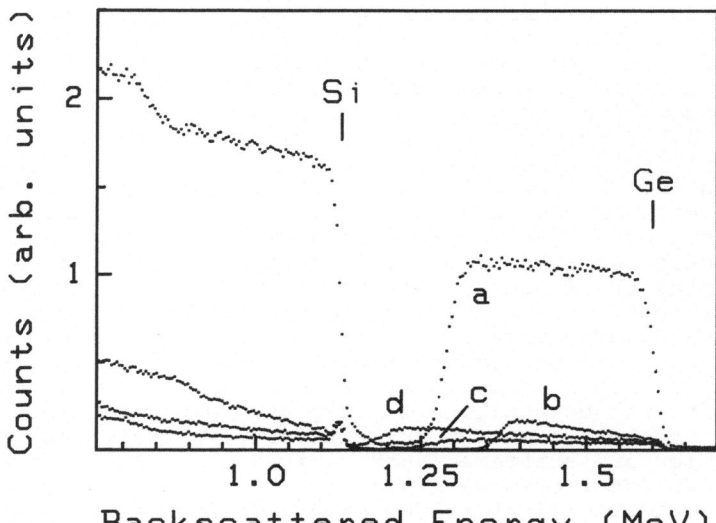

Fig. 2: RBS random and channeling spectra of a strained
layer Si/Ge superlattice with a period of 44 Å.
The total thickness is 4850 Å. (a) random spectrum, (b) <100> channeling direction, (c) <110>
channeling direction and (d) <111> channeling
direction.

ent directions of a SL with a total thickness of 4850 Å, the
individual layers of which are 4.9 Å and 39 Å for Ge and Si,
respectively. It is evident that the SL is of comparable
crystalline quality throughout the whole layer. The increase
of the channeling yield to lower energies is due to energy
straggling in the layer, an effect, which is lowest for the
<110> direction and somewhat higher for the other two direc-
tions. The minimum yields χ_{min} are generally measured directly
below the surface, where the signal is not yet influenced by
effects like energy straggling, which play a role when the
particles are passing through the material. χ_{min} in the <110>
direction is as low as 3.4%, indicating an excellent crystallinity. The minimum yields of our SL's in this direction are
typically 5% or less. The other directions display somewhat
higher yields, which is in agreement with theoretical calculations (5). For the SL in fig. 2, the values for the <111> and
<100> directions are 4.2% and 5.8%, respectively. Due to the
different angle between the incident beam and the channeling
direction, the energy width of the Ge signal is not the same
for different orientations. The random yield shown in fig. 2
is taken near the <110> direction and all channeling spectra
are normalized to the random yield.

4. RAMAN AND X-RAY MEASUREMENTS
Whereas RBS and channeling measurements describe the crystalline quality of a material we have used X-ray diffraction

and Raman scattering to confirm the quality of the interfaces
and the regularity of the superlattice periodicity. The new
periodicity introduced by the superlattice leads to a reduction of the Brillouin zone in the growth direction. The new
phonon dispersion relation can be described to a first approximation by the folding back of the dispersion curves of the
constituent materials (9). The optical branches of the dispersion of Ge and Si do not overlap. Only phonons, which are
generated by the acoustic branches, can propagate throughout
the entire SL. As the period of our SL's is equal or less than
70 Å the photon momentum q of the incident laser light is
small compared with the dimensions of the Brillouin zone.
Therefore, the folded acoustic phonons can be seen in our

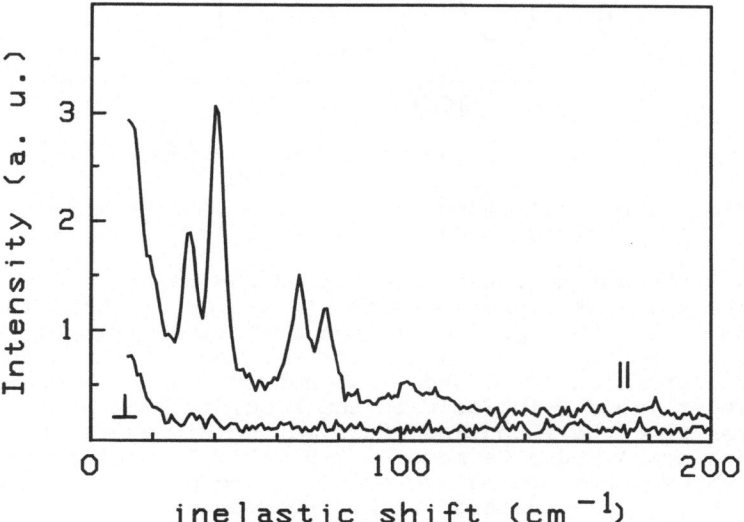

Fig. 3: Acoustic region in the Raman spectrum from a SL
with a period of 70 Å. Thickness of the individual Ge layers: 8 Å. ($\parallel$) spectrum taken with
parallel polarisation directions of the incident
and scattered light and ($\perp$) with crossed
polarisation.

Raman spectra as doubletts. Fig. 3 shows the acoustic region
in the spectrum of a SL with a period of 70 Å. Three orders of
doubletts can be seen, two of them clearly resolved. From the
energies of these modes, the thickness of one period of the SL
can be deduced by applying the elastic continuum model (10).
It corresponds well with the values obtained from RBS and X-
ray diffraction. The separation of one doublett is no longer
possible for the SL's with periods below 30 Å (11). From our
growth control (no RHEED oscillation controlled growth) we
have to expect an interface roughness and fluctuations of the
periodicity of the order of one monolayer at least. This
roughness suffices to explain the impossibility to resolve the

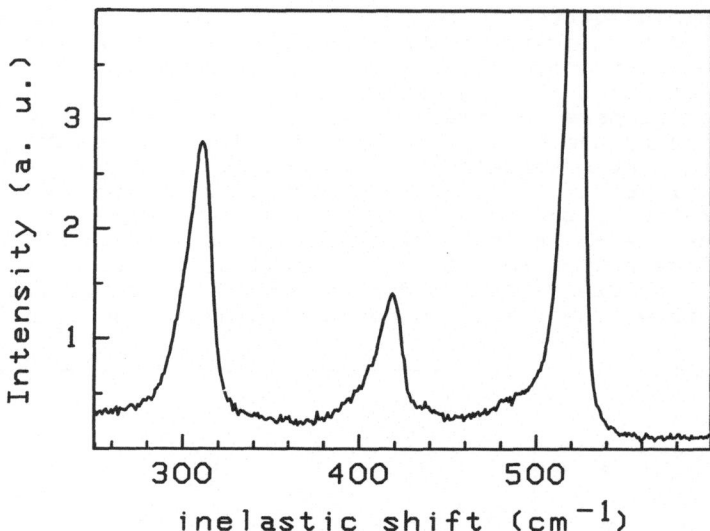

Fig. 4: Optical phonon region of a SL with a period of 25 Å.
Individual Ge layers: 4 Å.

doublets due to the broadening of the peaks.

The peaks at higher energies (> 250 cm^{-1}) shown in fig. 4
from a SL with a period of 25 Å are closely related with the
optical modes of the bulk material. The optical branches of
the dispersion of Ge and Si do not overlap, the one of Ge,
however, does coincide with the acoustic branch of Si. This
gives rise to a confined mode in the Si layers at 521 cm^{-1}
and a nearly pure Ge mode at 315 cm^{-1}. In addition a new mode
in between the optical modes of Si and Ge can be observed,
which can be associated with Si-Ge bonds at the interface (9).

Whereas the optical Si mode lies at the same energy as in
the case of bulk silicon, the Ge mode is shifted to higher
energies with respect to the bulk mode. This shift is of the
order of 15 cm^{-1} and can be attributed to the biaxial strain
in the Ge layers of the superlattices. Under the assumption of
pseudomorphic growth (strain of 4%), it can be calculated to
be 16 cm^{-1} by using parameters determined by uniaxial strain
experiments (12). The position of the Ge mode is the same for
all our strained layer superlattices. The persistence of
pseudomorphic growth has also been verified by X-ray diffraction experiments. In fig. 5 a X-ray measurement of the SL from
fig. 2 is displayed. Besides the zero order superlattice
reflex, the first and second order reflexes can also be seen.
From the distance of these orders, the period of the superlattice can be calculated. With the period from the rocking
curves and the relative thicknesses of the Ge and Si layers
from RBS, the strain in the SL can be deduced. A more detailed
study of our SL's by means of high resolution X-ray diffraction will be given elsewhere (13). Suffice it to say, that the
thicknesses and the amount of strain incorporated in the SL's

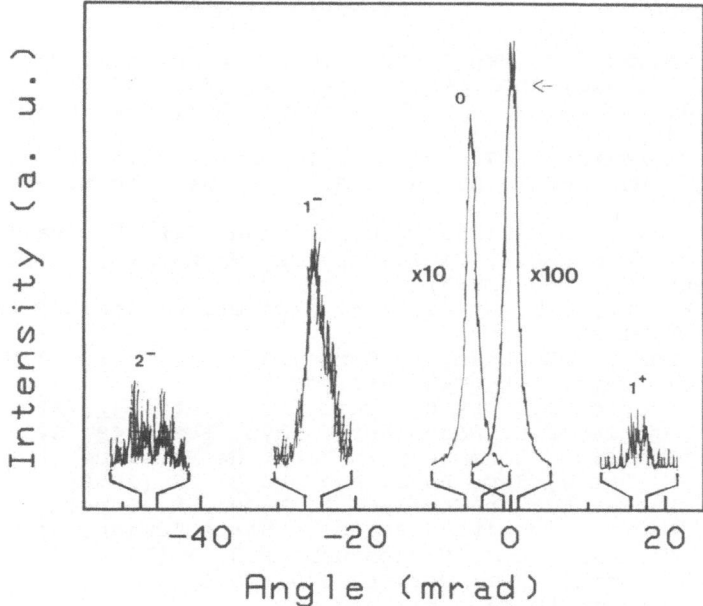

Fig. 5: X-ray rocking curves of th (400) substrate reflex
and the satellites of the Si/Ge superlattice.
The substrate reflex is marked with an arrow.

match the values from RBS and Raman measurement and, more
important, the SL's show the full tetragonal strain of 4% in
agreement with Raman data.

In conclusion, we have grown thick relaxed Ge layers up to
a thickness of 1 μm and strained-layer superlattices up to
5000 Å. The surfaces of both types of systems can be made to
be smooth, showing no roughness in Nomarski microscopy. The
large dislocation density at the Si-Ge interface of thick Ge
layers decreases with growing thickness. This leads to Ge
layers of excellent crystalline quality. The SL's are of equal
crystalline quality throughout the layers as shown by RBS.
X-ray and Raman measurements proove the pseudomorphic growth
of the SL's, which is evident by the measured strain of 4%. The
sharpness of the interfaces and the superlattice periodicity
lie in the range of one monolayer.

ACKNOWLEDGEMENT
We would like to thank very much Mr. H.-J. Gübeli for his
excellent technical support and Dr. R.E. Pixley from the
Physics Department of the University of Zürich for the
opportunity to use his RBS equipment and his help therewith.
Financial support by the Swiss National Science Foundation is
also greatfully acknowledged.

REFERENCES

1. J. Bevk, J.P. Mannaerts, L.C. Feldman, B.A. Davidson, A. Ourmazd, Appl. Phys. Lett. 49 (5), 286 (1986)
2. C.G. van de Walle, R.M. Martin, Phys. Rev. B 34, 5621 (1986)
3. T.P. Pearsall, J. Bevk, L.C. Feldman, J.M. Bonar, J.P. Mannaerts, A. Ourmazd, Phys. Rev. Lett. 58, 729 (1987)
4. J. Bevk, A. Ourmazd, L.C. Feldman, T.P. Pearsall, J.M. Bonar, B.A. Davidson, J.P. Mannaerts, Appl. Phys. Lett. 50, 760 (1987)
5. W.K. Chu, J.W. Mayer, M.A. Nicolet, Academic Press, New York, 1978
6. R. Hull, J.C. Bean, F. Cerdeira, A.T. Fiory, J.M. Gibson, Appl. Phys. Lett. 48, 56 (1986)
7. R. People, J.C. Bean, Appl. Phys. Lett. 47, 322 (1985)
8. R. People, J.C. Bean, Appl. Phys. Lett. 49, 229 (1986)
9. A. Fasolino, to appear in Jour. de Physique
10. S. M. Rytov, Sov. Phys. Acoust. 2, 67 (1956)
11 W. Bacsa, H. von Känel, K.A. Mäder, M. Ospelt, P. Wachter,submitted to Solid State Communications
12. F. Cerdeira, C.J. Buchenauer, F.H. Pollak, M. Cardona, Phys. Rev. 5, 580 (1972)
13. L. Tapfer, to be published

Direct Band-Gap Si-Based Semiconductors, Principles and Prospects

T. P. Pearsall

AT&T Bell Laboratories, 600 Mountain Avenue, Murray Hill, NJ 07974 USA

ABSTRACT

New direct optical transitions that occur at energies below the fundamental gap of Si have been measured in Si-based structures. The new energy levels have been induced in implanted Si as well as pseudo-morphic Ge-Si atomic layer superlattices. Important features that affect the bandstructure of strained-layer superlattice structures are: superlattice period and symmetry, substrate crystallographic orientation and strain. A strategy for manipulating these parameters to produce a direct bandgap semiconductor out of Ge and Si is discussed in terms of experimental data and theoretical calculations.

I. Introduction

A component of early research on Ge and Si was a persistent, but fruitless search to produce strong luminescence from these materials.[1,2] Although both Ge and Si have found applications in opto-electronics as photovoltaic detectors, so far no attempt to produce electro-luminescent sources from these materials has not been very successful. The motivation for developing solid-state light-sources is to obtain devices, grown on a Si substrate, for such applications as optical-fiber telecommunications, board to board computer back-plane links and intra-chip optical interconnects.

II. Getting Light Out of Si

Two approaches that are a subject of current research interest have addressed the challenge of obtaining useable electro-luminescence out of Si based structures. The first, and most successful of these has been the creation of a delocalized impurity band at 0.9 eV, about 200 meV below the indirect band edge in Si. Canham, Barraclough and Robbins at RSRE[3] have studied the possibility of doping Si heavily with an iso-electronic impurity to create a luminescence band in the same way that N-doping in GaP is used.[4]

Canham et al demonstrated that heavy electron bombardment, at the damage threshold of 13 eV, could be used to create an interstitial carbon-silicon complex known as a G-center. Using this technique, the electro-luminescence efficiency was raised by three orders of magnitude over a non-bombarded Si diode. Spectra comparing these two cases are shown in Fig. 1a and Fig. 1b. The carbon isovalent defect produces a luminescence band at 1.28μm whose spectral width is narrower at 77K than the corresponding width of an LED using III-V semiconductors[5]. Although the quantum efficiency is less than a few tenths of a percent, the narrow spectral linewidth compensates in part for this drawback. This technique appears to be quite promising and it is compatible with VLSI processing methods. It is too early to tell whether laser action in these diodes can be achieved, but these prospects appear dim because there is no simple way to make a double heterostructure, among other considerations. However, the demonstration alone of an LED operating at room temperature would be an impressive achievement.

At about the same time, a quite different approach was undertaken to modify the bandstructure of Ge-Si semiconductors grown on (110) GaAs by changing the fundamental symmetry of the underlying crystal structure[6]. Si molecular beam epitaxy can be used for

137

Y. I. Nissim and E. Rosencher (eds.), Heterostructures on Silicon: One Step Further with Silicon, 137–144.

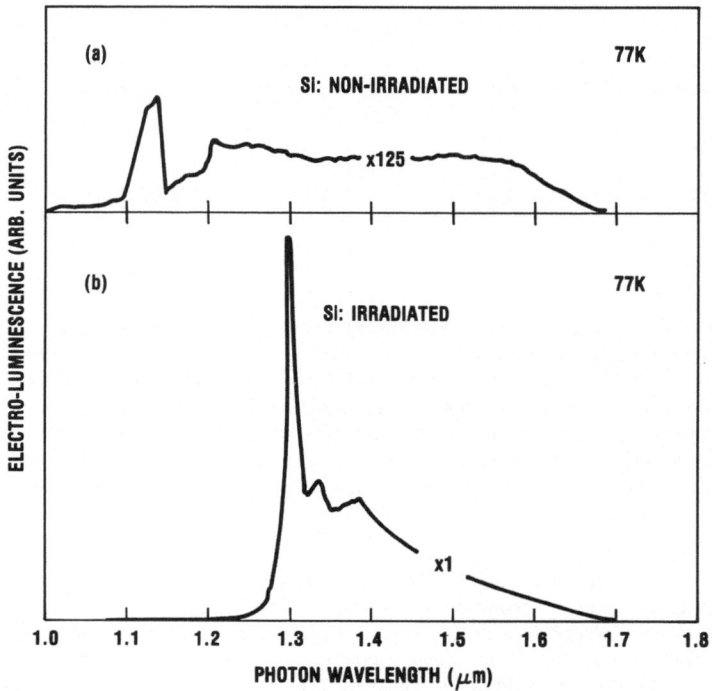

Fig. 1   a. 77K electroluminescence spectrum from a non-irradiated Si p-n junction.

b. 77K electroluminescence spectrum from an electron-irradiated Si p-n junction showing a sharp peak at X=1.28μm. The emission intensity in (b) is about 250 times stronger than in (a). These spectra are reproduced from Canham et al, Appl. Phys. **51**, 1509 (1987) with permission of the authors and the American Physical Society.

the deposition of alternating layers of Ge and Si only a few atoms in thickness. This procedure creates an artificial crystal structure with its own characteristic energy-level spectrum. This principle has been verified experimentally, and recent experimental results which show strong photoluminescence are promising evidence for the creation of a direct band-gap semiconductor. This kind of photoluminescence was first seen by Eberl et al at the Technical University of Munich, and a copy of their results is shown in Fig. 2.[7]

In the rest of this paper we will concentrate on this second approach and discuss the tools that can be used to pursue the goal of inducing a direct-gap Ge-Si superlattice.

III. Wavefunction Engineering

Initial spectroscopic experiments showed that new optical transitions in the infra-red below the Si bandgap were induced in 4-monolayer by 4-monolayer superlattices: Ge-Si (4:4).[8] Wavefunction mixing caused by the folding of the Brillouin zone was subsequently pointed out by People and Jackson.[9] Detailed bandstructure calculations[10-14] now show the importance of this effect on energy bands that lie along the direction of the superlattice axis. The creation of new energy levels with zone-center symmetry leading to a possible direct optical transition requires engineering on the electron wavefunction along the superlattice axis. The experimental parameters that can be modified are: The strain, the valence band offset, the orientation of the substrate, the number of Ge monolayers and the number of Si monolayers.

a. The Role of Strain

Pseudomorphic growth means that the lattice parameter of the superlattice in the plane of the substrate is the same as that of the substrate. Ge-Si strained-layer superlattices grown on Si are compressed by the smaller Si lattice parameter. The strain in the Ge layers is over 4%. If the structure is grown along the $<001>$ direction, then the $<100>$ and $<010>$ axes are in compression, while the unstressed $<001>$ axis is in extension. This strain geometry lifts the energy band directed along the $<001>$ superlattice axis in Ge and lowers the other two energy bands along the $<100>$ and $<010>$ directions. In the Si part of the structure there is no strain at all.

If the same superlattice is grown on $<001>$ Ge, then the Ge layers will be strain free. The Si layers will be in bi-axial extension and now the energy band along the $<001>$ direction will be lowered in energy, while the energy bands along the $<100>$ and $<010>$ will be raised. This feature is shown schematically in Fig. 3a and 3b.

b. The Role of Energy Band Offset

A heterojunction between Ge and Si perpendicular to $<001>$ has an energy offset that places the top of the Ge valence band about 0.5 eV higher than the valence band edge of Si. Strain imposed by the substrate will affect this offset further. This large band offset ensures that the conduction band edge of Si also lies below the conduction band edge of Ge. Since the conduction band edge of Si lies along the (100) directions, the band edge of superlattices of Ge and Si will tend to be Type II in nature with conduction bands formed from (100) X-like states. If the superlattice is grown on a Si substrate, the conduction band edge in Ge will be formed from (100) states lowered by the strain below the usual (111) minima. The conduction band of unstrained Si in the (100) directions will be still lower in energy. If the superlattice is grown on Ge, the conduction band edge will be formed from the strain-split $<001>$ conduction band in Si with the (111) minima in Ge slightly higher in energy. The fact that the Si band edge dominates the conduction edge bandstructure in Ge-Si superlattices is due almost entirely to the large valence-band offset.

140

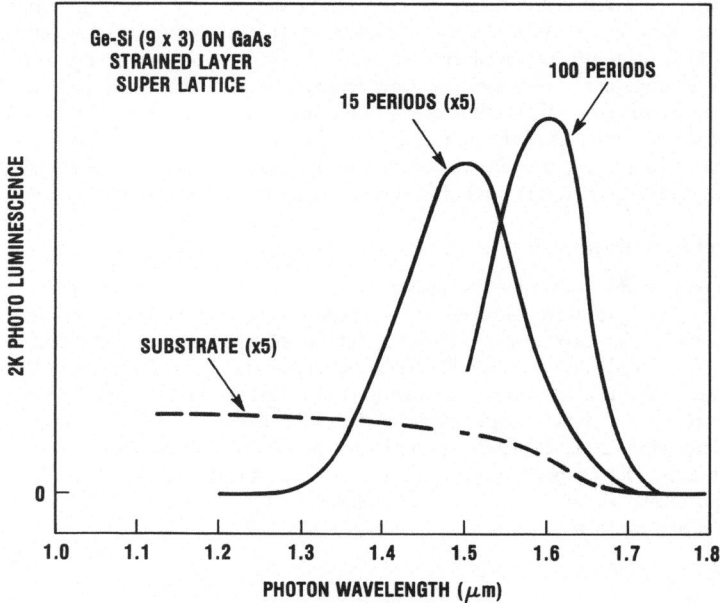

Fig. 2   2K photoluminescence spectrum for a strained Ge-Si atomic layer superlattice. These results support the notion of superlattice enhanced optical transitions. These spectra are reproduced with permission of the authors from K. Eberl, et al, Proc. MSS-III, Montpellier, France (1987).

c. Orientation of the Substrate

The substrate orientation determines how the perturbing effect of the large uniaxial strain along the growth direction affects the bandstructure. For example, if the Ge-Si superlattice were grown along the <111> direction on Si, then the Ge layers will be under strain. The Ge <111> direction will be raised in energy. The three remaining (111) directions will be lowered in energy. This strain splitting is greater than 300 meV, and the resulting superlattice becomes Type I with the lowest conduction band state residing in the Ge (111) minima.[15]

d. The Number of Monolayers of Ge and Si in the Superlattice Structure

There are fundamental limits that the number of atomic layers imposes on the electronic properties of Ge-Si superlattices. The smallest possible superlattice consists of alternating single monolayers of Ge and Si. Experimental results show that this superlattice is indistinguishable from a random alloy of the same average composition. At the other extreme, there is a maximum number of atomic layers per period that can support superlattice formation. The requirement here is for conduction and valence band electrons to tunnel freely from one set of Ge layers to the next. If the period of the superlattice is too long, the constituent layers will tend to behave more like slabs of bulk material, or coupled quantum wells.

It has been shown that bulk crystal potentials are established in Si and in Ge within two atomic monolayers.[16] This result implies that an (8:8) superlattice, for example, will consist mostly of material with bulk-like properties. Mechanical strain considerations limit the total layer thickness to fewer than 6 monolayers if the thickness of the Si and Ge regions are equal. This requirement coincides with the upper limit based on superlattice formation as well.

The above considerations suggest that superlattices of Ge_n and Si_m can be formed and grown on Si or on Ge if the number of monolayers n and m in each slab is between 2 and 5.

$$2 \leq n \leq 5$$

$$2 \leq m \leq 5 \qquad (1)$$

This restriction means that the slab thickness is large enough for the electrons not to wash out differences in potential, and that the number of monolayers per slab remains small enough for the electrons to "sense" more than one slab.

IV. Trends in Superlattice Formation

A picture of the complete effect of a superlattice on the energy band structure requires a true bandstructure calculation.[9-16] The simplified description of symmetry-induced changes that can be obtained by invoking energy band folding and the Krönig-Penney approximation can be useful for understanding certain trends related principally to the number of atomic monolayers in each superlattice slab.

In Fig. 4a we show, using band-folding ideas, how a bandstructure similar to that of Si and Ge would evolve as the superlattice dimensions go from 1 folding (2:2) to 2 foldings (4:4) to 3 foldings (6:6). The (6:6) structure is the maximum thickness allowed for the strained layer growth of 6 layers of Ge on Si. Because the minimum of the conduction band does not lie at the zone boundary, each successive folding makes the bandstructure look more direct, but each successive folding weakens the optical matrix element as the superlattice period becomes longer and longer.

(001) PSEUDO-MORPHIC EPITAXY

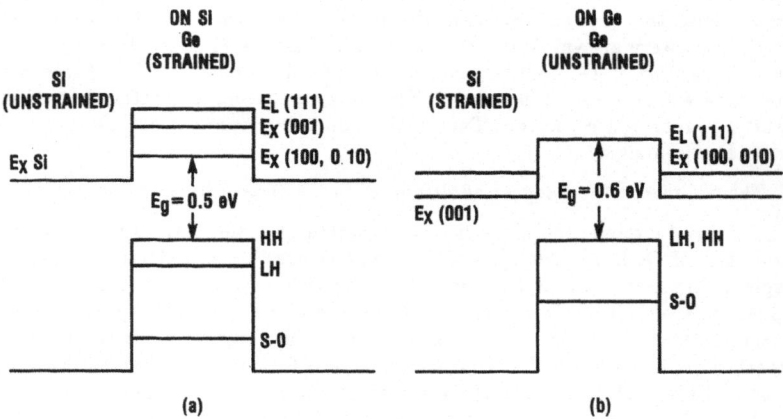

(a) (b)

Fig. 3   Conduction and valence band structure for a) Ge-Si grown on Si and (b) Ge-Si grown on Ge. These diagrams have been simplified by ignoring quantum confinement and zone folding effects. The large valence-band offset for (001)-oriented hetero-interfaces predisposes the superlattice to be Type II in nature with the Si conduction band states forming the conduction band edge.

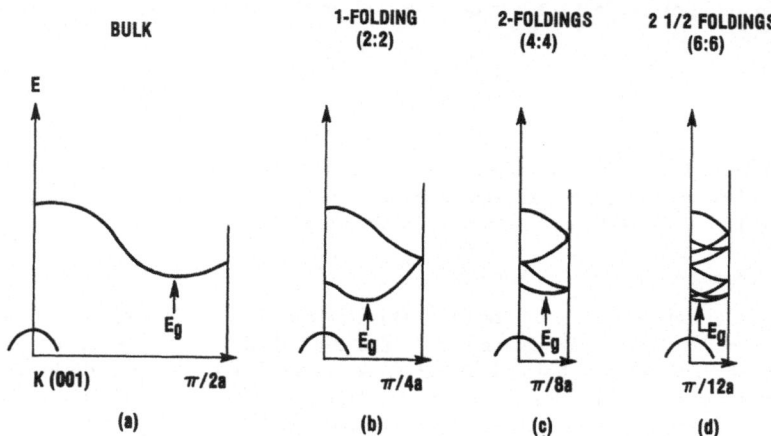

Fig. 4   Effects of multiple band-foldings are shown here for a series of superlattices with a Si-like bandstructure. As the number of layers per superlattice slab increases, the bandgap approaches the center of the Brillouin Zone. As the number of layers per slab increases, however, the effect of the superlattice potential diminishes exponentially.

Figure 4 shows also the importance of orientation of the superlattice with respect to the bandstructure of the semiconductor. In the case of (001)-oriented Si or Ge, the energy minimum along the (001) does not occur at a major symmetry point. Mixing zone edge and zone center states will create new levels at the zone center, but these will lie higher in energy than the minimum which does not get folded back to the zone center.

In general, one should try to pick the growth direction to coincide with a direction in k-space where the energy bands have a minimum at the boundary of the Brillouin zone. In the case of Si and Ge, this occurs along the <111> directions. In this manner, one can create a real minimum at the zone center with a single zone folding and thus maximize the effect of the superlattice on the optical matrix element. In order to create a pseudo-direct bandgap material, this new level must lie lower than any others in the conduction band. This condition may be filled if the sign of the strain imposed by the substrate lowers the energy of the conduction bands oriented along the axis of the superlattice. In the case of Si and Ge, this implies that the substrate have a lattice-constant larger than that of Si.

V. Conclusions

Some recent experimental results have extended some promise that Si-based semiconductor structures can be used to make electro-luminescent devices. Both iso-electronic doping and superlattice-induced mixing of conduction-band states have been used with some degree of success to increase optical transition rates. Much additional research must be done before it can be shown that either of these approaches is a technology for making practical light-emitting diodes or lasers.

REFERENCES

[1]   A. G. Chynoweth, in **Semiconductors and Semimetals**, Vol. 4, ed. R. W. Willardson and A. E. Bier (New York Academic Press) p. 307.

[2]   M. G. Bernard and G. Duraffourg, Phys. Stat. Solidi 1, 699 (1961).

[3]   L. T. Canham, K. G. Barraclough and D. J. Robbins, Appl. Phys. Lett. 51, 1509 (1987).

[4]   A. A. Bergh and P. J. Dean, **Light-Emitting Diodes** (Oxford, Clarendon Press, 1976).

[5]   T. P. Pearsall, L. Eaves, J. C. Portal, J. Appl. Phys. 54, 1037 (1983).

[6]   J. Bevk, J. P. Mannaerts, L. C. Feldman, B. A. Davidson and A. Ourmazd, Appl. Phys. Lett. 49, 286 (1986).

[7]   K. Eberl, G. Krötz, R. Zachai and G. Abstreiter. Proc. Conf. on Microstructures and Superlattices, Montpellier, 1987, J. de Physique 48 Supp. 11, C5-329 (1987).

[8]   T. P. Pearsall, J. Bevk, L. C. Feldman, J. M. Bonar, J. P. Mannaerts and A. Ourmazd. Phys. Rev. Lett. 58, 729 (1987).

[9]   R. People and S. Jackson, Phys. Rev. B15, 36, 1310 (1987).

[10] S. Froyen, S. M. Wood and A. Zunger, Phys. Rev. B15, 36, 4547 (1987).

[11] S. Ciraci and I. P. Batra, Phys. Rev. Lett. 58, 2114 (1987).

[12] M. S. Hybertsen and M. Schlüter, Phys. Rev. B15, 36, 9683 (1987).

[13] S. Satpathy, R. M. Martin and C. G. Van de Walle, submitted to Phys. Rev. (1988).

[14] I. Morrison, M. Jaros and K. B. Wong, Phys. Rev. B15, 35, 9693 (1987).

[15] R. People, J. C. Bean and D. V. Lang, Proc. Phys. of Semiconductors XVIII, p. 767 (1986).

[16] C. V. Van de Walle and R. M. Martin, J. Vac. Sci. Tech., B4, 1055 (1986).

GROWTH AND CHARACTERIZATION OF Si-Ge MULTILAYER
STRUCTURES ON Si(100)

J.-M. BARIBEAU, D.J. LOCKWOOD, M.W.C. DHARMA-WARDANA,
G.C. AERS AND D.C. HOUGHTON

Division of Physics, National Research Council of Canada, Ottawa, Canada, K1A 0R6

1. INTRODUCTION

Heteroepitaxy of $Si_{1-x}Ge_x$ on Si by molecular beam epitaxy (MBE) has been a subject of increasing interest in recent years (1). In spite of the large lattice mismatch between Si and Ge, high quality films of $Si_{1-x}Ge_x$ covering the entire range of alloy compositions have been grown utilizing strained-layer epitaxy methods. In this heteroepitaxial system, the misfit strain has dramatic effects on the band structure of the materials and can be used for tailoring their optoelectronic properties. This has been exploited for production of novel Si-based devices such as infrared photodetectors (2) and modulation-doped field-effect transistors (3). For these applications alternating layers of Si and $Si_{1-x}Ge_x$ were grown to form a superlattice structure. This has the advantage of increasing the total thickness of strained material attainable before plastic relaxation occurs. As the Si-Ge MBE technology matures, more complex heterostructures are being synthesized. For example, growth of $Si_{1-x}Ge_x$/Si Fibonacci superlattices has been reported (4). Such a superlattice, built according to the Fibonacci sequence is quasi-periodic and exhibits unusual structural properties (5). Resonant tunneling of holes across a double-barrier of Si buried in a $Si_{1-x}Ge_x$ layer has also been reported (6). Recently, growth of *artificial* crystals of Si_mGe_n made of alternating layers of pure Si and Ge has been demonstrated (7). The study of such materials is of much interest both from a basic physics point of view and for designing devices with unusual optical properties (8). In this paper, we report the growth of various $Si/Si_{1-x}Ge_x$ heterostructures including *thick* (~0.5 μm) periodic and Fibonacci strained-layer superlattices and *ultra-thin* (~10 nm) multilayers of Si_mGe_n (m, $n < 6$). The structural properties of these films are investigated by X-ray diffraction and Raman scattering spectroscopy.

2. EXPERIMENTAL

All the epitaxial layers discussed here were produced in a Vacuum Generators V80 MBE system. Both Si and Ge fluxes were obtained by e-gun evaporators. Deposition rates were carefully controlled by Sentinel III (9) optical sensors and flux shutters were computer controlled. The epitaxial films were grown at 500°C on 100 mm (100) Si wafers. Details on the substrate preparation and growth methodology can be found elsewhere (10). Deposition rates of 0.2-0.5 nm/s were used for the growth of the *thick* $Si_{1-x}Ge_x$/Si strained-layer superlattices. Rates were reduced to about 0.02 nm/s for the synthesis of the *thin* Si_mGe_n structures. Also, for the latter, a portion of the wafers was capped by a 10-15 nm Si protective layer.

The structural properties of the various heterostructures were studied using cross-sectional transmission electron microscopy (XTEM), double-crystal X-ray diffraction (DCD) and Raman spectroscopy . In DCD, 400 rocking curves were recorded in a non-dispersive geometry using a (100) Si first crystal and Cu K_α radiation (λ=0.154nm). The Raman spectra were measured in a 90° geometry with the sample (100) surface inclined at an angle of 12.3° to the incident light. The large refractive index of these materials makes this effectively a backscattering experiment inside the crystal. The samples were placed in an helium atmosphere at 295 K and the spectra were excited with 300 mW of 457.9 nm argon laser light and analyzed with a Spex 14108 double monochromator. Further details on the Raman measurements can be found elsewhere (11).

145

Y. I. Nissim and E. Rosencher (eds.), Heterostructures on Silicon: One Step Further with Silicon, 145–152.
© *1989 by Kluwer Academic Publishers.*

3. THEORY

3.1. X-ray diffraction

In the present study, the kinematical theory of X-ray diffraction has been used to analyze the experimental results (12-13). In the kinematical approximation, the diffracted amplitude is proportional to the Fourier transform of the spatial distribution of layered materials. The diffracted intensity $I(\theta)$ (neglecting absorption) about the 400 Bragg reflection θ_B for a superlattice consisting of N periods of $Si_{1-x}Ge_x/$ Si of thickness t_{Si} and t_{Ge} is given by

$$I(\theta) = \frac{\sin^2 NK(t_{Si}\Delta\theta_{Si}+t_{Ge}\Delta\theta_{Ge})}{\sin^2 K(t_{Si}\Delta\theta_{Si}+t_{Ge}\Delta\theta_{Ge})} \left[\left\{ \frac{F_{Si}}{V_{Si}} \frac{\sin K t_{Si}\Delta\theta_{Si}}{\Delta\theta_{Si}} \right\}^2 + \left\{ \frac{F_{Ge}}{V_{Ge}} \frac{\sin K t_{Ge}\Delta\theta_{Ge}}{\Delta\theta_{Ge}} \right\}^2 \right.$$

$$\left. + \left\{ 2\cos K(t_{Si}\Delta\theta_{Si}+t_{Ge}\Delta\theta_{Ge}) \frac{F_{Si}F_{Ge}}{V_{Si}V_{Ge}} \frac{\sin K t_{Si}\Delta\theta_{Si}}{\Delta\theta_{Si}} \frac{\sin K t_{Ge}\Delta\theta_{Ge}}{\Delta\theta_{Ge}} \right\} \right] \quad , \qquad [1]$$

where $K = \pi\sin 2\theta_B/\lambda\sin\theta_B$, $\Delta\theta_{Si,Ge} = \theta-\theta_B+ \varepsilon_{Si,Ge}\tan\theta_B$; F_{Si}, V_{Si}, ε_{Si} and F_{Ge}, V_{Ge}, ε_{Ge} are the structure factor, volume of unit cell and perpendicular strain (with respect to bulk Si lattice constant) of the Si and $Si_{1-x}Ge_x$ layers, respectively. From Eq. [1], the diffracted spectrum from a superlattice consists of a series of satellite reflections of angular spacing $\Delta\theta_n$ given by

$$\Delta\theta_n = \frac{\lambda\sin\theta_B}{(t_{Si}+t_{Ge})\sin 2\theta_B} \quad , \qquad [2]$$

with the zero order reflection centered at

$$-\Delta\theta_0 = \frac{t_{Si}\varepsilon_{Si}+t_{Ge}\varepsilon_{Ge}}{t_{Si}+t_{Ge}} \tan\theta_B \quad . \qquad [3]$$

Due to the limited thickness of the superlattice, the satellites have a finite width $\Delta\theta_p = \Delta\theta_n/N$ and the spectrum also exhibits secondary peaks of spacing $\Delta\theta_p$. Satellite intensity simulations using Eq. [1] were performed to obtain the thickness and strain of individual layers (13). Eqs [1]-[3] can also be applied in the case of *thin* multilayers. However, because of the very short period of these structures, the DCD spectrum will normally only display the zero order reflection which can be used to obtain the average composition and total thickness of the structure.

The jth generation Fibonacci superlattice is obtained from the recurrence relationship $S^j=S^{j-2}S^{j-1}$ where $S^1= \{A\}$ and $S^2= \{BA\}$, with A and B two different layer units. For a $Si/Si_{1-x}Ge_x$ Fibonacci superlattice, the position of the satellites is obtained from the following expression:

$$\sin\theta_{m,n} = \frac{2\lambda}{<a>} + \frac{\lambda(m\tau + n)}{2t_F} \quad , \qquad [4]$$

where $\tau = (1 + \sqrt{5})/2$, $t_F = \tau t_{Si} + t_{Ge}$ and <a> is the average lattice constant . Eq. [4] shows that a Fibonacci superlattice exhibits a complex spectrum consisting of an infinity of peaks that are subsequently labeled (m,n).

3.2. Raman scattering

If light of wavelength λ and momentum κ_i is incident on the superlattice surface, the large refractive index η of these materials ensures that the momentum k_i inside the crystal is essentially normal to the surface. If the scattered light has momentum k_s inside the crystal, then the change in momentum $\mathbf{q}$ in the scattering event is essentially backscattering-like, with $\mathbf{q}=\mathbf{k}_s-\mathbf{k}_i$ a vector of modulus $q=(4\pi\eta/\lambda)[1-1/4\eta^2]$.

In the case of a periodic superlattice made up of slabs of length t_a and t_b, the effective superperiodicity in the growth direction is determined by the mini-Brillouin zone vector $q_B=\pm\pi/t$, with $t=t_a+t_b$. In the simplest approximation the acoustic phonons can be thought of as being just the *bulk phonons* folded onto a minizone of $0<q<\pi/t$. This leads to the well known picture of folded acoustic modes with Raman peaks at $\omega_n=<v>(q_m\pm q)$ where $q_m=2\pi m/t$ and m=0, 1, 2, etc, and $<v>$ is the average sound velocity in the superlattice. In a Fibonacci superlattice there is no real periodicity but it can be shown that the quasi-periodicity t_F, with $\tau_F=\tau t_a+t_b$, controls the spectrum. Peaks occur at $\omega_{mn}=<v>(q_{mn}\pm q)$ where $q_{mn}=2\pi(m\tau+n)/t_F$. If m, n are any two Fibonacci numbers (ie, numbers in the sequence 0,1,1,2,3,5, etc...) then the peaks with neighboring Fibonacci numbers and occurring earlier in the sequence are dominant in the spectrum (4).

When we consider *thin* superlattices, the simple theory outlined above breaks down. This is because the infinite periodicity assumed for a large system with many layers becomes untenable for these strained layers having only hundred monolayers or so, as dictated by materials stability criteria (7). The lattice dynamics of the system inclusive of substrate and cap or surface oxide layer have to be considered explicitly. The *thin* multilayer system is analogous to a system of discrete states (of the multilayer) interacting with the continuum states (of the substrate) and can lead to localized states, band gap modes or resonant phonon modes. Our calculations using a linear chain model show that the intense broad peaks observed in the low frequency Raman spectra are indeed resonant phonon modes (14). Their position and width are sensitive to the modeling of the surface boundary conditions. The surface boundary condition is imposed via the parameter σ (see Fig. 9) which is zero for a freely vibrating surface, and unity for a fully anchored surface. This opens the possibility of using Raman studies for testing modes of the surface layers of these materials.

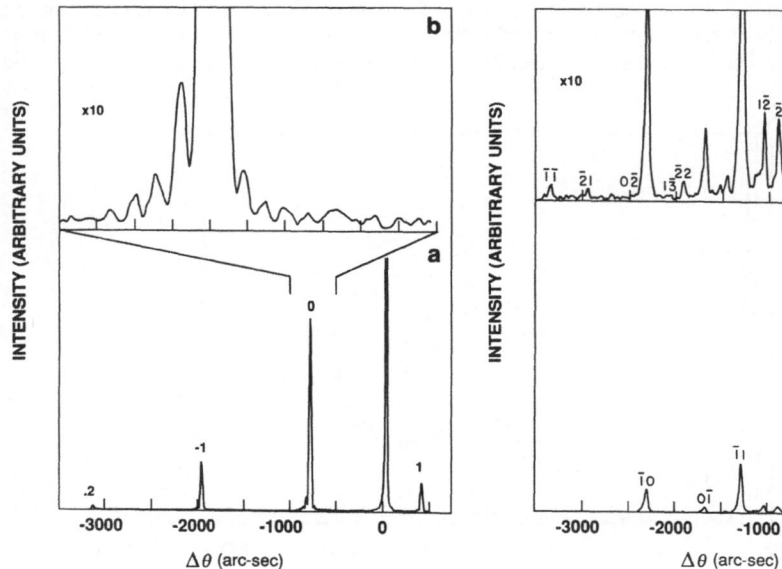

FIGURE 1. 400 rocking curve from a periodic Si/Si$_{1-x}$Ge$_x$ superlattice.

FIGURE 2. 400 rocking curve from a Fibonacci superlattice.

148

4. RESULTS AND DISCUSSION

4.1. $Si_{1-x}Ge_x$ periodic and quasi-periodic structures

Fig. 1 displays the 400 rocking curve from a 40 period superlattice made of alternating layers of Si and $Si_{1-x}Ge_x$. The spectrum exhibits strong satellite reflections with a spacing corresponding to a superlattice period of 16.5 nm. The total thickness of the superlattice is directly obtained from the secondary fringes seen each side of the satellites and magnified in Fig. 1(b). The spacing $\Delta\theta_p$ is 30±1" and corresponds to a total thickness of 640±20 nm. The sharpness of the superlattice reflections and the observation of the secondary peaks indicate that there are very little fluctuations of thickness or composition in this superlattice. Structural parameters were also obtained using Eq. [1]. For this superlattice the best agreement between experiment and theory is obtained using t_{Si}=10.3 nm, t_{Ge}=6.2 nm, x=0.17, $\varepsilon_{Si} = 1.1 \times 10^{-3}$ and $\varepsilon_{Ge}= 1.36 \times 10^{-2}$ (13). This result shows that most of the strain is accommodated by a lattice distortion of the alloy layers. Small residual strain in the Si layers is found and may result from the difference in thermal expansion coefficient of the two materials.

Fig. 2 shows the 400 rocking curve of a 11th generation Fibonacci superlattice build up with elementary units A and B consisting of Si and $Si_{1-x}Ge_x$ (89 layers A and 55 layers B). The spectrum exhibits a large number of peaks that can be labelled using Eq. [4]. From the position of the various reflections, values of $<a>$ = 0.544nm and t_F = 19.6 nm are obtained. No intensity calculation was performed for that structure. However, from the deposition rates used in this experiment, the thickness of individual layers is estimated at t_{Si} =8.0±0.5 nm and t_{Ge} = 5.0 ±0.5 nm corresponding to x = 0.2±0.02. Notice that the most intense peaks of the spectrum correspond to neighboring Fibonacci numbers in agreement with theory (4).

The Raman spectrum of the same periodic superlattice is shown in Fig. 3. Three strong lines at 287, 407 and 513 cm^{-1} are associated with the Ge-Ge, Ge-Si and Si-Si longitudinal optic (LO) phonons (15) in the alloy layers of the superlattice, while the intense line at 519 cm^{-1} corresponds to the LO phonon in the Si layers. As the bulk Si LO frequency is 520 cm^{-1}, a slight expansive strain is indicated confirming the X-ray results. The alloy layer phonon frequencies are shifted up in frequency from bulk alloy case (16) by an amount consistent with the strain value determined from X-ray study. The prominent features near 257 and 434 cm^{-1} are indicative of weak long-range order of Si and Ge atoms in the alloy layers (15).

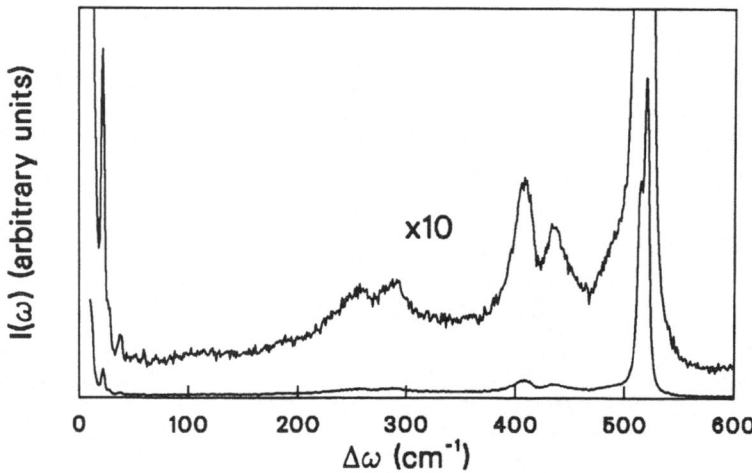

FIGURE 3. The unpolarized Raman spectrum of a periodic Si/$Si_{1-x}Ge_x$ superlattice.

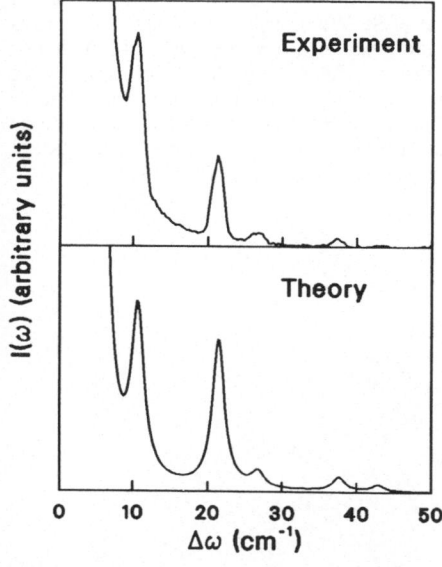

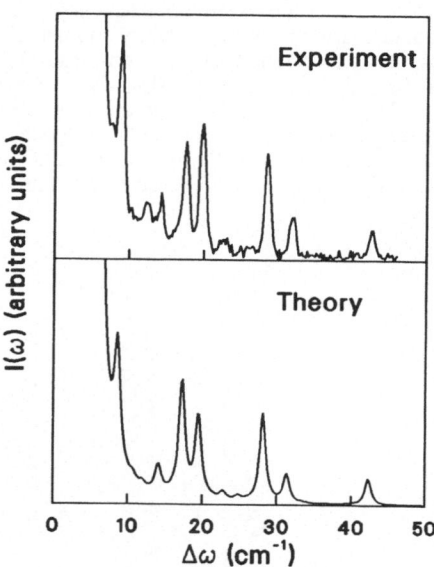

FIGURE 4. The unpolarized Raman spectrum of a periodic Si/Si$_{1-x}$Ge$_x$ superlattice.

FIGURE 5. The unpolarized Raman spectrum of a Fibonacci superlattice recorded at resolution of 0.8 cm^{-1}.

The sharp low frequency lines in Fig. 3 are shown in detail in Fig. 4. These lines are associated with the folded longitudinal acoustic (LA) phonons discussed earlier. These phonon lines are relatively intense, indicating the layer interfaces are sharp to within one atom layer. An analysis of the phonon LA frequencies and intensities (11) gives t_{Si}=10.5 nm and t_{Ge}=6.5 nm that are consistent with the values determined from X-ray data. The calculated spectrum is in good agreement with experiment (see Fig. 4).

The low frequency Raman spectrum of the Fibonacci superlattice is given in Fig. 5. The acoustic phonon spectrum exhibits a large number of intense sharp lines whose frequencies are in accord with the computed spectrum obtained from the theory outlined above using the structural parameters obtained from X-ray. The intensity of the lines are in agreement with experiment except for the lines near 20 cm^{-1}. This is most possibly due to the fact that the superlattice interfaces are not infinitely sharp, as required in the theory, and partly because a continuum model (4) has been used ignoring surface boundary and substrate effects.

4.2. Si$_n$Ge$_m$ multilayer structures

Several multilayer structures of Si$_m$Ge$_n$ (m,n<6) have been grown. The total thickness of these structures was less than 10 nm and do not exceed the critical thickness for plastic relaxation (7). Fig. 6 shows the XTEM lattice image of a multilayer structure consisting of 8 periods of Si$_4$Ge$_4$ with a 14 nm Si capping layer. In this micrograph the alternating four monolayers of Si and Ge are clearly seen. The waviness in the structure is believed to originate from the TEM sample preparation. The 400 DCD profile of the same structure is displayed in Fig. 7. Despite the reduced thickness of the epilayer, a very weak peak is systematically observed at –4200 ". The position and width of that reflection are consistent with a structure of thickness 14 nm and average Ge content $x = 0.41$ which suggests that the Si layers are actually slightly thicker than the Ge.

Fig. 8 shows the LA phonon spectrum of the Si$_4$Ge$_4$ structure. Although the optic phonon spectra of the uncapped and Si-capped superlattice are very similar, their respective acoustic phonon spectra are very different. A number of broad intense lines are observed at low frequencies, and their position and width are sensitive to the presence of a capping layer.

150

These features are too low in frequency to be associated with the folded modes of the Si$_4$Ge$_4$ structure, which has a periodicity comparable to that of Si. We interpret these new features as resonant phonon modes (14) arising from the interaction between the continuum of LA modes of the substrate and quasi-localized modes in the overlayer, which is, for these long wavelength modes, effectively a thin uniform *defect* layer attached to the substrate. The Si capping layer double the defect layer thickness and has the effect of decreasing the separations between the peaks of the resonances (14) as observed experimentally. A one-dimensional lattice dynamics model (14) accurately reproduces the observed spectrum (see Fig. 9).

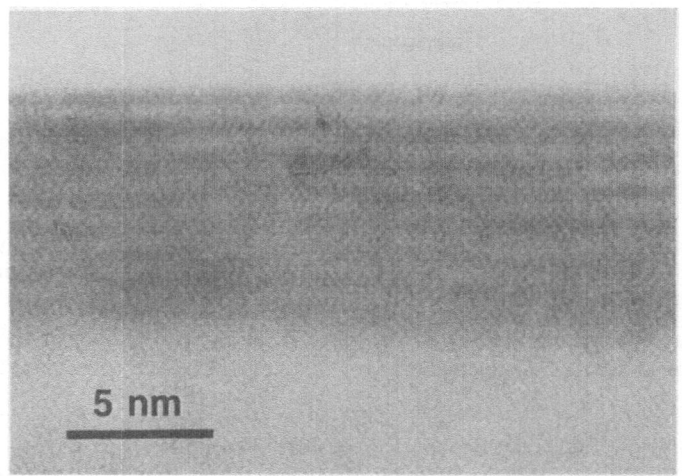

FIGURE 6. XTEM lattice image of a structure consisting of 8 periods of Si$_4$Ge$_4$.

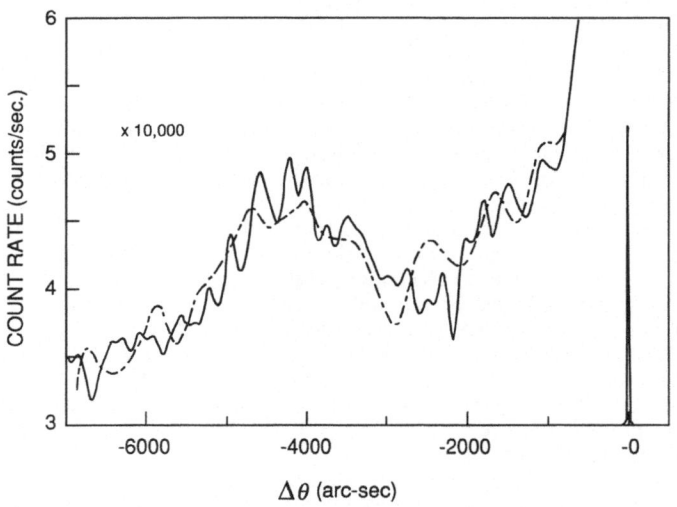

FIGURE 7. Two 400 rocking curves of the Si$_4$Ge$_4$ structure.

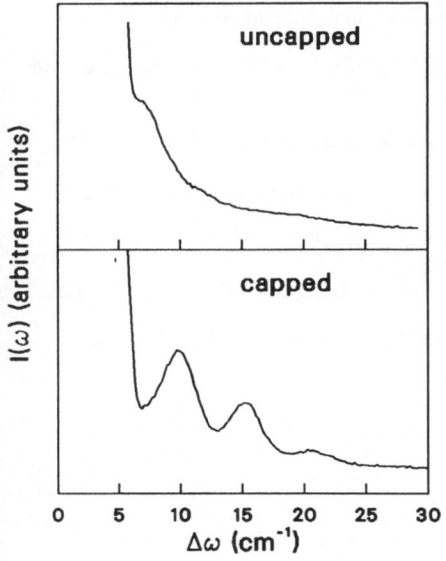
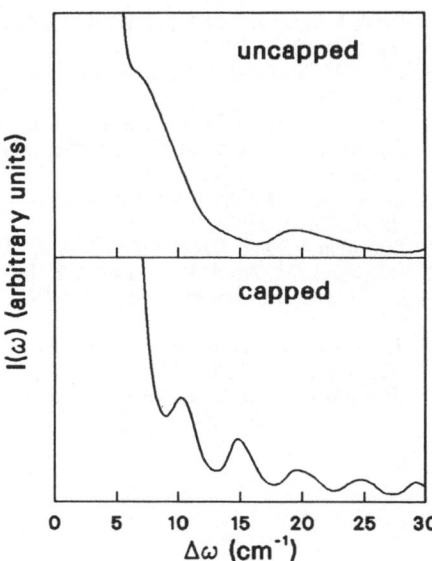

FIGURE 8. The experimental Raman spectrum of the Si_4Ge_4 structure, with and without a Si capping layer.

FIGURE 9. The theoretical Raman spectrum of the Si_4Ge_4 structure obtained for $\sigma=0.1$.

5. SUMMARY

We have presented an X-Ray diffraction and Raman scattering spectroscopy study of various Si-Ge multilayer heterostructures grown on (100) Si by MBE. The combination of these two techniques has proven very useful for the structural characterization of the *thick* $Si/Si_{1-x}Ge_x$ superlattices. Accurate determination of the dimension, composition and strain of the structures was obtained from a comparison between experimental and calculated spectra. The analysis confirmed the excellent crystalline quality of the grown material. We have also applied these techniques to the study of *ultrathin* Si_mGe_n multilayers. On these structures we have identified new resonant phonon modes arising from interaction between the continuum of acoustic modes of the substrate and the quasilocalized modes in the multilayer. The analysis of these modes has demonstrated the significant contribution of the substrate and boundary conditions at the surfaces on the long wavelength phonon spectrum.

REFERENCES

1. Bean JC, Phys. Today **39**, 36, 1986.
2. Pearsall TP, Bean JC, People R and Fiory AT, in *Proceedings of the 1st International Symposium on Silicon Molecular Beam Epitaxy* (Pennington, NJ, 1985), p. 633.
3. Daembkes H, Herzog HJ, Jorke H, Kibbel H and Kasper E, IEEE Trans Electron Devices **ED-33**, 633, 1986.
4. Dharma-wardana MWC, MacDonald AH, Lockwood DJ, Baribeau JM and Houghton DC, Phys. Rev. Lett. **58**, 1761, 1987.
5. Merlin R, Bajema K, Clarke R, Juang F-Y and Bhattacharya PK, Phys. Rev. Lett. **55**, 1768, 1985.
6. Liu HC, Landheer D, Buchanan M and Houghton DC, Appl. Phys. Lett., *in press*.

7. Bevk J, Mannaerts JP, Feldman LC, Davidson BA and Ourmazd A, Appl. Phys. Lett. **49**, 286, 1986.
8. Pearsall TP, Bevk J, Feldman LC, Davidson BA and Ourmazd A, Phys. Rev. Lett. **58**, 729, 1987.
9. Inficon Leybold-Heraeus, Schenectady, N.Y.
10. Baribeau JM, Jackman TE, Maigne P, Houghton DC and Denhoff MW, J. Vac. Sci. Technol. A **5**, 1898, 1987.
11. Lockwood DJ, Dharma-wardana MWC, Baribeau JM and Houghton DC, Phys. Rev. B **35**, 2243, 1987.
12. Speriosu VS and Vreeland T, J. Appl. Phys. **56**, 1591, 1984
13. Baribeau JM, Appl. Phys. Lett. **52**, 105, 1987.
14. Lockwood DJ, Dharma-wardana MWC, Aers GC and Baribeau JM, Appl. Phys. Lett., *in press* .
15. Lockwood DJ, Rajan K, Fenton EW, Baribeau JM and Denhoff MW, Solid State Commun. **61**, 465, 1987.
16. Brya WJ, Solid State Commun. **12**, 253, 1973.

REALIZATION OF SHORT PERIOD SI/GE STRAINED-LAYER SUPERLATTICES

K. EBERL, W. WEGSCHEIDER, E. FRIESS, AND G. ABSTREITER

Walter Schottky Institut
Technische Universität München, D-8046 Garching

ABSTRACT

We present experimental studies of growth, structural, and phonon properties of ultrashort period Si/Ge superlattices. The samples are grown by molecular beam epitaxy both on Si and Ge substrates with different orientations and strain distributions. Surface quality and lateral lattice constants are studied by LEED, interface sharpness by Auger Electron Spectroscopy and Raman measurements. Raman spectroscopy also gives information on Brillonin zone-folding effects and confined optical modes in thin layers.

1. INTRODUCTION

High quality growth of strained heterostructures based on epitaxial films of pure Si and Ge is a challange for both, technology and physics[1]. It is hoped that new artificial semiconductor structures can be achieved, which have a direct energy gap. Theoretical studies on pseudomorphic i.e. lattice matched, short period Si/Ge strained-layer superlattices (SLS) predict essential reduction of the minimum direct gap. Especially if the lateral lattice constant of the structure is larger or equal to the average of Si and Ge the direct gap is expected to become dominent on (100) substrates[2-4]. The conduction band minima of Si and Ge and also of Si_xGe_{1-x} alloys have Δ or L character depending on the strain conditions[5]. Thus zone folding induced modifications of the superlattice bandstructure using [100] and [111] growth directions are of special interest. First experimental evidence of new optical transitions were reported recently in Si/Ge SLS grown on Si(100)[6] and Ge(110)[7] buffer-layers. There are several possibilities to adjust the strain distribution in these superlattices. One can introduce a Si_xGe_{1-x} alloy buffer-layer[8] or grow an asymmetrically strained superlattice thick enough to relax it to an equilibrium lateral lattice constant[7,9]. Asymmetrically strained superlattices with a lattice constant close to a_{Ge} or a_{Si} are grown directly on Ge or Si substrates. In this article we report on studies of pseudomorphic growth, structural and dynamical properties of ultrashort period Si/Ge SLS grown on Si(100), Si(111), and Ge(100).

2. EXPERIMENTAL

The considerable lattice mismatch of about 4% provides an

153

Y. I. Nissim and E. Rosencher (eds.), Heterostructures on Silicon: One Step Further with Silicon, 153–160.
© 1989 by Kluwer Academic Publishers.

additional degree of flexibility in band structure engineering but also demands unconventional techniques for sample fabrication. We have grown our samples in a specially designed, partly homemade MBE-system equipped with Low Energy Electron Diffraction (LEED) and Auger Electron Spectroscopy (AES). As Si source we used a direct heated high purity Si bow connected to Ta electrical contacts. The flux-rate was kept constant within 5% by a careful current regulation. Ge was evaporated from a Knudson type effusion cell (PBN-crucible). The evaporation rates were calibrated with a quartz crystal microbalance at the sample position and kept below 2A/min for Si and 10A/min for Ge. Base pressure and pressure during growth are $< 4*10^{-11}$ mbar and $\approx 1*10^{-10}$ mbar respectively. The substrate (wafer of about 11x11mm) is heated by radiation. The substrate temperature (T_g) was monitored by a WRh thermocouple at the back calibrated with a second one that could be pressed against the surface and a pyrometer down to 450°C . For further description of the system see Ref. 10. Clean Si surfaces (AES detection limit is $\approx 1\%$ of a monolayer) have been obtained after slowly annealing the sample to 900°C for at least 5min without additional chemical pretreatment. It turned out to be important to remove any carbon contamination on top of the native oxide layer before this heattreatment. A Si buffer-layer of 200A was grown at T_g=550°C to reduce surface roughness. For Ge the situation is more difficult. The native oxide film can be removed already at a temperature of $\approx$ 500°C. It was not possible however to remove a carbon contamination of about 0.2 monolayers as detected by Auger measurements. Therefore we have grown a thick Ge buffer-layer of $\approx$2000A at T_g=440°C before starting with the super-lattice. Fig. 1 shows a typical layer sequence of the grown samples.

FIGURE 1.
Schematic diagram of
a typical sample.

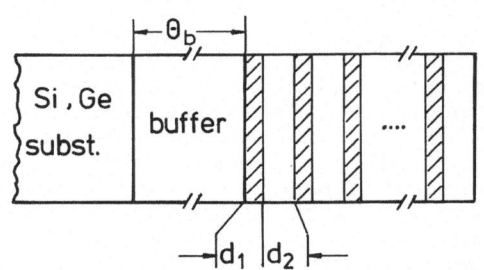

To study crystalline quality, lateral lattice constant and chemical composition of the structures the growth was interrupted repeatedly for coverage dependent LEED and AES measurements. LEED patterns have been obtained with electron energies of 20 to 400eV at a primary beam current of about 1μA. We use a vidicon camera to detect the distance and the energy dependance of profiles of the LEED spots. AES has been performed at a primary energy of 3keV and a beam current of 1μA. For quantitative analysis we evaluated the peak to peak height in the derivative spectra and normalized it to the values measured on corresponding clean buffer-layers.

3. RESULTS & DISCUSSION
There is a principal difference between epitaxial growth of Si

along [100] and [111] direction. Epilayers can be grown on Si(100) down to very low temperatures of about T_g=350°C in contrast to Si(111) where one needs at least 500°C for fairly perfect crystallinity. This is probably a consequence of the different atomic structures. On Si(111) there is a characteristic closely packed double-layer arrangement which leads most likely to nucleation in a double-layer mode. By means of LEED and RHEED intensity oscillations in Si coverages have been observed with double-layer and monolayer periodicity for Si(111) and Si(100) respectively[11,12]. Corresponding atomic steps on Si surfaces have been reported[13-15]. For Ge on Ge(100) we found good epitaxial growth down to T_g≈250°C.

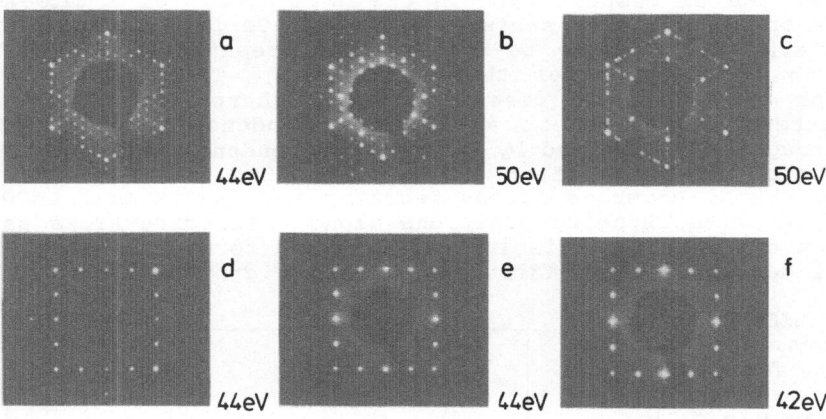

FIRURE 2. LEED pattern of: a) Si(111) buffer layer; b) 2ML Ge in the 30th period of a Si_6Ge_2 SLS on Si(111); c) 6ML Si on top of the surface providing Fig.2b; d) 9ML Ge in the 20th period of a Si_3Ge_9 SLS on Ge(100); e) 3ML Si on top of the surface providing Fig.2d; f) 4ML Si on Ge(100).

Preparing the Si(111) buffer as described above results in the well known, sharp (7x7) pattern (see Fig. 2a). The critical thickness for two-dimensional lattice matched growth of Ge on Si(111) is only ≈4ML [16]. AES measurements indicate considerable intermixing between a 4ML thick Ge film and the Si surface during deposition at T_g>350°C. Consequently for growing Si/Ge SLS on Si(111) it is necessary to switch the substrate temperature during growth. In Fig. 2b one sees the (5x5) reconstruction of 2ML Ge deposited at T_g=350°C after annealing to 500°C. It shows the surface of the completed Ge layer in the 30th period in a superlattice composed of a sequence of alternating 2ML Ge and 6ML Si (Si_6Ge_2 SLS) on Si(111) using T_g=350°C and 500°C for the Ge and Si layers respectively. Subsequent annealing of the Ge film to 500°C causes no intensity decrease of the Ge(47eV) Auger line and no increase of the Si(92eV) Auger line. The LEED pattern in Fig. 2c belongs to 6ML Si on top of the surface shown in Fig. 2b. It's a superposition of (7x7) and (5x5). In this superlattice the LEED pattern switches between Fig.2b and 2c up to the overall thickness of 30 periods. At least 14ML Si on a 2ML Ge film are

necessary to regain the pattern shown in Fig. 2a. These LEED
results and coverage dependent AES measurements within the Si
layers indicate that there is still some exchange between the
impinging Si atoms and the last Ge layer. Finally it can be
said that only within a very small window in the growth
parameter space it is possible to realize ultrashort period
Si/Ge SLS on Si(111). We did not observe any shift in the
distance of the LEED spots for a Si_6Ge_2 and $Si_{12}Ge_4$ SLS up to
240ML indicating full accomodation to the substrate. This
overall thickness corresponds to ≈380Å encounting tetragonal
deformation of the lateraly compressed Ge layers.
Fig. 2d, and 2e shows (2x1) LEED patterns from the surface of
9ML Ge and 3ML Si respectively in the 20th period of a Si_3Ge_9
SLS grown on Ge(100). The surface of the 9ML Ge film provides a
LEED pattern very similar to that of the prepared Ge buffer.
The growth temperature for this sample was T_g=310°C. After
deposition of 3ML Si we observe still a sharp and intense
(2x1) pattern. But there is a beginning tendency to enhanced
surface roughness indicated by an energy dependent inhomogenous
broadening of the spots after 4ML Si on Ge(100) (see Fig. 2f).
Exceeding 4ML Si coverage causes faceting i.e. additional LEED
spots appear with changing positions along {111} directions as
a function of energy[15]. This indicates island formation after 3
to 4ML Si depending on T_g (Stransky Krastanov growth mode).

FIGURE 3. LEED intensity
profiles across (11) and
(1$\bar{1}$) beams for the Ge
buffer and different Si
and Ge layers in a Si_3Ge_9
SLS. a) 30th period, b)
120th period.

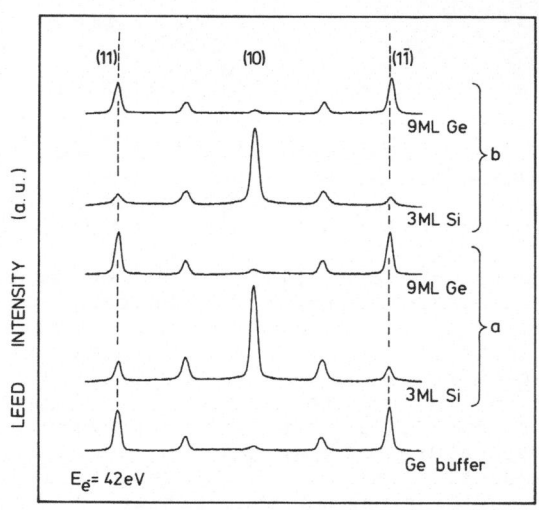

Thus the critical thickness (h_c) for pseudomorphic growth of Si
on Ge(100) is only 4ML. To determine the critical overall
thickness (h_c^{SLS}) of asymmetrically strained superlattices, we
evaluated the separation of the main LEED beams as a function
of coverage[7]. Fig. 3 shows profiles across the (11) and (1$\bar{1}$)
LEED spots of different surfaces in a Si_3Ge_9 SLS. The dashed
lines mark the spot separation of the Ge buffer. The profiles
in Fig. 3b belong to the 120th period of this superlattice and
indicate a lateral contraction of the lattice constant of a

little less than 1%. That means that the overall thickness is already beyond h_c^{SLS} consequently the Si layers relax partly to about 3% extention whereas the Ge layers are about 1% biaxially compressed. This change in the strain distribution is close to the expected equilibrium lattice constant. The critical thickness h_c^{SLS} of a Si_3Ge_9 SLS grown on Ge(100) at $T_g=310°C$ is found to be about 1200A. 120 periods of this superlattice correspond to a thickness of nearly 2000A.

The LEED pattern of a Si(100) substrate is uniform to that of Ge(100) (Fig. 2d) and it is due to a (2x1) reconstruction with scattering from two domains rotated by 90 degree[14]. Epitaxial growth of thin Ge films on Si(100) is at least in the initial stage (1-2ML) similar to the inverse situation on Ge discussed above. The first 3ML Ge grow in a two-dimensional mode followed by island formation with increasing coverage[17] It should be noticed that h_c is probably a little bit smaller than for Si on Ge(100).

FIGURE 4. Ratio of the AES intensities of the Si(92eV) and Ge(47eV) line as a function of the coverage in the 20th period of Si_3Ge_9 SLS's grown at different substrate temperatures.

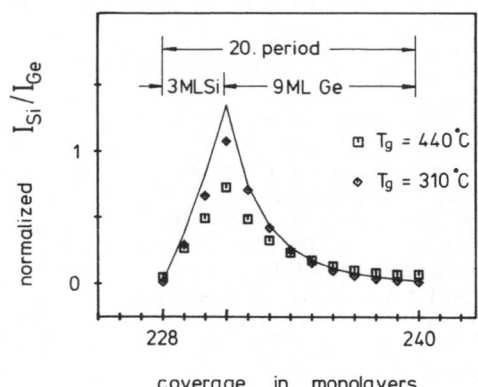

coverage in monolayers

We also determined the intermixing of the individual layers in a Si/Ge superlattice during deposition by means of AES performed between the growth intervalls. Beside the high energy Auger transitions of Si and Ge at 1147eV and 1619eV respectively we mainly evaluated the $Si(L_{23}VV)$ transition at 92eV and the $Ge(M_1M_2M_3$ Coster Kronig) lines at 47 and 52eV. The escape depth of these Auger electrons have been determined for deposition at room temperature and are about 5.5A for Si(92eV) and 7A for Ge(47eV). Fig. 4 shows the measured ratios between the intensity of this low energy Si and Ge Auger-lines normalized to the bulk intesities as a function of the coverage for two different growth temperatures. The full line reflects the calculated ideal ratio for sharp interfaces. Samples grown at $T_g=310°C$ are well described by this curve at least within the Ge layer . Significant deviations of the full line is observed for the experimental points obtained with $T_g=440°C$ in both the Si and Ge layers. There is considerable intermixing between the impinging Si or Ge atoms on the surface of the topmost layer during growth. LEED measurements on this superlattice show energy dependent spot broadening referring to enhanced surface roughness but not yet to three-dimensional growth which would lead to similar AES results. This assumption is confirmed by Raman spectroscopical measurements which will be discussed later.

Experimental results for pseudomorphic Si/Ge SLS grown on Si(100) substrates show that they are more difficult to realize than on Ge(100) substrates. The strain distribution in such a structure is inverse to the situation on a Ge substrate i. e. the Ge layers are 4% lateraly compressed. LEED and AES measurements on a Si_9Ge_3 SLS grown at $T_g=400°C$ on Si(100) indicate even more intermixing and reduced crystalline quality than the sample depicted in Fig. 4 prepared at $T_g=440°C$ on Ge(100). These results are in agreement with theoretical considerations predicting that for negative misfit ($a_{substr.} > a_{adsorb.}$) it is easier to achieve pseudomorphic growth due to anharmonic adsorbate interatomic forces[18]. Monte carlo simulations of epitaxial growth[18] also predict defect formation caused by strain induced diffusion for Si_xGe_{1-x} alloys on Si(100) for $x \leq 0.5$. The diffusion length calculated from bulk diffusion

FIGURE 5. Raman spectra of
a) a $Si_{0.25}Ge_{0.75}$ alloy of 240ML thickness;
b) Si_3Ge_9 SLS grown at $T_g=440°C$;
c) Si_3Ge_9 SLS grown at $T_g=310°C$.
All structures are grown on Ge(100) substrates.
The dashed lines indicate the position of "folded–LA" phonons appearing especially in the superlattice grown at the lower temperature.

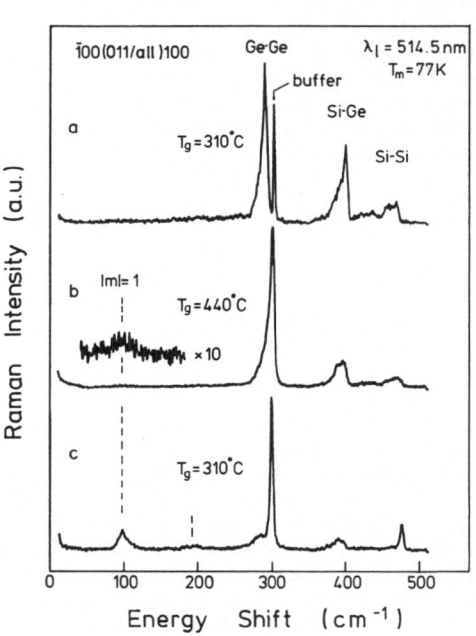

constants is negligible in the temperature range used. This is also demonstrated by recent stability experiments carried out on a symmetrically strained Si/Ge superlattice[19]. Finally we assume that intermixing dominantly takes place during growth, and the strain situation is an important parameter.

A versatile method to investigate growth and properties of such SLS structures is Raman spectroscopy[20]. Information on crystalline quality, composition, and strain distribution can be extracted from the energy positions, intensities, and line shapes of the various optical phonon modes. Additional modes appear for superlattices in the low energy region. They are due to Brillouin-zone "folding" of the longitudinal acoustic (LA) mode induced by the artificial periodicity along the growth direction. Fig. 5 shows three Raman spectra measured in backscattering geometry of a $Si_{0.25}Ge_{0.75}$ alloy and two Si_3Ge_9 SLS's grown at different substrate temperatures. The two superlattices are the same one as used for the AES studies shown in Fig. 4. The spectra b) and c) demonstrate the effects of intermixing between the individual layers. An indication of the excellent quality of the SLS grown at $T_g=310°C$ is the narrow linewidth especially of the Ge-Ge and Si-Si optical phonon mode.

FIGURE 6.   Raman spectra of a Si_4Ge_{12} SLS of 15 periods grown at $T_g=310°C$.

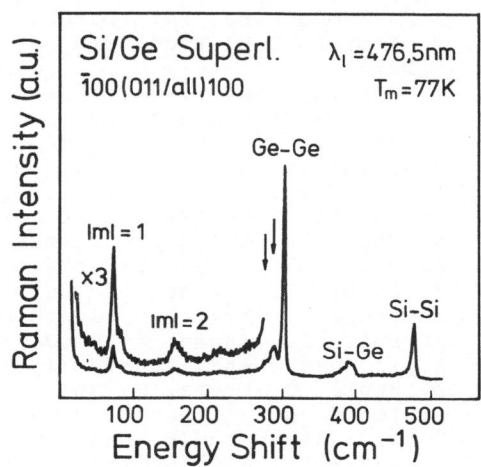

The satellite peak at the low energy side of this line corresponds to a "confined" optical mode[21] which is observed clearly in the spectrum shown in Fig. 6 using Si_4Ge_{12} SLS. Another characteristic feature for the formation of sharp interfaces is the small intensity of the Si-Ge interface mode. The shift to higher energy of the Si mode in the spectrum shown in Fig. 5c is due to the partly relaxed Si layers. The thickness of this sample is with 120 periods beyond h_c^{SLS} in contrast to the samples a) and b) which are both 240ML thick. "Folded LA" phonons (dashed lines) are due to the periodic concentration profile which still exists in the sample grown at $T_g=440°C$. For a superlattice with sharp interfaces i.e. rectangular function of composition a series of folded peaks is expected. The decay in intensity is related to the fourier coefficients of the concentration profile and thus provides additional information on the interface sharpness[22]. The energy position of the first "folded" peak ($|m|=1$) for d=12ML (Fig. 5c) and d=16ML (Fig. 6) is in good agreement with an elastic continuum theory using an average sound velocity. For the higher energy peaks deviations are observed which can be understood by taking into account the correct dispersion of the existing modes. Corresponding calculations appeared recently in the literature[23]. A quantitative analysis of the new experimental data is under way.

160

ACKNOWLEDGEMENTS
We thank Dr. Glasow from Siemens AG Erlangen for kindly
providing the Ge substrates. The work was supported financially
by the Deutsche Forschungsgemeinschaft and by Siemens AG.

REFERENCES

1a) See, e.g., Springer Series in Solid State Sciences, edited
    by G. Bauer, F. Kucher, and H. Heinrich, Springer Verlag
    Berlin Heidelberg, Vol. 67, (1986);
 b) Mat. Res. Soc. Proc. edited by M. Gibson, G.C. Osbourn and
    R.M. Tromp, MRS, Pittsburgh, Vol. 56, (1986)
2a) S. Froyen, D.M. Wood, and A. Zunger, Phys. Rev. B.36, 4547
    (1987)
 b) I. Morrison and M. Jaros, Phys. Rev. B.37, 916 (1988)
3. M.S. Hybertsen and M.Schlüter, Phys. Rev. B.36, 9683 (1987)
4. R. People and   S.A. Jackson, Phys. Rev. B.36, 1310 (1987).
5. C.G. Van de Walle, R.M.Martin, Phys. Rev. B.34, 5621 (1986)
6. T.P. Pearsall, J. Bevk, L. Feldmann, J.M. Bonar, and J.P.
    Mannaerts, Phys. Rev. Lett. 58, 729 (1986)
7. K. Eberl, G. Krötz, R. Zachai, and G.Abstreiter, Journal de
    Physique C5 no 11, 329 (1987)
8. E. Kasper, H. Kibbel, H. Jorke, H.Brugger, E.Friess, and G.
    Abstreiter, Phys. Rev. B (to be published)
9. R. Hull, J.C. Bean, f. Cerdeira, A.T. Fiory, and J.M.Gibson
    Appl. Phys. Lett. 48, 56 (1986)
10. K. Eberl, G. Krötz, T. Wolf, F. Schäffler, and G.Abstreiter
    Semicond. Sci. Technol. 2, 561 (1987)
11. K.D. Gronwald and M. Henzler, Surf. Sci. 117, 180 (1982)
12. T. Sakamoto, N.J. Kawai, T. Nakagawa, K. Ohta, and T.Kojima
    Appl. Phys. Lett. 47, 617 (1985)
13. J.A. Martin, C.E. Aumann, D.E. Savage, M.C. Tringides, M.G.
    Lagally, W. Moritz, and F. Kretschmar, J. Vac.Sci. Technol.
    A 5, 615 (1987)
14. W. Weiss, D. Schmeisser, and W. Göpel, Phys. Rev. Lett. 60,
    1326 (1988)
15. M. Henzler, Surf. Sci. 152/153, 963 (1985)
16. P.M.J. Marée, F.M. Mulders, K. Nakagawa, J.F. van der Veen,
    K.L.Kavanagh, Surf. Sci. (to be published) and Ref. therein
17. Y. Kataoka, H. Ueba, and C. Tatsuyama, J. Appl. Phys. 63,
    749 (1988)
18. A. Kobayashi and S. Das Sarma, Journal de Physique C5 no
    11, 329 (1987); S. Das Sarma, S.M. Paik, K.E. Khor, and A.
    Kobayashi, (to be published)
19. H. Brugger, E. Friess, and G. Abstreiter, E. Kasper, and H.
    Kibbel, (to be published)
20. G. Abstreiter, and H. Brugger, Proc. 18th Int. Conf. Phys.
    of Semicond., Stockholm 1986, edited by O. Engström, Wourld
    Scientific, Singapure, Vol. 1, 739 (1987)
21. E. Friess, H. Brugger, K. Eberl, and G. Abstreiter, Phys.
    Rev. B, (to be published)
22. C. Colvard, T.A, Gant, M.V. Klein, R.Merlin, R.Fischer,
    H. Morkoc, and A.C. Gossard, Phys. Rev. B.31, 2080 (1985)
23. A. Fasolino and E. Molinari, Journal de Physique C5 no 11,
    569 (1987)

DOPANT SEGREGATION AND INCORPORATION IN MOLECULAR BEAM EPITAXY

S. Andrieu*, F. Arnaud d'Avitaya, J.C.Pfister.

Centre National d'Etudes des Telecommunications, 38243, Meylan, France.
**I.S.A. Riber*, Bd national, 92503 , Rueil-Malmaison, France.

1-INTRODUCTION.

In Molecular Beam Epitaxy (MBE), doping is performed by coevaporation of the semiconductor material(s) and the dopant, opening and shutting the dopant shutter source. Since the growth temperature is quite low (typically 500°C for GaAs and 700°C for Si), the lack of bulk diffusion should lead to abrupt profiles. However, a smearing of the profile as well as a dopant surface enrichment is observed for many dopants both in Si (Sb^{1-9}, $Ga^{7,8,10}$, In^{11-13}, $As^{14,15}$) and GaAs ($Sn^{16,17,34,35}$, $Mg^{33,18}$, Mn^{18}). Moreover, incomplete incorporation of the incident dopant flux is reported for dopants where desorption is significant (Mg in $GaAs^{33}$ and In^{13}, Sb^6, Ga^8, in Si).

The desorption phenomenon is not sufficient to explain this behaviour. Incomplete incorporation can be due to a very low sticking coefficient, i.e. for a desorption lifetime shorter than the time taken to complete an epitaxial layer (assuming a two dimensional growth). However, this is never the case with the all dopants used. Moreover, a very low desorption lifetime cannot lead to doping profile transients and dopant surface enrichment. Consequently, this behaviour can only be explained by the occurrence of another phenomenon. In fact, during the time taken to complete a layer, the incident dopant flux is not sufficient to constitute the dopant surface concentration. Consequently, this surface enrichment can only be explained by a segregation of dopant atoms. Several models have been proposed.

Barnett and Greene[19,20] have assumed a segregation mechanism by a backward diffusion of incorporated dopant atoms from a several nanometer thick bulk layer below the surface. Therefore, they proposed a higher dopant diffusion coefficient near the surface. Their calculated profiles were then compared to experimental ones by fitting this diffusion coefficient, and a good agreement between experiments and theory was obtained. However, MBE being a non equilibrium phenomenon, the application of thermodynamics is not justified. Indeed, incorporation enthalpy is found to be negative for Sb and Ga[19], contrary to In in Si[12].

Tabe and Kajiyama[5] have considered a system of three phases (vapour, adsorbed and bulk phases) with exchanges between them. The segregation phenomenon was taken into account by the exchange between the bulk and adsorbed phases. In the case of Sb in Si, a good agreement between theory and experimental data is obtained. However, such an approach is also based on thermodynamic considerations.

Iyer, Metzger and Allen[8] have proposed a phenomenological equation concerning the dopant surface concentration. They did not take into consideration the segregation phenomenon explicitly but they drew up a review of the dopant fluxes on the surface, the incident flux, the desorbing flux, and the incorporated atom flux. They consequently introduced an arbitrary incorporation time. In spite of the absence of an explanation of the segregation phenomenon, the experimental results, and especially the high surface concentration, were well described by such an equation. In fact, the segregation phenomenon is implicitly taken into account in this incorporation time. Besides, many workers have successfully applied this phenomenological equation to explain their observations[6,22,33].

161

Y. I. Nissim and E. Rosencher (eds.), Heterostructures on Silicon: One Step Further with Silicon, 161–168.
© *1989 by Kluwer Academic Publishers.*

Consequently, we think that any kind of model has to predict such an equation.

In this paper, we propose a segregation phenomenon occurring on the surface[21]. By solving the system in a continuous way, we come accross the Iyer, Metzger and Allen equation. But the incorporation time is thus made explicit.

2-SEGREGATION PHENOMENON HYPOTHESIS.

2-1-Model foundations and predictions.

To explain the epitaxial growth, Burton, Cabrera and Frank [23] (B.C.F.) have proposed a two dimensional (2D growth) process : under certain conditions, the crystal grows layer by layer, and before starting another layer, the underlayer has to be completed. This growth type occurs in MBE for appropriate substrate temperatures T_{sub}. In the case of Si growth, the improved verification is probably given by recent LEED observation[24] and RHEED oscillation results[25-27]. However, certain discrepancies in this 2D growth process seem to appear in GaAs[28,29].

In B.C.F. theory, the growth layer is delimited by steps, which contain a certain amount of kinks. The atoms constituting the epitaxial layer are adsorbed on the surface, and diffused towards the steps and kinks. Consequently, steps progress in order to complete the layer. The dopant atoms impinging on the surface follow the same course. In the model, we suppose that an impurity on a kink can climb over the step. Therefore, this atom is not necessarily incorporated but stays on the surface. Such a mechanism is perfectly possible[30] since dopants diffuse on the surface and a kink or a step represents only a surface potential perturbation. According to this hypothesis, two kinetic equations can be written for a growth layer : one concerning the number of dopant atoms on the underlayer, the other on the layer in the process of formation (figure 1), as :

$$\frac{dn^k_0}{dt}(t) = (F_D - F_{Ddes}).S^k_0(t) - n_{Dinc}(t) - n_{Dclimb}(t) \qquad (1)$$

$$\frac{dn^k_h}{dt}(t) = (F_D - F_{Ddes}).S^k_h(t) + n_{Dclimb}(t) \qquad (2)$$

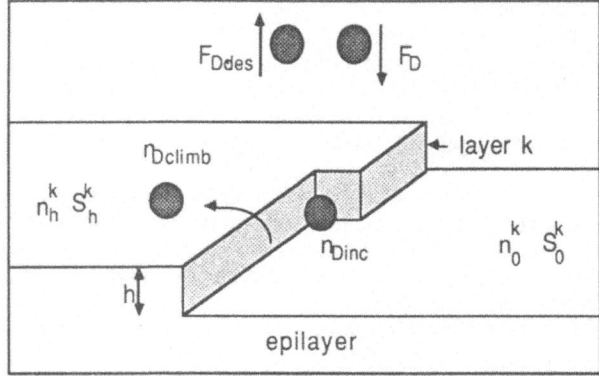

figure 1 : Schema of the surface during growth. F_D is the incident dopant flux, F_{Ddes} the desorbing flux, S^k_0 the surface of the underlayer, n^k_0 the number of dopant atoms on this surface, S^k_h the surface of the growth layer with n^k_h dopant atoms. n_{Dinc} and n_{Dclimb} are respectively the numbers of incorporated and climbed dopant atoms per unit time.

Since the incident epitaxial material flux is constant during growth and its desorption is insignificant (in the case of Si MBE see ref. 31), the surfaces S^k_0 and S^k_h are linearly dependent on time[21]. Concerning the desorbing flux, a first order law is taken into account. In fact, this generally seems to be the case[32]. Furthermore, we consider that steps move towards dopant atoms (the contrary hypothesis should give the same final results). Consequently, we have :

$$n_{Dinc}(t) = \alpha \cdot \frac{n^k_0}{S^k_0} \cdot \left| \frac{dS^k_0}{dt} \right| \tag{3}$$

$$n_{Dclimb}(t) = \gamma \cdot \frac{n^k_0}{S^k_0} \cdot \left| \frac{dS^k_0}{dt} \right| \tag{4}$$

where α and γ are respectively the impurity probabilities of being incorporated or climbing over the step. Consequently, $\alpha + \gamma = 1$ (since the desorption is taken into account at any moment), so we will only use α in equations. Such a system with boundary conditions allows for analytical solutions[21]. Moreover, the total surface concentration ρ_s can be written upon n^k_0 and n^k_h. By treating the system in a continuous way[21], we then come back to the phenomenological equation of Iyer, Metzger and Allen[8], as :

$$\frac{d\rho_s}{dt} = F_D - \frac{\rho_s}{\tau_{des}} - K_i \cdot \rho_s \tag{5}$$

and the incorporation coefficient K_i is now made explicit using the incorporation coefficient α, and the time to complete a layer T (related to the growth rate v_g, and the height of a step h), as :

$$K_i = \frac{\alpha}{T} = \frac{\alpha \cdot v_g}{h} \tag{6}$$

By solving equation 5, different cases appear and can then be compared with experiments.

2-1-Comparison with experiments.

As the doping level N_{DB} is given by the ratio of the incorporated flux $K_i \cdot \rho_s$ to the growth rate, the following expression is obtained at steady state :

$$N_{DB} = \frac{K_i}{K_i + 1/\tau_{des}} \cdot \frac{F_D}{v_g} = \frac{\alpha}{\alpha + q} \cdot \frac{F_D}{v_g} \quad \text{with } q = T/\tau_{des} \tag{7}$$

When the desorption lifetime is so short that the desorbing flux is much higher than the incorporated flux (i.e. $K_i \ll 1/\tau_{des}$, occurring at high substrate temperatures), the experiments show that the doping does not depend on the growth rate (Sb in Si[5] and Mg in GaAs[33]). Such a result is proof of the dependence of K_i on v_g. Wood et al[33] have followed the same reasoning in order to understand the Mg doping results in GaAs. Moreover, the solution to equation 5 leads to three cases (Figure 2) which actually describe the observations especially with B[36,37], Sb[6], In[11-13], Ga[8] in Si, and Mg[33], Sn[34] in GaAs.

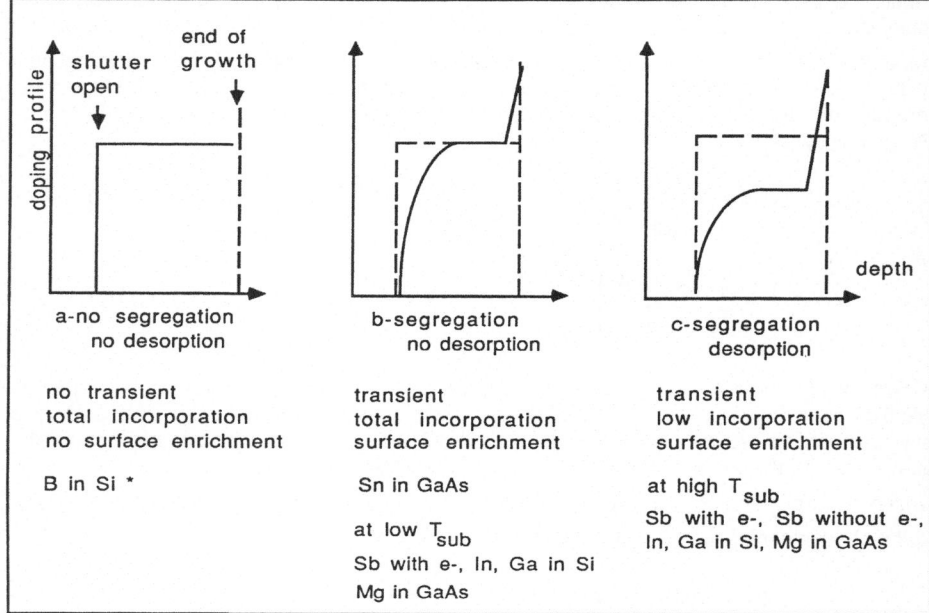

Figure 2 : Influence of desorption and segregation phenomenon on the doping profile, illustrated by experimental examples.(*In fact, B segregates but its segregation effect is small and difficult to observe[36,37,21]).

In the second stage of the study , the relation between the incorporation coefficient α and the substrate temperature T_{sub} is examined. In order to characterize a dopant, the incorporation probability σ, defined as the ratio of the measured to the ideal doping level (considering total incorporation), is usually drawn versus T_{sub}, for a fixed dopant incident flux and growth rate. According to the model, σ is expressed as :

$$\sigma = \frac{1}{1 + \dfrac{q}{\alpha}} \tag{8}$$

Experimentally, σ is an exponential function of $1/ T_{sub}$ at high temperatures. Moreover, the desorption lifetime is described by an Arrhenius' law (Ga[8], sb[6] in Si, Mg[33] in GaAs, see ref. 32 in the general case), the activation energy being the dopant desorption energy on the crystal. Consequently, the incorporation coefficient α is also described by such a law, as :

$$\alpha = \alpha_0 . \exp E_i/ T_{sub} \tag{9}$$

From knowledge of σ at high temperatures and τ_{des}, energy E_i and coefficient α_0 can be calculated[21], as shown in Table 1 for several dopants in Si and GaAs. Consequently, incorporation energy E_i is found to be positive, except for Ga and Sb without electron irradiation in Si. In fact, these two dopants are particular. Firstly, Sb is evaporated as Sb_4 clusters : a dissociation is then necessary, implying a dissociation activation energy (if the dissociation is thermally activated, which is the case at high temperatures[38]), which is not taken into account here. Moreover, the Sb doping behaviour at low temperatures[6] is not explained by this model, as well as the surface concentration evolution with the doping level[5,9,21]. Secondly, concerning Ga, it is not so clear, but this dopant presents some anomalous behaviour[8,39].

dopants	crystals	E_σ (eV)	E_d (eV)	α_0	E_i (eV)	ref.
Sn	GaAs (100)	no desorption*		5E-12	1.35	34,21
Mg	GaAs (100)	5.2	2.4	7E-22	2.8	33,21
In	Si (100)	2.7	1.5		1.2	13,21
Sb with e-	Si (111)	3.9	2.4	8E-11	1.5	9,21
B	Si (111)	no desorption			1.8**	21
Sb without e-	Si (111)	2.0	2.4		-0.4	6
Ga	Si (100)	1.6	2.9		-1.3	8

TABLE 1 : Activation energies for several dopants. E_σ is the incorporation probability activation energy, E_d the desorption energy, E_i the incorporation energy. *In fact, the desorption effect can be seen at high temperatures[35]. **Calculated upon the estimation of ref.21 where $\alpha \approx 7.10^{-2}$ at Tsub = 650°C, taking $\alpha_0 = 10^{-11}$.

Consequently, the incorporation energy has a positive value in normal cases. Moreover, such an assumption is confirmed by the increase in surface concentration with increasing temperature when only segregation occurs (case of B in Si[21] and Sn in GaAs[35]). It should also be noted that the E_i calculated values are consistent with a step climbing process : for example, the energy to take over a Si atom on a kink is about 2.2 eV [31,40], and a little less was found for In, Sb with e-, and B in Si. In the case of Mg in GaAs, the calculated incorporation energy is quite high. In fact, α_{Mg} can be written as the square of the regular incorporation coefficient found ($\sqrt{\alpha}_{0Mg}$ n 10^{-11} and $E_{iMg}/2$ =1.4 eV). Such a result leads us to suppose that a Mg atom should be incorporated with another impurity, but it is not clear at this stage of investigation. In fact, this model is probably not sufficient to explain the doping in GaAs in all cases, since several phenomena are in competition : imperfect 2D growth[29], desorption affected by As_4 pressure[33], two possible incorporation sites, Ga or As, formation of complex...

3-DOPING PROFILE CONTROL

This study shows that profile smearing is a consequence of both segregation and desorption. In fact, the transients are due to the time needed for the dopant surface concentration to reach a steady state. Consequently, following the surface concentration during growth will allow control of the doping level. This is why we have performed Sb surface concentration measurements on Si in the monolayer range using an in situ spectroscopic ellipsometry technique[41]. We have shown that the ellipsometric signal cosΔ is linearly dependent on the Sb coverage at room temperature, as reported earlier[42-44], and is sensitive to less than 0.1 monolayer. Indeed, it is possible to follow in real time the adsorption and desorption kinetics of Sb on Si at higher temperatures. Figure 3 shows an adsorption kinetics at low temperature, where desorption is insignificant. This result is in agreement with Metzger and Allen observations[6], who have measured a desorption lifetime of

10^5 seconds at this temperature. Consequently, the adsorption kinetics must theoretically increase linearly with time, which is actually observed. Moreover, a saturation of the surface concentration occurs. In fact, when the surface is covered with antimony, corresponding to 0.8 monolayer[5,9] (since Sb atoms are larger than Si ones), antimony does not stick on this new surface as the temperature is too high. Indeed, we have observed the desorption of submonolayers (after room temperature deposition) at a temperature as low as 200°C. Such behaviour allows the incident dopant fluxes impinging on the surface to be calculated. On the contrary, Figure 4 shows the adsorption-desorption kinetics for a substrate temperature of 700°C. Knowing the incident dopant fluxes and using the surface concentration saturation, the desorption lifetime can be calculated, as at steady state $\rho_s = F_D \tau_{des}$. However, we have observed variations depending on the pre-cleaning. Particularly, if an oxide remains on the surface, no antimony adsorption is observed. Therefore, further investigations using Si deposition[47] are in progress.

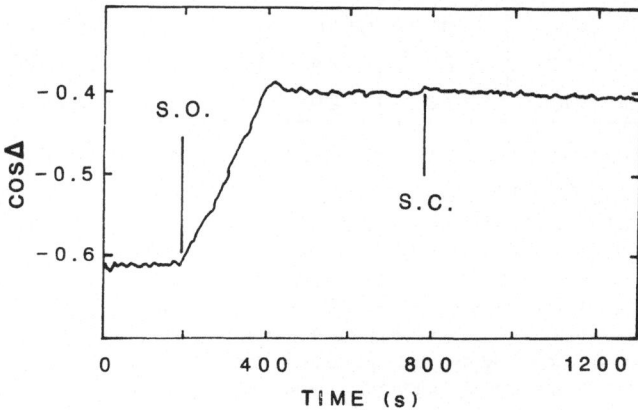

Figure 3 : Adsorption kinetics of Sb on Si(111) at T_{sub}=600°C , for a temperature cell T_{cell}=275°C. (S.O. : shutter open, S.C. : shutter closed).

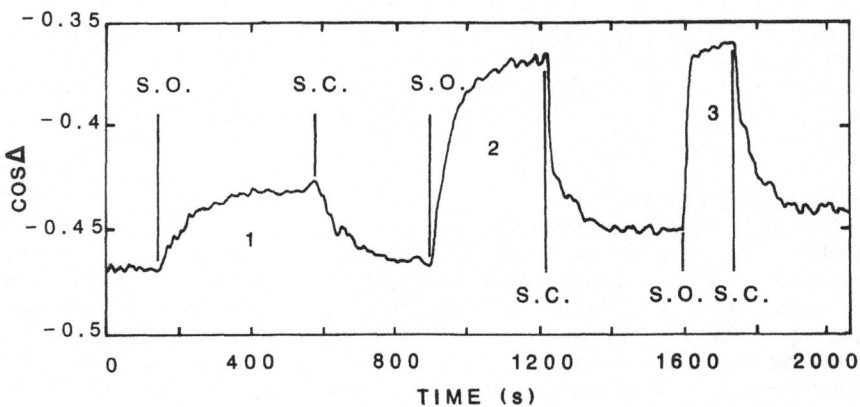

Figure 4 : Adsorption and desorption kinetics of Sb on Si(111) at T_{sub}=700°C for several dopant fluxes : 1- T_{cell}=250°C, 2- T_{cell} =275°C, 3- T_{cell} =300°C.

4-CONCLUSION

A new hypothesis for dopant segregation on a crystal surface during two dimensional growth has been presented. Such a mechanism provides a good explanation for the dopant surface enrichment, the low incorporation when desorption occurs and the profile transients. The incorporation coefficient of a dopant on a kink is described by an Arrhenius' law. The calculated activation energies are consistent with the energy necessary to "move "a dopant atom from a kink to the surface. However, the role of surface reconstruction has not been taken into account, and certain questions on this point remain unanswered [45,46]. Moreover, even if the calculated activation energies of the incorporation coefficient are consistent with the segregation phenomenon hypothesis, the value of constant α_0 is not justified. Further investigations are in progress in order to explain this α_0 value. This study also shows that the doping profile can be controlled by knowledge of the dopant surface concentration during growth. We have presented surface concentration measurements in the monolayer range by in situ spectroscopic ellipsometry for Sb in Si, showing that such measurements are possible. To conclude, such surface analysis, as well as a good choice of dopant, should help us to obtain devices with abrupt profiles, like doping superlattices or IMPATT diodes.

Acknowledgement

The work on spectroscopic ellipsometry analysis is supported by the european reaserch programm (E.S.P.R.I.T. porject n° 305).

Bibliography

1 - J.C. Bean, Appl. Phys. Lett., **33**, (1978), 654.
2 - Y. Ota, J. Electrochem. Soc., (1979), 1761.
3 - U. König, H. Kibbel and E. Kasper, J. Vac. Sci. Technol., **16**, (1979), 985.
4 - H. Siugiura, J. appl. Phys., **51**, (1980), 2630.
5 - M. Tabe and K. Kajiyama, Jap. J. Appl. Phys., **22**, (1983), 423.
6 - R.A. Metzger and F.G. Allen, J. Appl. Phys., **55**, (1984), 931.
7 - D. Streit, R.A. Metzger and F.G. Allen, Appl. Phys. Lett., **44**, (1984), 234.
8 - S.S. Iyer, R.A. Metzger and F.G. Allen, J. Appl. Phys., **52**, (1981),5608.
9 - S. Delage, Y. Campidelli, F. Arnaud d'Avitaya, and S. Tatarenko, J. Appl. Phys., **61**, (1987), 1404.
1 0 - G.E. Becker and J.C. Bean, J. Appl. Phys., **48**, (1977), 3395.
1 1 - J. Knall, J.E. Sundgren, J.E. Greene, A. Rockett and S.A. Barnett, Appl. Phys. Lett., **45**, (1984), 689.
1 2 - A. Rockett, J.J. Drummond, J.E. Greene, J. Knall and S.E. Sundgren, J. Vac. Sci. Technol. A, **3**, (1985), 855.
1 3 - M.A. Hasan, J. Knall, S.A. Barnett, A. Rockett, J.E. Sundgren and J.E. Greene, J. Vac. Sci. Technol. B, **5**, (1987), 1332.
1 4 - Y. Ota, J. Appl. Phys., **51**, (1980), 1102.
1 5 - G. Bajor and J.E. Greene, J. Appl. Phys., **54**, (1983), 1579.
1 6 - C.E.C. Wood and B.A. Joyce, J. Appl. Phys., **49**, (1978), 4854.
1 7 - A. Rockett, J.J. Drummond, J.E. Greene, and H. Morkoc, J. Appl. Phys., **53**, (1982), 7085.
1 8 - C.E.C. Wood, *Molecular Beam Epitaxy and Heterostructures*, L.L.Chang and K.Ploog, NATO ASI series; Applied sciences n°87 (1985) 149.
1 9 - S.A. Barnett and J.E. Greene, Surf. Sci., **151**, (1985), 67.
2 0 - J.E. Greene, S.A. Barnett and A. Rockett, Appl. Surf. Sci., **22/23**, (1985), 520.
2 1 - S. Andrieu, F. Arnaud d'Avitaya, and J.C. Pfister, to be published.
2 2 - E. Kasper, *Proc. 1st Int. Symp. on Si MBE*, Toronto, (1985), 102.

168

23- W.K. Burton, N. Cabrera and F.C. Frank, Trans. Roy. Soc. London, **243A**, (1951), 299.
24- K.D. Gronwald and M. Henzler, Surf. Sci., **117**, (1982), 180.
25- T. Sakamoto, N.S. Kawai, T. Nakagawa, K. Ohta and T. Kosima, Appl. Phys. Lett., **47**, (1985), 617.
26- T. Sakamoto, T. Kawamura and G. Hashiguchi, Appl. Phys. Lett., **48**, (1986), 1612 .
27- M. Ichikawa and T. Doi, Appl. Phys. Lett., **50**, (1987), 27.
28- J.H. Neave, B.A. Joyce, P.J. Dobson and N. Norton, Appl. Phys. A, **31**, (1983), 1.
29- C.T. Foxon, J. Vac. Sci. Technol. B, **4**, (1986), 867.
30- J.A. Venables, J. Vac. Sci. Technol. B, **4**, (1986), 870.
31- E. Kasper, Appl. Phys. A, **28**, (1982), 129.
32- P.J. Pagni, J. Chem. Phys., **58**, (1973), 2940.
33- C.E.C. Wood, D. Desimone, K. Singer and G.W. Wicks, J. Appl. Phys., **53**, (1982), 4230.
34- C.E.C. Wood, D. Desimone and S. Judaprawira, J. Appl. Phys., **51**, (1980), 2074.
35- F. Alexandre, C. Raisin, M.L. Abdalla, A. Bernac and J.M. Masson, J. Appl. Phys., **51**, (1980), 4296.
36- R.A.A. Kubiak, W.Y. Leong and E.H.C. Parker, J. Vac. Sci. Technol. B, **3**, (1985), 592.
37- S. Andrieu, J.A. Chroboczek, E. Andre, Y. Campidelli and F. Arnaud d'Avitaya, to be published in J. Vac. Sci. Technol B, (1988).
38- S.A. Barnett, H.F. Winters and J.E. Greene, Surf. Sci., **165**, (1986), 303.
39- R.A.A. Kubiak, S.M. Newstead, W.Y. Leong, R. Houghton, E.H.C. Parker and T.E. Whall, Appl. Phys. A, **42**, (1987), 197.
40- H.C Habbink, R.M. Broudy, G.P. McCarthy, J. Appl. Phys., **39**, (1968), 4673.
41- S. Andrieu, F. Ferrieu, F. Arnaud d'Avitaya, (to be published).
42- T. Smith, J. Opt. Soc. Am., **58**, (1968), 1069.
43- F.H.P.M. Habraken and G.A. Bootsma, Act. Elect., **24**, (1981/82), 167.
44- R.M.A. Azzam and N.M. Bashara, *Ellipsometry and polarized light*, North-Holland Publ. Comp., Amsterdam, (1977).
45- R.J. Culbertson, L.C. Feldman, and P.J. Silverman, Phys. Rev. Lett., **45**, (1980), 2043.
46- H.J. Gossmann and L.C. Feldman, Phys. Rev. B, **32**, (1985), 6.
47- D.C. Streit and F.G. Allen, J. Appl. Phys., **61**, (1987), 2894.

HIGH Tc SUPERCONDUCTING INTERCONNECTIONS IN SEMICONDUCTOR - BASED ELECTRONIC SYSTEMS

ROBERT C. FRYE

Recent materials advances have demonstrated superconductivity at temperatures well above 77K. It is no longer unreasonable to consider electronic systems based on superconducting interconnections and semiconductor devices. The interconnections between devices occur on several levels, ranging from the large scale, printed-wiring-board sized structures, down to the very small lines that connect devices on a single chip. An important question to begin to examine at this stage is where, within the hierarchy of device interconnections, it is most (or least) advantageous to consider using superconductors. This paper will present analytical results that address two important issues: the performance benefits of a resistance-free interconnection technology and the levels of critical current density that the superconductors must exhibit in order to realize these benefits.

INTRODUCTION

It is a well recognized fact that circuits based on silicon field-effect transistors can realize substantial speed improvements if operated at reduced temperatures. Recent breakthroughs in high-Tc ceramic superconducting materials have raised the interesting possibility of building systems that combine the advantages of this increased device speed with those of a resistance-free interconnecting medium. Because superconductivity can now be achieved at temperatures well above that of liquid nitrogen, silicon based electronics and superconducting wires can now be operated in a mutually compatible temperature range. The benefits of cryogenic operation have been well studied from the point of view of the transistor. What we will consider in this paper, instead, are the performance advantages that can be obtained from superconducting interconnections between the transistors.

If we focus on the contraints imposed by interconnections in particular, physical dimensions play a key role in determining their electrical properties. A major step in the revolutionary growth of VLSI speed was the recognition that semiconductor devices obey certain scaling rules. Exploiting this behavior has lead to unprecedented, sustained improvement in device performance. Interconnections obey similar, but distinctly different scaling rules. The nature of electronic systems obviously requires that we use interconnections spanning a large range of physical sizes from the very large ones found, for example, on printed wiring boards down to the smallest ones on the chips themselves. Each of these has its own set of constraints, and the improvements that we can expect to result from superconductors will depend on where, within the

169

Y. I. Nissim and E. Rosencher (eds.), Heterostructures on Silicon: One Step Further with Silicon, 169–185.
© *1989 by Kluwer Academic Publishers.*

hierarchy of interconnections we use them.

We will examine interconnections at three different levels - printed wiring boards, thin-film hybrids and integrated circuits. These levels represent a progression from very large to very small structures, and from low to high wiring densities. One feature that all three have in common is that they are inherently planar interconnection technologies. In high performance, high speed applications, the transmission-line properties of these structures become important. It is through analysis of these properties that we can understand not only the ways that the finite conductivity of conventional materials impacts the performance of electronic systems, but also, and move importantly, how these properties are impacted by scaling down the dimensions of such planar interconnections.

RESISTANCE EFFECTS IN SIGNAL TRANSMISSION

Figure 1 shows a schematic diagram of a typical microstrip transmission line structure. It consists of a rectangular cross section conductor overlying a uniform ground plane, with an intervening dielectric. Such lines are characterized by their resistance, inductance, and capacitance per unit length, R, L, and C. Resistance per unit length is a familiar concept, and is simply determined by the conductor resistivity, ρ, and the cross sectional area, A.

$$R = \frac{\rho}{A} \qquad (1)$$

The inverse relationship between resistance and cross sectional area is a particularly straightforward example of a scaling rule. Inductance per unit length is a quantitative measure of the net magnetic flux per unit current that passes between the line and the ground plane. Similarly, capacitance per unit length expresses the net electric field flux (or charge) per unit voltage. These

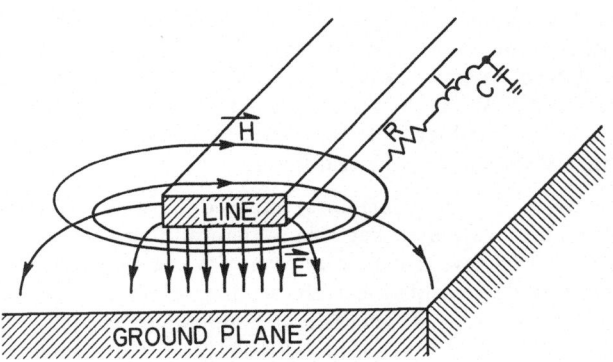

Figure 1. Basic microstrip transmission line structure.

two parameters describe one of the fundamental links between circuit theory and Maxwell's equations. If we consider, for simplicity, a simplified case in which the conducting line is very wide and thin, so that we can neglect fringing fields and flux penetration into the metal, the L and C are approximately given by

$$L = \frac{\mu d}{W} \tag{2}$$

and

$$C = \frac{\epsilon W}{d} \quad , \tag{3}$$

where W is the line width, d is the separation from the ground plane, and μ and ϵ are the dielectric's permeability and permittivity, respectively. For most practical line geometries, Equations 2 and 3 are not very good estimates of L and C. These expressions, however, illustrate a very important point regarding the way that line parameters are influenced by scale changes. Note that both L and C are determined by the **aspect ratio** of the line width and dielectric layer thickness. This is a very general principle; it is a consequence of the scale invariance of Laplace's equation in the dielectric. If we scale down the line dimensions by some arbitrary factor X, then the capacitance and inductance per unit length are unchanged. The decrease in size is offset by the fact that both the electric and magnetic fields become more intense. The resistance per unit length, on the other hand, increases by a factor of X^2. It is this quadratic increase in line resistance that limits the minimum useful size of interconnections.

A more general model of a line would also include a conductance in parallel with the line capacitance to account for losses in the dielectric. Since our concern in this paper is to compare the effects of line resistance, dielectric losses are only of tangential interest. It is also important to be aware that they exist, particularly at very high frequencies. In conventional interconnecting lines which have been pushed to high wiring densities, losses in the metal are so dominant that dielectric losses are entirely negligible. In very low resistance or superconducting lines, particularly very long ones, dielectric loss may, however, become the limiting factor. It is also important to consider resistive losses in the ground plane. For the case of a superconductor, we will assume that the ground plane is also superconducting. For conventional metal lines and groundplanes, the model could be corrected so that the variable R also included ground plane resistance, which would slightly alter Equation 1 but **not** its scale dependence. For clarity, we will neglect this effect in the analysis that follows.

Given the distributed parameters R, L and C, the analysis of wave propagation in the z direction along the line is given by the well-known coupled wave equations[1]

$$\frac{\partial V}{\partial z} = -RI - L\frac{\partial I}{\partial t} \tag{4}$$

and

$$\frac{\partial I}{\partial z} = -C\frac{\partial V}{\partial t} \tag{5}$$

Where V and I represent the voltage and current propagating down the line. Decoupling the above equations leads to the familiar traveling wave equations

$$V(z,t) = V_+ e^{i(\omega t - kz)} + V_- e^{i(\omega t + kz)} \tag{6}$$

and

$$I(z,t) = I_+ e^{i(\omega t - kz)} + I_- e^{i(\omega t + kz)} \tag{7}$$

where

$$k = \omega \left[LC(1 - \frac{iR}{\omega L}) \right]^{1/2}. \tag{8}$$

Here, ω is the angular frequency ($\omega = 2\pi f$) and k is the propagation coefficient. In both Equations 6 and 7, the first term on the right describes a wave traveling in the +z direction, and the second describes a wave traveling in the -z direction. By substituting the solutions 6, 7 and 8 back into the coupled wave equations, it can be shown that

$$\frac{V_+}{I_+} = \frac{-V_-}{I_-} = \left[\frac{L}{C}(1 - \frac{iR}{\omega L}) \right]^{1/2}. \tag{9}$$

The ratio of voltage to current for both the forward and reverse traveling waves on the line, as expressed by Equation 9, is the quantity conventionally known as the characteristic impedance, Z_o. For cases in which resistance, R, is negligible, Equation 9 resolves to its well known form

$$Z_o = (\frac{L}{C})^{1/2} \quad , R < \omega L. \tag{10}$$

At sufficiently low frequencies, where this condition is no longer met, all lines with finite resistance exhibit RC, rather than LC, type behavior.

The characteristic impedance is a key parameter for transmission lines. The remaining steps in the solution of the coupled wave Equations, 4 and 5, require that the boundary conditions be taken into account. If the line is terminated at a length ℓ with a load impedance Z_L, then the ratio of the magnitudes of the reverse to the forward traveling voltages is given by

$$\frac{V_-}{V_+} = e^{-2ik\ell} \frac{Z_L - Z_o}{Z_L + Z_o}. \tag{11}$$

If the line is terminated by an impedance that is equal to the characteristic impedance of the line, then the reverse component vanishes. The ratio given by Equation 11 is the reflection coefficient, Γ. In most high frequency signal applications it is imperative to terminate the line at either the load or the source end with an impedance that matches Z_o. If Γ is nonzero, the transmitted and reflected signals will resonate or interfere, resulting in very non-ideal signal transmission. If we consider the characteristic impedance for a general lossy line

$$Z_o = \left[\frac{L}{C}(1 - \frac{iR}{\omega L}) \right]^{1/2}, \tag{12}$$

then one of the important effects of line loss becomes obvious. In a regime where the resistance is appreciable, the magnitude of the impedance varies inversely with the square-root of the frequency. It is not possible to provide a terminating element, like a resistor or capacitor, that will properly terminate the line over all frequencies.

Another of the problems associated with resistive lines is that they are dispersive. A pulse traveling down the line propagates with a group velocity given by

$$v_g = \frac{d\omega}{d(Re[k])} \tag{13}$$

As Equation 8 shows, for the special case of zero resistance, the group velocity is a constant. For the more general case, however, velocity is frequency dependent. This causes distortions in broadband signals as some of their frequency components outrace others. Note also in Equation 8 that non-zero resistance leads to an imaginary component in the propagation coefficient. This means that the voltage and current will suffer exponential decay as they travel along the line.

Although the effects of line resistance discussed above can, in some particular cases, have significant negative impact on system performance, the most severe problem associated with lossy transmission lines is that they limit the bandwidth of the interconnection. In order to examine the extent to which this is important, it is necessary to consider the lines in greater detail. The maximum speed at which such lines will operate is critically dependent on their size, length and terminations.

BANDWIDTH VS. WIRING DENSITY

The previous section discussed some of the issues that arise in scaling the size of interconnection structures. In order to be more specific, let us consider the example structure, shown in Figure 2, which shows a cross section of two lines

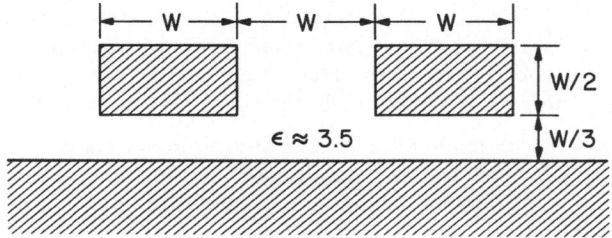

Figure 2. Example line geometry.

overlying a ground plane. The aspect ratio of the structure and the dielectric constant of the insulator are fairly representative of real interconnection lines. This choice of dimensions results in a capacitance of 1.2pF/cm, inductance of 3.1nH/cm, leading to a nominal characteristic impedance of approximately 50 Ω. The exact choice of the shape of the line is not really important, as well shall see. Choosing a particular structure like this, however, allows us to examine the effects of scaling on interconnection performance, and to see how superconductivity changes the rules for the lines. For example, it relates the line resistance per unit length -

$$R = 2\rho/W^2 \tag{14}$$

to the wire pitch, ie the number of lines per unit width

$$P = 1/2W. \tag{15}$$

At this point it is clear that superconductors represent a special case for which $\rho = 0$. This breaks the linkage between resistance and wiring density. Similarly, conventional conductors, unlike superconductors, will also exhibit the effects of accumulated resistance as the lines become longer.

Strictly speaking, superconducting interconnections only have zero resistance for dc signals. Signals with nonzero frequency components will cause a voltage drop, arising from inductance, along the superconducting line. This inductive drop, in turn, causes normal current to flow, leading to finite resistance losses. In traditional, low temperature superconductors, these losses would be negligible for the line dimensions and signal frequencies that we will be concerned with.[2,3] For the newer, high T_c materials, similar measurements are complicated by the large variability in material structure and properties that can result from different process technologies. There is, however, no reason at present to suspect that good high frequency properties in multilayer thin film structures can not be achieved in these new materials. For the analysis that follows, we will simply model the superconducting line as being resistance free.

At high frequencies, the skin effect begins to increase the resistance of metal lines. Current flows in a surface layer of thickness δ given by

$$\delta = (\frac{\rho}{\pi f \mu_o})^{1/2}. \tag{16}$$

At high frequencies, where δ becomes comparable to the metal thickness, the resistance begins to rise. In superconducting lines, current flow is generally confined to the surface, even at very low frequencies. An effect analogous to the skin effect occurs only at very high frequencies (determined by the superconductor's energy gap) well outside the range of this analysis.

Figure 3 shows the basic circuit elements used to model signal propagation in the system. In addition to the distributed elements R, L, and C that characterize the line, this model includes an ideal voltage driver with an output impedance R_s, and a load impedance Z_L. Between the driver and the line and between the load and the line, provision has been made to model the parasitic series inductance and shunt capacitance of any connectors that may be present at either end of the line.

CIRCUIT ELEMENTS

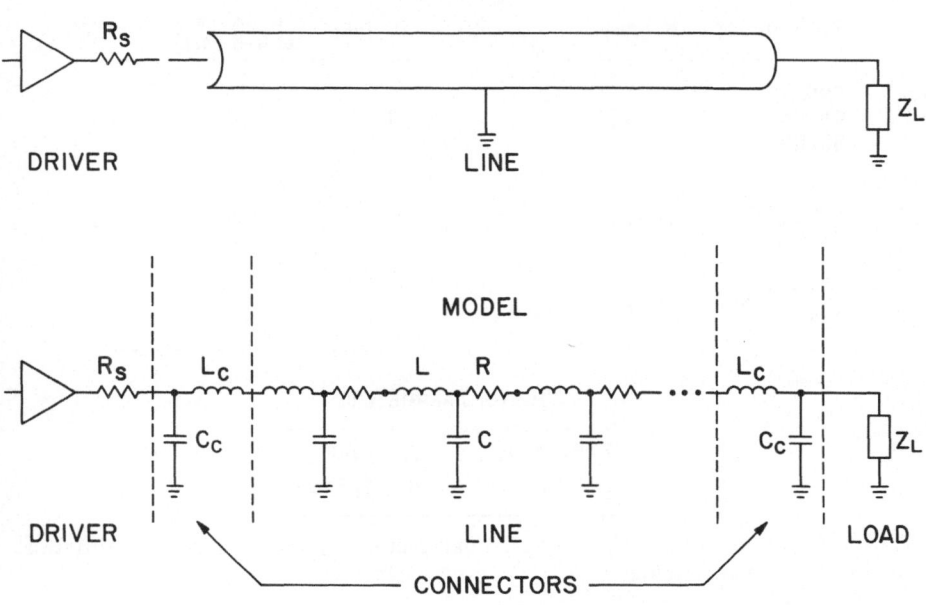

Figure 3. Circuit element model.

There are two basic methods that are commonly used to avoid problems of signal reflections in the transmission line. One is to use a matched load, $Z_L = \sqrt{L/C}$. For propagation of digital signals, this approach usually requires making R_s small. Since the load is resistive, during those times in which the output voltage is nonzero, current flows down the line and static power is dissipated at the load resistor. For large systems, with many interconnections, the power requirements can become prohibitive. An alternative, and the one that we will consider here, is to use a matched source, ie to make $R_s = \sqrt{L/C}$, and to terminate the line with a high input impedance receiver. For digital systems this approach has an added benefit, since the effect of line resistance is to reduce the speed, but not to attenuate the output, as it does for a resistive load.

Table 1 lists parameters used in this model for three different interconnection technologies. The maximum interconnection lengths are representative of the physical size of the overall structures themselves. The common parameters are those that are invariant with scale. For printed wiring boards, it is assumed that the driver and receiver are mounted in packages, and the values of L_c and C_c are chosen to the representative of high performance packages.[4,5] The values of these parasitic elements in thin film interconnections are much less well defined because the technology, is not yet mature, and a

PARAMETERS

TECHNOLOGY	W (μm)	L_c (nH)	C_c (pF)	MAXIMUM LENGTH (cm)	METAL
PRINTED WIRING BOARD	50–200	2	1	50	Cu
THIN FILM HYBRID	10–50	0.1	0.25	10	Cu
INTEGRATED CIRCUIT	2–10	0	0	1	Al

COMMON PARAMETERS

Z_0, R_s: 50 Ω	Z_L: .01pF
L: 3.1nH/cm	C: 1.2pF/cm

Table 1. Values of the model parameters used for three different interconnection technologies.

variety of chip attachement and interconnection techniques are being employed.[6-8] Some possible methods range from tape automated bonding, which has a reported inductance of 1.2nH[9] to direct solder attachment with an inductance less than 0.05nH[8]. For this analysis we have used 0.1nH as an intermediate value. The capacitance of 0.25pF represents a typical 100μm x 100μm pad on a chip. Finally, for integrated circuits, the parasitics are essentially absent. For such structures the line often is simply a direct extension of the driver itself.

The right side of Figure 4 shows the magnitude and phase of the frequency response for the 50cm long, printed wiring board level interconnection. In this plot, and the ones to follow, an equal propagation delay has been subtracted from all the curves to make the phase plots clearer. An interesting feature to notice in this figure is that the frequency response, even for the superconducting line, begins to fall off above 3GHz. This is a result of the limitation imposed by the connector parasitics associated with packages. In addition, these parasitic circuit elements cause an impedance mismatch at the driver end. The resulting signal reflections resonate in the line, causing the periodic ripples in the frequency response. For copper lines, the frequency response falls off more quickly as the line dimensions are scaled down. This shows clearly one of the fundamental trade-offs in conventional interconnection design - wiring density increases can only be obtained at the expense of bandwidth. Superconductors, by virtue of having zero resistance for all values of W, are not subject to the same trade-off.

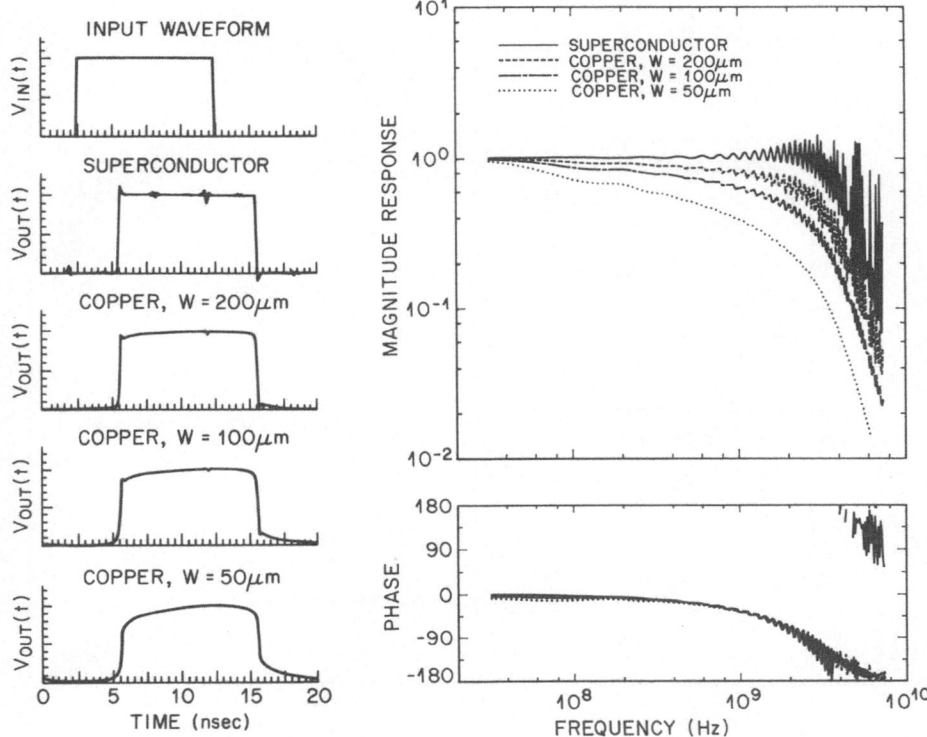

Figure 4. Performance comparisons for 50 cm long, printed wiring board level interconnections.

The waveforms on the left of Figure 4 show equivalent information, but in the time domain. For all of the lines, the overall time delay for the signal to propagate is equivalent. It is determined by the length of the line and by the speed-of-light travel of the electromagnetic wave through the dielectric, which for a resistance free line is given by

$$v = 1/(LC)^{1/2} \simeq 1/(\mu_o \epsilon)^{1/2}. \qquad (17)$$

The bandwidth reductions that occur for the smaller sized wires show up in the time domain as a degradation in the rise-time of the pulse. At low levels of wiring density, it is possible to make conventional transmission lines that perform almost as well as superconducting ones. It is only when we push the dimensions of the lines beyond those that are typically used that we begin to see significant reductions in the performance.

Figure 5 shows similar calculations for thin film hybrid interconnections. In contrast to the previous figure, these structures are much less constrained by their connectors. Note that the frequency range for this figure is somewhat higher than for Figure 4. Shorter lines are inherently faster and, consequently,

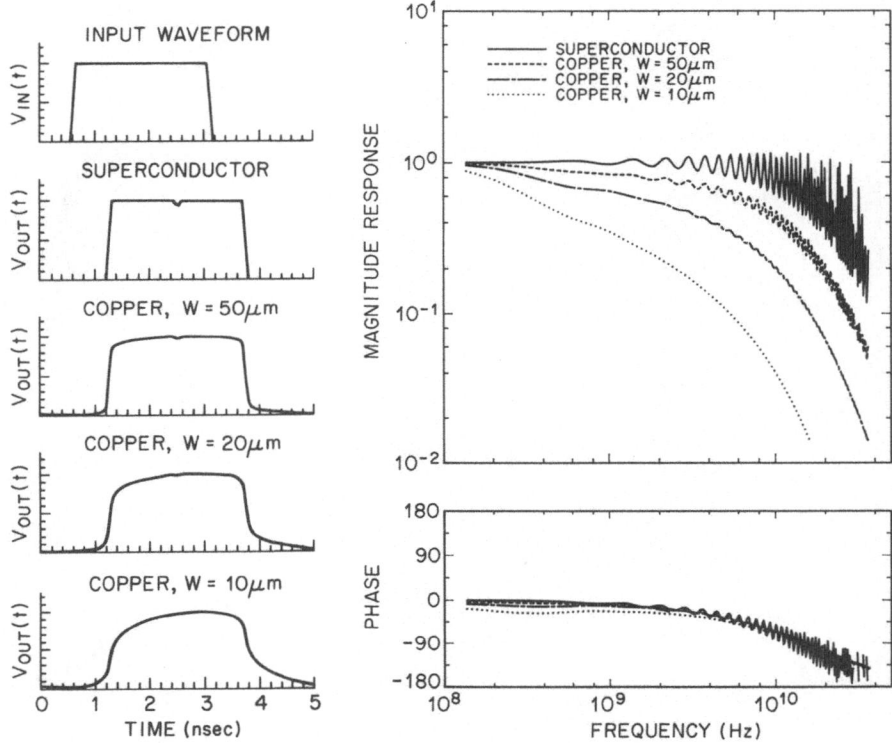

Figure 5. Performance comparisons for 10 cm long, thin film hybrid level interconnections.

can be scaled down to smaller dimensions before their resistance once again begins to dominate their bandwidth. As in the previous calculations, at the lower levels of wiring density conventional interconnections can perform almost as well as superconducting ones.

Chip level interconnections, shown in Figure 6, continue the same trend. Again, because they are even shorter they are also faster. Because there are no parasitics at all, the response for the superconducting line is nearly flat. The small ripples are caused by the small load capacitor which slightly influences the signal reflection. Frequencies of several tens of gigahertz probably approach the limits of semiconductor device performance.

These sets of calculations demonstrate clearly the problems in scaling down interconnecting structures. More importantly, however, they show that for most purposes, it is generally possible to make adequate transmission lines using normal conductors. It is only when we attempt to scale these lines down to small dimensions that they show severe bandwidth degradation. Superconducting transmission lines are generally considered to be important for

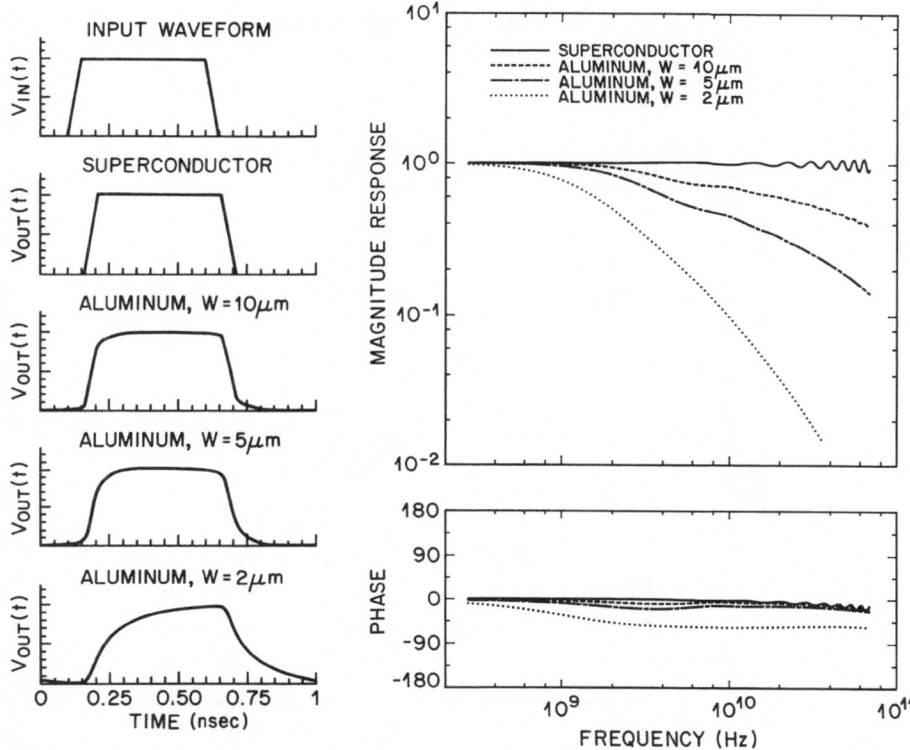

Figure 6. Performance comparisons for 1 cm long, integrated circuit level interconnections.

their high bandwidth, but an alternative view, and one that is not so generally recognized, is that they offer wiring density improvements. The vast majority of semiconductor devices are incapable of utilizing multi-gigahertz speed interconnections. In many instances where such speeds are necessary, the number of wires needed is low enough that larger, conventional lines are adequate.

The question to pose, then, is at what point does resistance begin to significantly affect the performance of the interconnections? If we examine closely the equations that describe the transfer functions depicted in Figures 4-6, we find that resistance is negligible if the total line resistance is small compared to the characteristic impedance of the line. Figure 7 shows the relationship between line length and wiring density for several metal systems, allowing a maximum accumulated resistance of 5Ω (ie 10% of Z_o). These curves use the same example line aspect ratio as the example calculations, except for both copper and aluminum films. Below a pitch of about 250 lines/cm (ie for W $> 20\mu$m) it becomes impractical to make deposited films much thicker, so these

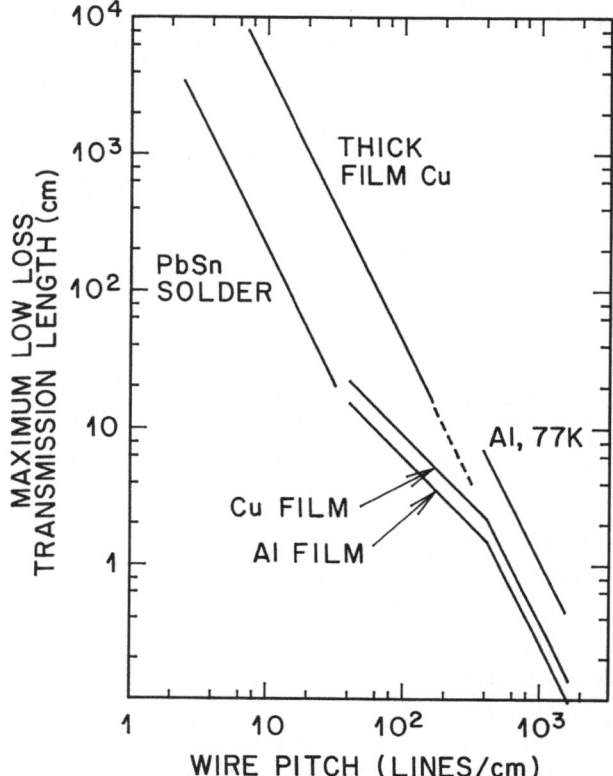

Figure 7. Maximum low-loss transmission line length vs wire pitch for various metals.

curves take on linear rather than quadratic behavior. Aluminum at 77K is also included, since it may be used in cryogenic systems. Depending on the demands placed on the signal bandwidth, the choice of 5Ω for maximum tolerable line resistance may be revised either upward or downward.

Superconductors are subject to different constraints on their maximum wiring density. They are only capable of supporting a limited current density in the superconducting state. Above this critical current density, they exhibit resistance. Figure 8 shows a plot of the voltage input and the current at the driver and load ends of the line using a matched impedance driver. The current in this figure has been multiplied by the characteristic impedance, so its dimensions are the same as the input voltage. For a matched source, the maximum current in the line occurs at the input end, and is given by

$$I_{max} = V_{sig} / 2Z_o, \tag{18}$$

where V_{sig} is the amplitude of the input voltage. At the load end, the small current pulse charges the load capacitance.

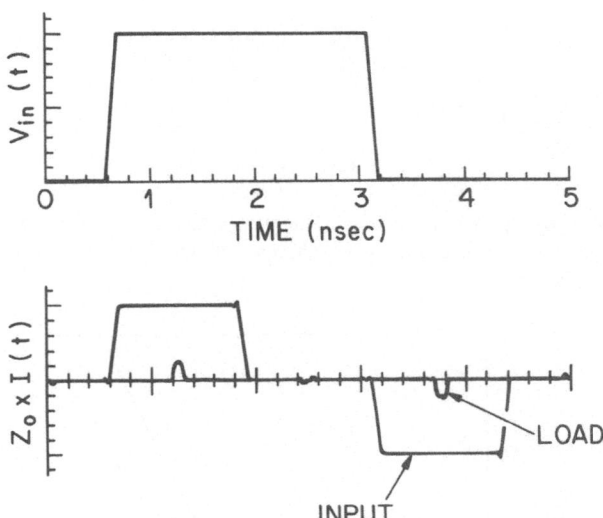

Figure 8. Input voltage and normalized current at both ends of the line for a matched impedance driver.

As lines are scaled down to smaller dimensions, the current density needed to transmit the signal grows proportionally. Figure 9 shows the peak current density as a function of wire pitch for two commonly used values of V_{sig}. For superconducting interconnections, these curves specify the limits of obtainable wiring density.

Based on these criteria, we can evaluate relative merits of superconducting and conventional interconnections. Generalizing these results beyond specific line aspect ratio that we used in the examples, if we consider **any** arbitrary shape, since Zo is scale invariant, the wiring density limits can be expressed, instead, as limits to the minimum cross sectional area of the wire. For conventional metals,

$$A_{min} = \frac{\rho \ell}{0.1 Z_o} \tag{19}$$

and for the superconductor

$$A_{min} = \frac{V_{sig}}{J_{crit} Z_o} \tag{20}$$

Equation 19 is calculated by requiring the total line resistance to be less than 10% of the characteristic impedance, while Equation 20 restricts the current density in the superconducting line to be less than half of the critical current.

Equations 19 and 20 provide a way to determine, for a particular set of parameters, whether or not superconductors offer an advantage. The two interconnection approaches are equilavent when both can achieve the same minimum cross-sectional area, ie when

$$\frac{\rho\ell}{0.1Z_o} = \frac{V_{sig}}{J_{crit}Z_o} \tag{21}$$

Superconductors are superior if

$$J_{crit} > \frac{V_{sig}}{10\rho\ell} \tag{22}$$

Note that this is **independent** of Z_o. The important point to remember is that the factor of 10 in the denominator may change, depending on the exact amount of resistance that is tolerable.

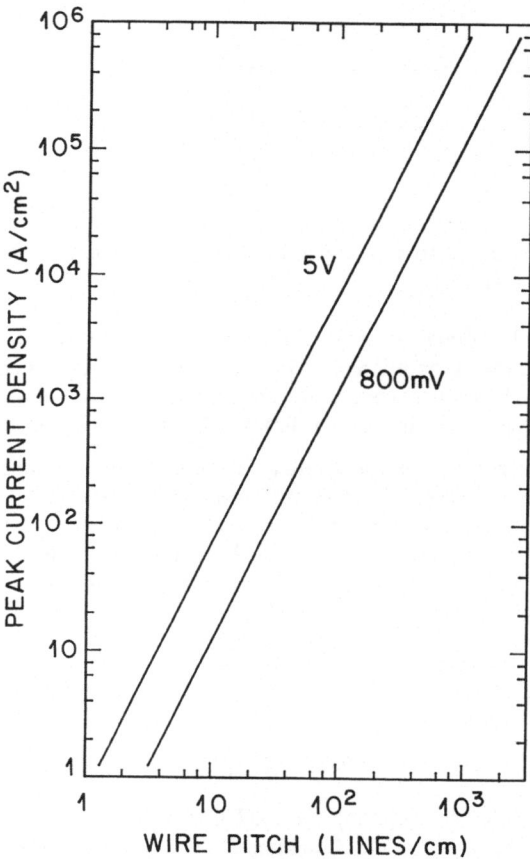

Figure 9. Peak current density vs wire pitch for two commonly used signal levels.

Figure 10 shows the break-even limits for superconducting lines compared with several different metal lines. These curves are shown for signal levels of 1 volt, and should be sifted upward proportionally for higher voltages. For values of signal path length and critical current density that lie above the appropriate curve, superconducting interconnections provide greater interconnection density. For shorter interconnections, resistance in metal lines is less important, so we place more stringent demands on the superconductor.

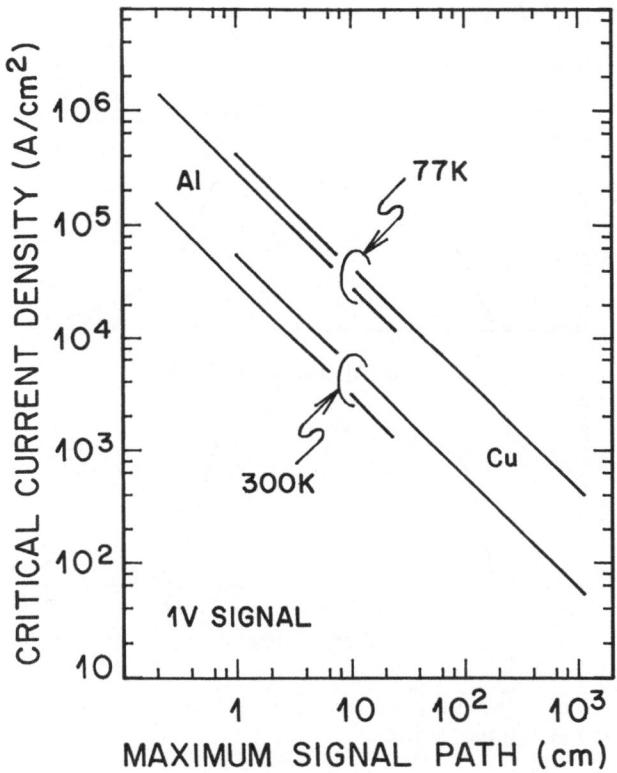

Figure 10. Break even limits for superconducting and conventional interconnections. For points above these limits, supreconductors offer wiring density advantages.

DISCUSSION AND CONCLUSION

Results for metals at low temperatures are included in Figure 10 because superconductors, if they are to find application in eventual system designs, really need to outperform more conventional materials at the lower temperatures. This is particularly true for very small structures like integrated circuits. Figure 11 shows the improvement in speed that results from cooling a 1cm long, 2μm x 1μm aluminum line down to 77K. Even if high T_c superconductors are eventually developed with the high critical current densities needed ($\sim 10^6 A/cm^2$) to carry signals at such reduced dimensions, they must also meet a host of process compatibility requirements before

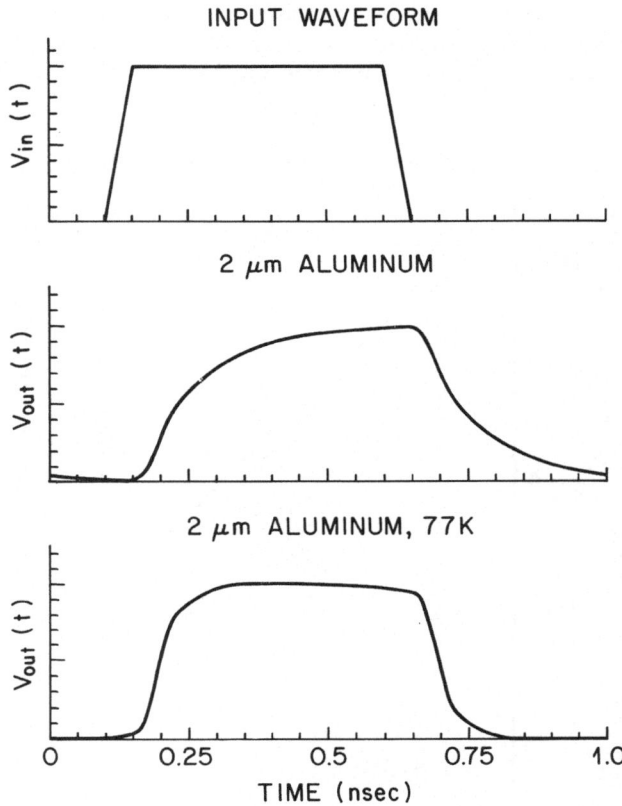

Figure 11. Improved pulse propagation in 1cm long, integrated circuit level aluminum lines at low temperatures.

they can be used on the chips. Given the level of performance that can be obtained using conventional aluminum lines, it is unlikely that superconductors will find widespread applications as interconnections on the chips themselves.

At the other extreme in signal path length, superconductors would appear to offer more promise. For printed wiring boards, however, packages are a major limiting element in overall system performance. This is an area that is receiving increased attention, and future developments may remove this constraint. Until packages are improved, superconductors are not likely to find use at the printed wiring board level for the simple reason that they offer no appreciable benefits for the system as a whole.

The most promising area for the use of superconducting interconnections is in thin film hybrids. Precisely because the technology is immature, it is also more flexible. Superconducting materials do not need to fit into an existing fabrication scheme. More importantly, multichip modules can be made with performance on a par with the chips themselves.

REFERENCES

1. T. C. Edwards "Foundations for Microstrip Circuit Design" John Wiley and Sons, New York (1981).

2. W. D. McCaa and N. S. Nahman, J. Appl. Phys. **39** 2592 (1968).

3. W. D. McCaa and N. S. Nahman, J. Appl. Phys. **40** 2098 (1969).

4. A. J. Rainal, AT&T Bell Laboratories Tech. J. **63** 177 (1984).

5. C. J. Stranghan and B. M. MacDonald, IEEE Proc. 35th Electronic Components Conf. 361 (1985).

6. E. T. Lewis, IEEE Trans. Components, Hybrids and Manuf. Tech. **CHMT-2** 441 (1979).

7. C. W. Ho, D. A. Chance, C. H. Bajorek and R. E. Acosta, IBM J. Res. Dev. **26** 268 (1982).

8. C. J. Bartlett, J. M. Segelken and N. A. Teneketges, IEEE Trans. Components, Hybrids and Manuf. Tech. **CHMT-10** 647 (1987).

9. D. Herell and D. Carey, IEEE Trans. Components, Hybrids and Manuf. Tech. **CHMT-10** 99 (1987).

SUPERCONDUCTOR-SILICON HETEROSTRUCTURES

A.W. KLEINSASSER

IBM T.J. Watson Research Center, P.O. Box 218, Yorktown Heights, New York 10598, U.S.A.

1. INTRODUCTION

The dramatic recent discoveries of oxide superconducting materials having transition temperatures (T_c) exceeding 120 K have led to great expectations for electronic applications (1,2). There is now an apparent convergence of semiconductor technology, with growing low temperature applications such as liquid nitrogen cooled high speed computers, and superconductor technology, which now offers the possibility of superconductors which function at or above 77 K. Although significant technical difficulties must be overcome before any hybrid technology can be practical, there is already significant interest in the use of superconducting interconnects between semiconductor circuits (1,3). Applications for high T_c superconducting devices, taking advantage of the convenience of higher operating temperature or improvements in parameters such as the energy gap, are also expected. However, the lack of generally useful three-terminal superconducting devices (4) should lead to an increased interest in new devices, including superconductor-semiconductor hybrids (5).

This paper deals with hybrid superconductor-semiconductor structures and devices, an area of growing interest (4-8) even before the advent of high temperature superconductivity. Silicon is the principal semiconductor material considered, with discussion its limitations where appropriate. Most of the discussion will necessarily center on conventional (low T_c) superconductors, since there has been virtually no experimental work to date on high T_c hybrid structures. The prospects for high T_c devices will be addressed where possible. However, the excitement over the new materials should not obscure the need for basic hybrid device physics research, which is the major concern here. Much of this discussion will deal with the physics of superconductor-semiconductor interfaces (Section 2). Section 3 discusses devices based upon such heterojunctions.

2. PHYSICS OF SUPERCONDUCTOR-SEMICONDUCTOR INTERFACES

2.1 Basic interface properties

Figure 1 is a band diagram illustrating an SuSm contact for the case of a degenerate n-type semiconductor. (In this paper, N signifies normal metal, I insulator, Sm semiconductor, and Su superconductor. Thus a thin film metal-oxide-metal tunnel junction is NIN, and a Schottky contact is NSm.) The behavior of contacts to p-type material is, as in the case of normal metal Schottky contacts, essentially similar. As in most NSm contacts, the Fermi level is located within the semiconductor energy gap, E_g at the interface, resulting in a Schottky barrier of height E_B and depletion width w. Note that the superconducting energy gap, $2\Delta \ll E_g$, E_B (typical superconducting gap energies are of order a few meV). The structure behaves as a SuIN tunnel junction, with the Schottky barrier as the insulator and the neutral semiconductor as the normal metal. With the new high T_c superconductors, 2Δ (and possibly kT) may be an order of magnitude or so larger, but the above picture is still basically correct.

Under bias, current transport is dominated by tunneling, since $kT \ll E_B$. For small applied bias, the contact is ohmic and the current density is $J = gV$, where V is the applied voltage. Using the WKB approximation, the specific conductance, g, is given by (9):

Y. I. Nissim and E. Rosencher (eds.), Heterostructures on Silicon: One Step Further with Silicon, 187–201.
© 1989 by Kluwer Academic Publishers.

$$g \simeq (4\pi e^2 m^* E_{00}/h^3) \exp[-E_B/E_{00}]. \tag{1}$$

The factor $\exp[-E_B/E_{00}]$ is the WKB tunneling probability. The exponent is proportional to barrier width (which varies as the square root of barrier height) and square root of barrier height. Here $E_{00} = (eh/4\pi)(N_D/\varepsilon m^*)^{1/2}$, with m^*, ε, and N_D the effective mass, dielectric constant, and dopant concentration on the semiconductor, and $kT \ll E_{00} \ll E_B$.

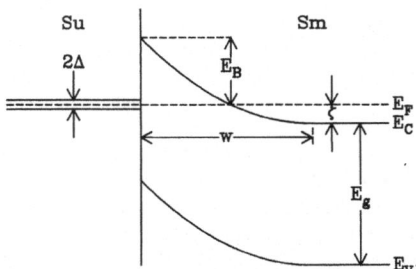

FIGURE 1. Energy bands near the interface between a superconductor and an n-type semiconductor.

2.2 Proximity effect

2.2.1 Introduction. At the boundary between a superconductor and a normal metal (SuN interface), Cooper pairs leak out of Su and into N, inducing superconductivity in N. This is known as the proximity effect (10), as illustrated in Figure 2a, a qualitative plot of the spatial behavior of the amplitude for finding a Cooper pair near the interface. The pair density is reduced in the superconductor over a distance of order the superconducting coherence length, ξ_s. Penetration into the normal region is limited to a depth of order the normal coherence length, ξ_n.

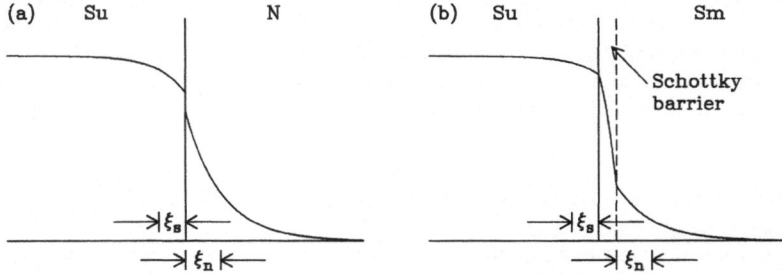

FIGURE 2. Schematic of the variation of the Cooper pair amplitude at (a) an SuN contact and (b) an SuSm contact. In (b), tunneling through the Schottky barrier reduces the penetration of pairs into the semiconductor (a typical reduction would be orders of magnitude).

In principle the proximity effect occurs at an SuIN interface as well, however, as illustrated schematically in Figure 2(b), pairs must tunnel through the interfacial potential barrier in order to cause a proximity effect, resulting in a tremendous reduction of the effect (8). Since an SuSm contact is an SuIN tunnel junction, the pair density in the semiconductor layer should be reduced by a factor of order the tunneling probability $\exp[-E_B/E_{00}]$ due to the Schottky barrier (i.e. by many orders of magnitude). Thus, although the characteristic lengths for the decay of the pair density in the normal and superconducting materials are the same as for the SuN case, it is somewhat surprising that proximity effects are seen at all in SuSm junctions. However, ample evidence for such effects has been reported, as discussed below.

2.2.2 <u>Length scale.</u> The Cooper pair density asymptotically approaches its bulk value in the Su layer and drops off exponentially with distance into the N layer near an SuN interface (8). The characteristic lengths in Su and N are the superconducting and normal coherence lengths, ξ_s and ξ_n, roughly the distance an electron travels in time $\hbar/\Delta$ in Su or $\hbar/kT$ in N. In the clean limit $(\ell \gg \xi_n)$, $\xi_{sc} = \hbar v_F/\pi\Delta$ and $\xi_{nc} = \hbar v_F/2\pi kT$, where ℓ, v_F, and D are the electron elastic mean free path, Fermi velocity, and diffusion constant ($D = v_F \ell/d$ in a d-dimensional material). In the dirty limit $(\ell \ll \xi_n)$, $\xi_{sd} = (\hbar D/\pi\Delta)^{1/2}$ and $\xi_{nd} = (\hbar D/2\pi kT)^{1/2}$. In general (11), $\xi_n \simeq (\xi_{nc}^{-2} + \xi_{nd}^{-2})^{-1/2}$. As an example, the coherence length in the usual case of experimental interest, a dirty three dimensional semiconductor with carrier density n, effective mass m· and mobility μ is:

$$\xi_{nd} = (\hbar^3\mu/6\pi m^* e k_B T)^{1/2} (3\pi^2 n)^{1/3}. \qquad (2)$$

The normal coherence length depends on carrier mobility as shown, for various carrier concentrations in n-Si (m· = 0.26 m_e) at 4.2 K, by the solid curves in Figure 3(a). The straight lines show the asymptotic behavior in the clean and dirty limits. For bulk Si, ξ_n increases, even though μ decreases, with increasing n. For n = 3×10^{19} cm^{-3} and μ = 100 cm^2 /V-s, ξ_n = 16 nm. Larger values, up to ~100 nm, are obtainable in a two dimensional electron gas (2DEG). Figure 3 is numerically correct for the 2D case if the identification $N_s = n^{2/3}$ is made for the sheet carrier concentration. Note that ξ_n is independent of μ in the clean limit. ξ_n (T) is plotted in Figure 3(b). The lower curve is for 3×10^{19} cm^{-3} and 100 cm^2/V − s, showing a drop to only a few nm at temperatures of order 77 K. The upper curve represents the clean limit, in principle obtainable in 2DEG devices, for the same vale of n (or for $N_s = 10^{13}$ cm^{-2} in 2D).

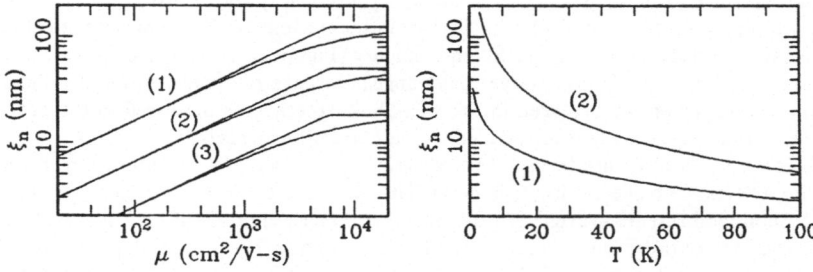

FIGURE 3. (a) Dependence of normal coherence length in n-Si on mobility in n-Si at 4.2 K, with n = 3×10^{19}, 2×10^{18}, and 1×10^{17} cm^{-3} in curves 1, 2, and 3. (b) Temperature dependence. Curve 1 is for 3×10^{19} cm^{-3} and 100 cm^2 /V-s, while curve 2 shows the clean limit value for the same value of n (or, in 2D, $N_s = 10^{13}$ cm^{-2}).

Devices such as the Josephson weak links described below must not exceed a few ξ_n in length. In other words, a few hundred nm represents an upper limit for the length of Si links at 4.2K. At higher temperatures, the limits are more stringent. The smallness of the coherence length in Si is one reason that low effective mass materials are attractive in SuSm structures. In InAs (m· = 0.023 m_e), for example, for a given carrier density, ξ_n is an order of magnitude larger than in Si in the clean limit. The much larger mobilities possible in such materials, as well as the small effective masses, are important in the the dirty limit.

2.2.3 <u>Boundary conditions.</u>

Ohmic contacts between semiconductors and metals are very important to device technology. Good contact resistances in Si and GaAs technologies are ~$10^{-7} - 10^{-6}$ Ω − cm^2. A typical high performance Nb-based Josephson junction with a critical current density of the $10^3 - 10^4$ A/cm^2 has the same specific resistance in the normal state. So an active device in superconductor technology has a specific resistance corresponding to a low value of contact resistance in semiconductor technology. In the context of the proximity effect, contact resistance is related to the difficulty of getting Cooper pairs into the semiconductor, and an understanding of the effect of the Schottky barrier (or the contact resistance) on the boundary conditions for the pair density is of great importance.

It has been well established that the proximity effect can be observed in SuSm contacts, mostly through experiments involving semiconductor-coupled Josephson weak links (Section 2.3). However, the simplest demonstration of the effect is the measurement of the transition temperature of SuSm bilayer as a function of layer thicknesses, a widely-used technique in SuN systems (10). Leakage of pairs out of the superconductor into a normal region (Figure 1) causes an observable lowering of T_c if the superconductor thickness is of order a few coherence lengths or less. Such measurements have been reported in the Nb-pSi system (12), where they were used to extract the length ξ_n. This early experiment is important in directly verifying the proximity effect in SuSm systems through study of ξ_n. However, it is the number of pairs which penetrate into the normal layer, rather than how far they penetrate, which is most important for the proximity effect (11), a fact which has received little attention in the case of SuSm systems (aside from cursory references to Schottky barriers). What needs to be emphasized is that it is surprising that proximity effects are observed at all in most SuSm experiments. Study of the T_c dependence of thin superconductor films on thick semiconductor layers with varying doping can directly study the boundary conditions for the pair amplitude (the variation of barrier thickness or tunnel probability with doping is inherently more important than the variation of normal coherence length). This is qualitatively evident in the experiment of Hatano, et al. (12), which showed reductions of T_c from the bulk value of 7-23% for 40 nm Nb films on $4.5 - 30 \times 10^{18}$ cm^{-3} pSi. Using a value of 0.41 eV for E_B for Nb-pSi (13), the tunneling probability for a carrier through this barrier, $\exp [-E_B/E_{00}]$, ranges from $2 \times 10^{-7} - 2 \times 10^{-3}$ in these samples. This implies a great reduction of the pair amplitude in crossing the SuSm boundary (Figure 2b), which is inconsistent with such a large reduction of T_c. Thus, Cooper pairs appear to penetrate more easily into the semiconductor than expected, as also seen in the weak link experiments discussed in the next section. More experiments of this type are needed to study this effect in detail.

Bilayer T_c studies can also be used to study contact processes, for example chemical or sputter cleaning to remove oxides, since T_c is particularly sensitive to oxide (or other insulating) layers at the SuSm contact. However the experiments are generally difficult to carry out, since they require Su films of order one coherence length thick with bulk T_c's, a difficult proposition for most superconducting materials.

Tunneling experiments can be used to probe the superconducting order parameter in proximity systems. Recent experiments have measured the energy gap in both the Su and Sm sides of Pb-Si SuSm contacts by studying the conductance-voltage characteristics of MISu and MISm tunnel junctions grown on the Su and Sm layers, which were of order one coherence length thick (14).

Barrier-free metal-semiconductor contacts are possible on materials such as InAs, in which the Fermi level is pinned in the conduction band at surfaces and interfaces. The possibility of avoiding of one of the supposed limitations of Si, along with longer coherence lengths, has led to considerable interest in such materials for SuSm structures.

2.3 dc Properties of Josephson Weak Links.

2.3.1 Structures. Josephson effects occur in systems of weakly-coupled superconductors (15). Figure 4(a) illustrates a thin film sandwich junction structure. In a tunnel junction, the coupling between the superconductors is via an insulator (which may be a lightly-doped semiconductor) and normal conduction is by tunneling. The term weak link is often reserved for non-tunneling

Josephson devices, for example an SuNSu junction. SuSmSu junctions span the full range of behavior, from SuISu to SuNSu. Although there has been much interest in SuSmSu tunnel junctions
(8), we are concerned here with SuNSu-like SuSmSu weak links.

A bridge geometry is usually employed for weak link devices (Figure 4b). This allows the use
of single-crystal semiconductor substrates and a single metal evaporation. It also allows the
possibility of a superconducting field effect transistor (FET) (Figure 4c), in which a gate is used
to control critical current and resistance of a weak link (Section 3.3.2).

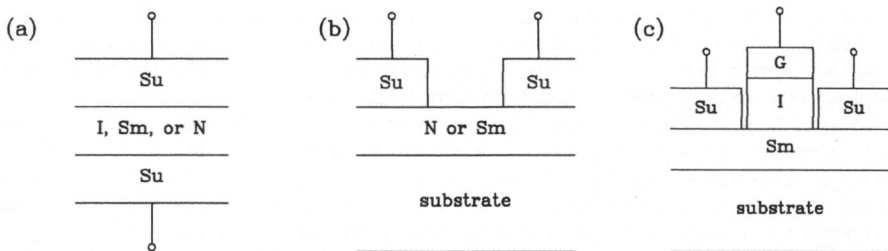

FIGURE 4. SuSmSu weak link structures. (a) Sandwich. (b) Bridge. (c) Superconducting field
effect transistor (gate-controlled bridge).

2.3.2 Theory. The basic experimentally observable properties of Josephson devices are the
critical current, I_c, (the maximum zero-voltage supercurrent) and the normal resistance, R_n (the
device resistance in absence of superconductivity or at voltages well in excess of the superconducting energy gap). The product $I_c R_n$ is a figure of merit, relating the strength of coupling of the
superconductors (i.e. I_c) to the normal conductance, setting the output voltage scale, and corresponding to an upper frequency limit for device response. This quantity has a maximum value
of $\sim\Delta$. It is rarely discussed in detail for SuSmSu junctions.

Most theoretical work on SuSmSu weak links has been based on the sandwich geometry
(16,17). For long links (L $\gg \xi_n$, where L is the normal layer thickness), the crical current is
exponential in device length due to the exponential decay of the Cooper pair amplitude in the
normal region, $I_c \propto (\Delta^2/kT_c)\xi_n^{-1} \exp[-L/\xi_n]$. The proportionality constant is related to the
boundary conditions for the pair amplitude. Essentially the same result is obtained for the case
in which the link is coupled via a two-dimensional electron gas (18,19).

Virtually all experiments involve bridge structures, and a thorough theoretical treatment of
bridge devices, both SuNSu and SuSmSu, would be desirable. Fortunately, the general form of
the critical current is similar to that in sandwich structures. The problem of an SuNSu bridge
structure was considered by van Dover, et al. (20), who argued that the structure can be treated
in two separate one-dimensional problems, that of finding the value of the pair amplitude in the
normal layer of an SuN bilayer (the normal layer is assumed to be thin compared with ξ_n), and
that of calculating the critical current of a homogeneous link composed entirely of N material.
(The current density at the contact must be small, a condition which is satisfied for long bridges.)
This treatment has the advantage that the boundary condition at the SuN interface is arbitrary;
the change in the pair amplitude in crossing the boundary is treated either as a fitting parameter
or as a parameter whose value can be determined by an independent experiment (eg. SuN bilayer
T_c or tunneling measurements).

The same treatment can be applied to the SuSm bridge (21). For a bridge of length L, width
w, thickness d_n, and resistivity ρ_n, the resistance is $R_n + 2R_c$, where $R_n = \rho_n L/wd_n$ is the resistance of the semiconductor bridge and $R_c = [(\rho_n r/d_n)^{1/2}/w]$ is the contact resistance (22). Here
r is specific resistance of the SuSm contact (the reciprocal of the conductance expressed in

Equation 1). The contact length is assumed to be large compared with the transfer length. Then (21):

$$I_cR_n = \frac{\pi\Delta^2}{2ekT_c}\frac{\rho_n m_s}{\rho_s m_n} A^2 f^2 \frac{L}{\xi_n} e^{-L/\xi_n}[1 + 2(rd_n/\rho_n)^{1/2}/L], \tag{3}$$

where Δ, ρ_s, and m_s are the energy gap, resistivity, and effective mass in Su, m_n is the effective mass in Sm, $f^2 \sim 1$ is the ratio of value of order parameter in Su at the interface to its bulk value, and A^2 is the factor by which the order parameter changes in crossing the SuSm interface. A is related to exp $[-E_B/E_{00}]$, or r, and can be determined by other experiments, as discussed above. The last factor in Equation (3) is due to the contact resistance. Note that this entire argument applies to the case of materials which form barrier-free contacts.

Measurement of I_c (T) for a single device can be used to extract $\xi_n(T_c)$, since the dominant dependence is exp $\{ -[L/\xi_n(T_c)](T/T_c)^{1/2}\}$, for temperatures not too close to T_c. Measurement of $I_c(L)$ for several devices on a single wafer is an even better way to determine ξ_n, since the only length dependence of I_c is in exp $[-L/\xi_n]$. The prefactor in both cases gives a value for the boundary condition parameter A. Measurement of $I_c(N_D)$ on several wafers, knowing ξ_n, is the best way to study the boundary conditions, which are dominated by the change in Schottky barrier width.

Josephson effects are observable only in devices which are shorter than several times ξ_n. The optimal I_cR_n product ($\sim\Delta$) is achieved in devices having $L \sim \xi_n$. Device length is thus maximized by maximizing ξ_n, so that the clean limit is of interest. The behavior of bridges in this case is not worked out. It is interesting to note that if $L \sim \xi_n \sim \ell$, transport becomes ballistic.

2.3.3 Experiment. The most notable SuSmSu sandwich weak link experiments involved single-crystal Si membranes (13,23). Detailed experiments on several such devices of known thickness and uniform doping would provide the best opportunity for comparison with theory, however there has been no such systematic work on this type of structure. We concentrate here on bridge results.

Bridge-type weak links were first reported by Schyfter, et al. (24). Most of the work to date has been on Si, the best and most complete study being that of Nishino, et al. (25), who studied the critical current of Pb alloy-pSi weak links as a function of temperature, device length, doping, and magnetic field. For 0.14-0.22 μm, 3×10^{19} cm^{-3} devices at 4.2 K, both temperature and length dependences gave 24.6 nm for $\xi_n(4.2$ K), in good agreement with the expected value of 21 nm (for some reason the numbers reported in Ref. 23 were roughly half these values).

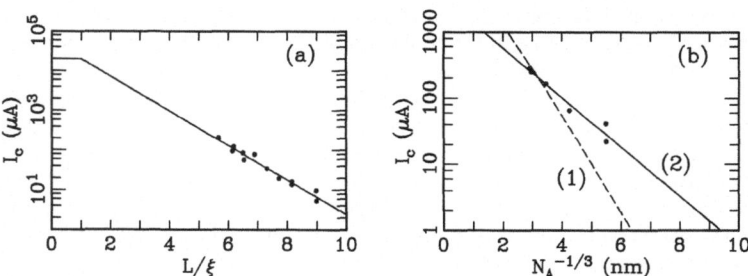

FIGURE 5. Critical current data from experiment of Ref. (25). (a) I_c vs. L/ξ_n for devices on 3×10^{19} cm^{-3} pSi. (b) I_c vs. $N_A^{-1/3}$. Line (1) was obtained using the value of I_c at $N_A = 3 \times 10^{19}$ cm^{-3} and the expected dependence on ξ_n, with $\xi_n \propto N_A^{1/3}$, ignoring the dependence of μ on N_A Line (2) is a fit to the actual data.

Figure 5(a) is a semilog plot of I_c vs. L/ξ_n at 4.2 K. The $I_c R_n$ values listed in Ref. (25) were 0.4-0.8 mV, the largest values presumably corresponding to the shortest devices. Contact resistance must be a major contributor to R_n for devices this short, but even ignoring this, the extrapolated value of $I_c R_n$ for short ($<$1-2$\times$ ξ_n) devices is many tens of mV, far in excess of Δ (I_c itself extrapolates to a value $\sim$100 times that of the shortest device). Equation 3 can account for a large prefactor to the $(\Delta^2/kT_c)(L/\xi_n)$ exp $[-L/\xi_n]$ dependence of $I_c R_n$, through the factor $\rho_n m_s/\rho_s m_n$ and the contact resistance. However, the required value of A^2 is still (unexpectedly) much larger than exp $[-E_B/E_{00}]$, which is $<<$ 1. (This does not explain the large extrapolated $I_c R_n$ for short links. This will require more theoretical work.)

Figure 5(b) is a semilog plot of the dependence of critical current on acceptor density (I_c vs. $N_A^{-1/3}$). Line (1) is fitted to the data using the values of ξ_n consistent the known value at $N_A = 3 \times 10^{19}$ cm^{-3}. Line (b) is a fit to the actual data. The discrepancy is in a direction opposite to that expected for changing tunnel probability (Schottky barrier transparency), and can be accounted for by the dependence of mobility on carrier concentration, which affects the value of ξ_n (In principle, values for μ and ξ_n can be measured for the individual samples.). Evidently, the boundary condition factor, A, does not change much over this range.

The weak link experiments performed to date have several shortcomings, including the use of diffused Sm layers which ar too thick ($d_n >> \xi_n$), have nonuniform carrier concentrations, and (presumably) contact-dominated resistances. Future experiments should include the use of well-characterized epitaxial layers, with careful measurement of μ, n, d_n, etc., and correlation of weak link results with the other types of experiments in order to independently obtain a value for A.

2.3.4 Weak link summary. Experiments have convincingly shown that the proximity effect occurs at interfaces between superconductors and doped semiconductors, with the length scale determined by the normal coherence length. Much less is known about the boundary conditions, but unexpectedly large values of $I_c R_n$ in weak links and T_c depression in SuSm bilayers indicate that pairs penetrate more easily into the semiconductor, at least under some conditions, than one would expect. (In weak links showing no supercurrent, due to large L or an applied magnetic field, the current-voltage characteristics are those expected for two series SuIN junctions. It is clear that transport across the interfaces is due to tunneling.)

Some of the experimental anomalies can be explained by invoking nonuniform dopant distributions (eg. higher concentration at the surface than assumed in the calculations) or alloying at the contacts. However, it is clear that the simple tunneling argument uused to describe the SuSm contact must break down. Effects such as image force lowering of the barrier make some difference, but the real problem the discrete nature of the dopants, which makes a large difference at high concentrations, with inter-dopant spacing exceeding the calculated depletion width. What is needed at this point is more experimental and theoretical work, with an emphasis on experimentally relevant geometries, well-defined structures, and fundamental work on the SuSm proximity effect emphasizing boundary conditions over length scale. Similar work on materials other than Si, especially those, such as InAs (26,27), that form barrier-free contacts, would aid in achieving an understanding how much of an advantage these materials really provide over silicon.

3. HYBRID SUPERCONDUCTOR-SEMICONDUCTOR DEVICES

3.1 Materials problems

The major problem in SuSm device structures is achieving a good SuSm contact, which is the basis for all devices. The cleaning process for the semiconductor surface is critical, particularly since achieving a good superconductor layer within a coherence length or so of the interface may preclude the use of alloyed contacts in some circumstances. The sensitivity of the proximity effect to contact properties may allow the phenomenon to be used as a probe of metal-semiconductor contact properties if the phenomenon becomes better understood. For example, measurement of contact resistance becomes difficult with low specific resistance contacts, while measurement of supercurrent is relatively straightforward, once the right device structure exists.

The known high temperature superconductors introduce new difficulties into potential hybrid device applications. They suffer from being very reactive with semiconductor materials, as well as from having exceptionally short coherence lengths. The situation is complicated by the anisotropic nature of these materials. The use of low temperature epitaxial growth techniques and barrier layers, both metallic and insulating, are obviously of great interest.

3.2 Two-terminal devices

3.2.1 Josephson devices. SuSmSu Josephson devices were discussed in Section 2. They are of technological interest because they offer usefully high impedances along with extremely low capacitances. They are of special interest for high T_c applications because of the difficulty of fabricating good tunnel junctions, which require superconductors with good properties within a coherence length or so on both sides of the interface, with the barrier only a few monolayers thick. These requirements are relaxed with SuNSu and SuSmSu devices, particularly bridges (which require only a single superconductor deposition), although the shortness of ξ_n at high temperatures (Section 2.2.2) may pose problems.

3.2.2 Super-Schottky diodes. Superconducting tunnel junctions are of interest as detectors at mm and sub-mm wavelengths. Fundamental to such detectors is a highly nonlinear current-voltage characteristic, such as that of a Schottky diode. In a Super-Schottky diode (6) the metal electrode is a superconductor, and the nonlinearity is greatly enhanced for voltages below its energy gap, resulting in improved performance. This SuSm diode is essentially an SuIN device, but the use of a semiconductor allows improved impedance matching at high frequencies (6).

3.3 Three-terminal devices

3.3.1 Introduction. There is significant general interest in the possibility of transistor-like (three-terminal) superconducting devices for a variety of applications (4). One approach to seeking such devices is to apply working principles from semiconductor technology to hybrid devices, with the hope of obtaining some advantage from the combination. Several such devices are described here.

The scale of operating voltages in superconducting devices is usually of order the superconducting energy gap, which is a few meV or less for conventional superconductors and, being proportional to T_c, is expected to be a few tens of meV for the new oxide materials (i.e. in a material with $T_c \simeq 90$ K). Typical semiconductor device voltages are of order 1 V, although scaling down of device dimensions in order to allow denser circuits reduces this to several hundred mV in submicrometer CMOS circuits. Despite this partial convergence of the voltage scales for superconducting and semiconducting devices, they are at present still an order of magnitude or so apart. This is an important consideration is considering the operating voltages of hybrid devices, since low power dissipation is often a major requirement.

3.3.2 Proximity effect superconducting FETs. The gated proximity effect Josephson weak link of Figure 4c is a superconducting FET. The gate controls the carrier concentration in the link/channel, thereby changing ξ_n and I_c. This device was first proposed about a decade ago (6,28). Experimental devices based on this principle were first reported for InAs (29) and Si (30). The Si device was constructed on a 70 nm thick pSi membrane, with the gate on the back. It exhibited a degree of gate control which was difficult to account for by the expected mechanism (31). No suitable insulator gate has been found for InAs, so the response of the MISFET devices was rather poor, and recent work has emphasized back-gated JFETs (32). Recently a GaAs/AlGaAs MODFET device was reported (33). In addition to the demonstrated FET devices, a self-aligned coplanar Si-coupled weak link structure has been demonstrated which should be directly applicable to coplanar FET devices (34), an improvement over the back-gated membrane structure. Unfortunately, none of the experimental work has been very systematic. Most of it has been on single working devices.

Given that these FETs are Josephson junctions, it is natural, but incorrect, to assume that device speed and power dissipation are essentially the same as for digital Josephson technology. The control mechanism is identical to that in conventional FETs, with the same gate voltage swing required in both cases to produce a given change in channel carrier concentration. Switching the device on and off will require the same CV^2 energy per cycle in both cases, as long as the same change in carrier density is required. Also, making the channel superconducting does not change carrier transit time, which represents a limit to the speed of FET devices. The role of superconductivity in a superconducting FET is to change the device characteristics, allowing a zero on-state resistance.

One potential misconception about JOFETs which is rather natural is that voltage gain is essentially impossible to achieve, since the natural output voltage scale of these devices is set by the $I_c R_n$ product of the weak link, and thus limited by the superconducting energy gap, while the input (gate) voltage is orders of magnitude larger. In fact, voltage gain is perfectly natural, but it originates not from superconductivity, but rather from control of channel resistance, as in an ordinary FET. This is illustrated in Figure 6. We define "on" and "off" states corresponding to large and negligible critical current. I_c responds exponentially to gate voltage ($I_c \sim \exp [-L/\xi_n]$, $\xi_n \propto n^{1/2}$, $\Delta n \propto \Delta V_{gate}$). So a large change in I_c can be made with only a small change in R, as experimentally observed (30). The "off" output voltage is of order $I_c R_n$ of the weak link in the "on" state, as shown in Figure 6(a), and the voltage gain is $<< 1$. However, an optimal link is of order one coherence length long, in which case the critical current response is linear, not exponential. Thus, in a well-designed device, the output resistance switches between small and large values, and large voltage gain is possible, as shown in Figure 6(b). Note, however, that the difference between the IV characteristics in the "on" state for superconducting and non-superconducting electrodes (solid and dashed curves) is not nearly as dramatic in the latter case.

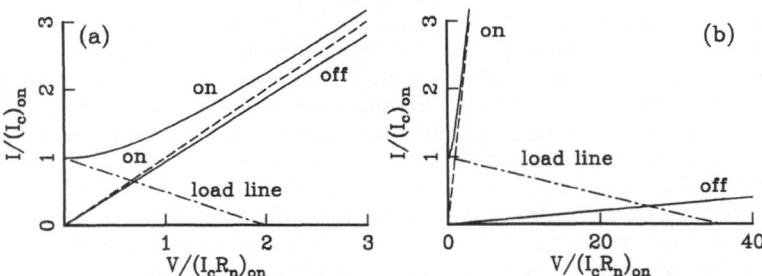

FIGURE 6. Schematic IV characteristics of superconducting FET. (a) If resistance change is small between "on" (large I_c) and "off" (small I_c). The dashed line applies to the same device in the absence of superconductivity. Output voltage is limited by the device $I_c R_n$ product, which is much smaller than the gate voltage swing. (b) If resistance change is large between "on" and "off," output voltage can be made arbitrarily large (i.e. the device can have large voltage gain).

The small scale of voltages for which superconductivity makes a significant difference in device characteristics makes it vital to make the gate voltage swing be as small as possible; this is limited by the ability to control the inversion threshold (or equivalent voltage) and by the gate capacitance (the number of carriers introduced by a given gate voltage). It is difficult to conceive of mV gate voltages, but tens of mV should be possible. Thus, although it is difficult to imagine operation at voltages corresponding to the energy gaps of conventional superconductors, device operation at larger source-drain voltages should be considered. Also, high T_c superconductors offer the possibility of gap voltages in the tens of mV range. The implications of this for applications, (i.e. whether or not a superconducting channel can make a useful difference in device characteristics) remain to be worked out.

3.3.3 <u>Metal base transistor.</u> Mead (35) first proposed the metal base transistor (MBT) in 1960 as a unipolar analog of the bipolar transistor. The original device had an NININ configuration, although various other conceptually equivalent devices, also called hot electron transistors (HET), have been proposed (SmNSm, NINSm, etc.). The device is based on the injection of electrons into the base film which have a large kinetic energy in the base , and are thus separate from the equilibrium carriers in the base. These hot carriers must be collected before they lose so much energy or normal component of momentum that they cannot surmount the base-collector barrier. The ultimate goal in an HET device is ballistic transport, in which electrons traverse the base region without scattering, and emerge from the base with their full injection energy, since this promises the ultimate in device speed (or at least transit time) and current gain.

The initial interest in hot electron transistors waned due to problems with achieving current gain. The ratio of collector current to emitter current α, must approach unity if the device is to have a useful common-emitter current gain $\beta = \alpha/(1-\alpha)$. But losses in the base and quantum mechanical reflection at the base-collector barrier tend to make α rather small. Experimental values of α too large to believe are usually attributed to pinholes, and measurement of α alone are of little value in demonstrating that transport is dominated by the MBT mechanism. Interest in hot electron transistors was revived recently (36-38), and experimental progress has been impressive, due to advances in semiconductor heterostructure technology which make it possible to form an entire multilayer device structure from epataxial, lattice-matched heterojunctions, with reduced scattering at interfaces, reduced electron-electron scattering in the base (due to the low carrier density) and the reduced quantum mechanical reflections, the latter being due to tailoring of barrier shapes, or to the small value of the Fermi energy (39,40).

A difficulty with these devices is the large base resistance. Two possible ways around this problem are the use of materials which have long ballistic mean free paths (41) or the use of a high mobility 2DEG in the base (40). A superconducting base would eliminate base resistance altogether, although this is a backward step along the path that HET work has followed in recent years, with the use of low doping or 2DEGs in the base to minimize both electron-electron and impurity scattering. The mean free path for ballistic motion should be very short in a metal due to the large carrier density, and superconductivity does not alter this fact. Of course it is the ratio of base thickness to effective mean free path which matters, and zero base resistance should allow the base layer to be extremely thin, down to tens of Angstroms.

The problem of quantum mechanical reflections is expected to be much larger with a metal than with a semiconductor base, due to the large value of the Fermi energy in a metal (39,40). Again, nothing changes if the metal is a superconductor. However, this argument overlooks the possibility of a wide difference in effective mass between collector and base. The use of a low effective mass material for the collector is claimed to greatly reduce the problem of quantum mechanical reflections (42) at the base-collector interface, as suggested by theory (43,44). Also, the reflection probability should be much smaller in the case of ballistic electrons (45). This may remove a major objection to metal base transistors in general.

Tonouchi, et al. have studied a metal base device, which they named Super-HET, both theoretically (42) and experimentally (46). The device has a SmSuSm structure, with a GaAs emitter, an InSb collector (InSb, which forms low Schottky barriers to metals, is deposited in polycrystalline form on the base film; GaAs is the substrate), and a Nb or NbN base. Electrical measurements of the common-base characteristics of Nb (20-40 nm thick) and NbN (60 nm thick)-based devices at 4.2 K were consistent with α values of $\simeq$0.6-0.8 and $\simeq$0.6, respectively. Assuming that $\alpha \propto \exp[-d/\lambda]$, where d is the base thickness, λ, the effective hot electron mean free path, must be of order 70 nm or greater in the Nb case (and even larger for NbN). This group measured the hot electron mean free path to be 110 nm in Nb at 4.2 K in a separate set of related experiments (47). Of course a simple measurement of common base characteristic does not differentiate between ballistic electrons and electrons which have undergone scattering but which still have sufficient forward momentum to surmount the collector barrier, and the claim of a measurement of a ballistic mean free path (47) was later withdrawn (42). For Nb, the same group calculated a ballistic mean free path (42) of $\simeq 14$ nm at low temperatures, limited by phonon emission (they claim that the component due to electron-electron scattering is in excess of 200 nm for all energies of interest).

It is difficult to establish the absence of pinholes which may allow permeable base transistor action. A similar situation exists in the case of epitaxial $Si - CoSi_2 - Si$ devices (48), where analysis of electrical measurements in conjunction with careful electron microscopy studies allowed some insight into the pinhole question. This sort of analysis or, even better, hot electron spectroscopy (38) are needed on the Super-HET structure. So more experimental work is needed in order to establish values for both ballistic and non-ballistic mean free paths. Also, other structures, such as NISuSm, perhaps based on Si, would be interesting.

The voltage scale for these devices is determined by the (Schottky) barrier height of the base-collector contact, and is of order hundreds of mV. This is far larger than the characteristic voltage associated with superconductivity, the energy gap. It is a limitation for low temperature applications, where a mV (or few tens of mV) scale is more appropriate. The base-collector barrier height is a more or less intrinsic property of the particular metal-semiconductor system used, and low barrier height systems are of interest.

It is questionable whether high T_c materials are of interest for this device. They would allow high operating temperatures, but the extremely short mean free paths would imply a base thickness which is too small to be of practical interest.

3.3.4 <u>SUBSIT</u>. A superconducting metal base transistor which, in principle, avoids the difficulties of ballistic mean free path and quantum mechanical reflections, and actually resembles a bipolar transistor, is called the SUperconducting-Base Semiconductor-Isolated Transistor, or SUBSIT (49). The device structure is illustrated in Figure 7.

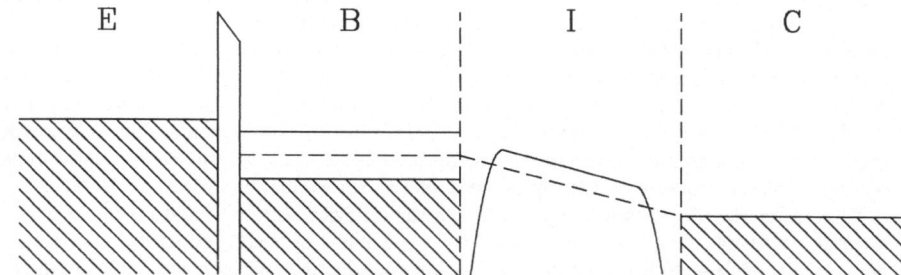

FIGURE 7. Schematic diagram of bands in SUBSIT. The NISu (or SuISu) Emitter-base (E-B) junction is biased for injection of quasiparticles, with energies just above the base energy gap, which diffuse across the base and are collected before recombining into pairs. No barrier is shown between base and isolator (I) , although an extremely transmissive one might be tolerated. The collector (C) is shown as a metal, but could be a degenerate semiconductor.

The device has a MISuSm structure, with the MISu being the emitter-base section (the emitter can also be superconducting). Quasiparticles (essentially unpaired electrons) are injected into the base film by tunneling. The two "fluids" in the base are the pairs (majority carriers) and quasiparticles (minority carriers) in the superconductor, in close analogy to a bipolar transistor, but with zero base resistance. Unlike the usual metal base transistor, these quasiparticles do not need to be "hot." They can be injected with energies just above the superconducting energy gap in the base, or they can relax down to this level by scattering. The injected quasiparticles diffuse across the base, and gain in the device depends upon their being collected before they recombine into pairs. Unlike hot electron transistors, however, they do not have to enter the collector on the first attempt. This improves the collection efficiency, but increases the base transit time (it is not a ballistic device). However initial estimates show that SUBSIT should be a fast device (49), with response in tne 10 ps range or better. In contrast to the Super-HET, the operating voltage scale for SUBSIT is set by the base energy gap voltage. Current gain depends on the base transit time, the effective base quasiparticle recombination time, and a transmission factor, and should be large (49).

As in other metal base transistors, the key to SUBSIT operation is the base-collector contact. Frank (49) refer to the semiconductor layer in contact with the base as the "isolator," the collector being the metal (or heavily-doped semiconductor) contact to this layer. Contacts to the isolator must be ohmic, and should be barrier-free (on the scale of the superconducting energy gap), since the presence of a barrier would tend to trap quasiparticles in the base, decreasing the current gain and slowing down the device. The isolator also serves the function of preventing direct pair-breaking tunneling between base and collector; this process represents a potential leakage current. The isolator must be frozen out, since thermally-generated carriers also represent a source of leakage. Of course, the operating temperature must be low enough that thermally-generated quasiparticles in the base do not contribute a significant leakage current.

A metal (superconducting base)-semiconductor (isolator) contact having a barrier height smaller than the superconducting energy gap is not easy to realize. A material in which the Fermi level pinning position is in the conduction (or valence) band forms contacts with negative barriers, so that an accumulation layer forms between base and collector. However the effect of such a layer, which may act as a two-dimensional proximity effect superconductor, has not been explored. (The role of the proximity effect in this device has not been worked out. However, any reduction of the base energy gap would be deleterious.) A contact having no barrier (either positive or negative) is possible in principle using ternary semiconductors (50) such as $In_xGa_{1-x}As$. However, the barrier height change with composition (x) is of order 10 meV/at. %, so that requirements on the ability to control composition (spatial distribution on a wafer and between fabrication runs) are stringent. Also, given the discrete nature of dopants, and the relatively large local variations of potential on the meV scale of the superconducting gap, it is not really permissible to use the usual continuum picture for the spatial variation in semiconductor bands. High T_c superconductors allow some relaxation of requirements, due to the order of magnitude or so increase in the gap, but the problem is still present, and experimental studies of the base-isolator structure are clearly of great interest for evaluating this device (50).

Tamura, et al. (51) proposed essentially the same device structure, and made several experimental attempts at realizing an it. Their original structure consisted of $Nb - Al_2O_3 - Nb - nInSb$ (note the use of a superconducting emitter), and had a common base current transfer factor α of order 10^{-4} (there was no reference to an isolator in their original proposal). (The energy band diagram which appears in Ref. 51 should include a substantial Schottky barrier, much larger than the superconducting energy gap, between base and collector.) A later design (52), which included an isolator, replaced the InSb collector with a heterostructure based on nInGaAs and InAlGaAs, lattice-matched to an InP substrate. An α of ≈ 0.3 was reported for this device. The philosophy followed by this group has been to accept a small, and therefore very transmissive, barrier at the base-collector interface, rather than to aim for a barrierless contact. As mentioned above, the barrier should affect speed and current gain, which will presumably be studied in future experiments. This group has also done modelling of the device (53) also predicting achievable large gain and ≈ 10 ps response. The acceptance of a barrier at the base-isolator interface would allow Si to be considered as a semiconductor material for SUBSIT.

4. CONCLUSIONS

There is considerable interest in hybrid superconductor-semiconductor structures and devices, which should grow as both technologies develop. Proximity effects occcur at interfaces between superconductors and semiconductors, often in situations in which simple arguments indicate they should be suppressed. The physics of these effects at SuSm interfaces is very interesting, and understanding it is important for hybrid devices, most of which are in a primitive state of realization. High T_c superconductivity brings both attractive and deleterious factors into the device picture. Hybrid devices require the use of both advanced superconducting and semiconducting technologies. Early work points to an important role for exotic technologies, such as III-V heterostructures, MBE, small bandgap materials, etc. However, the extent to which Si can be used is unknown, and the experience with simple proximity effect physics suggests that simple materials and structures should not be forgotten in future experiments and devices.

The author wishes to acknowledge valuable conversations with A. Davidson. D.J. Frank, W.J. Gallagher, T.N. Jackson, P. Santhanam, and J.M. Woodall. Partial support for this work was provided by the U.S. Office of Naval Research under contract N00014-85-C-0361.

REFERENCES

1. A.P. Malozemoff, W.J. Gallagher, and R.E. Schwall, in Chemistry of High Temperature Superconductors, D.L. Nelson, M.S. Whittingham, and T.F. George, Eds., Am. Chem. Soc., Washington, D.C., 1987, page 280.
2. M. Nisenoff, in Low Temperature Electronics and High Temperature Superconductors, S.I. Raider, R. Kirschman, H. Hayakawa, and H. Ohta, Eds., Electrochem. Soc., Inc., Pennington, N.J., 1988, page 344.
3. O.K. Kwon, B.W. Langley, R.F.W. Pease, and M.R. Beasley, IEEE Electron Dev. Lett., EDL-8, 582 (1987).
4. W.J. Gallagher, IEEE Trans. Magnetics, MAG-21, 709 (1985).
5. T. van Duzer, in Low Temperature Electronics and High Temperature Superconductors, S.I. Raider, R. Kirschman, H. Hayakawa, and H. Ohta, Eds., Electrochem. Soc., Inc., Pennington, N.J., 1988, page 352.
6. A.H. Silver, A.B. Chase, M. McColl, and M.F. Millea, in Future Trends in Superconductive Electronics, B.S. Deaver, C.M. Falco, J.H. Harris, and S.A. Wolf, Eds., American Institute of Physics, New York, 1978, page 364.
7. H. Kroger, IEEE Trans. Electron Dev. ED-27, 2016 (1980).
8. M. Russo in Josephson Effect-Achievements and Trends, A. Barone, Ed., World Scientific, Singapore, 1986, page 216.
9. F. A. Padovani, in Semiconductors and Semimetals, Vol. 7, Part A; R.K. Willardson and Albert C. Beer, eds., Academic Press, New York, 1971, page 75.
10. G. Deutscher and P.G. de Gennes in Superconductivity, R.D. Parks, ed., Marcel Dekker, New York, 1969, Vol. 2, page 1005.
11. W. Silvert, J. Low Temp. Phys., 20, 439 (1974).
12. M. Hatano, T. Nishino, and U. Kawabe, Appl. Phys. Lett. 50, 52 (1987).
13. L.B. Roth, J.A. Roth, and P.M. Schwartz, in Future Trends in Superconductive Electronics, B.S. Deaver, C.M. Falco, J.H. Harris, and S.A. Wolf, Eds., American Institute of Physics, New York, 1978, page 384.
14. T. Nishino, M. Hatano, and U. Kawabe, Jpn. J. Appl. Phys. 26, Suppl. 26-3, 1543 (1987).
15. K.K. Likharev, Revs. Mod. Phys. 51, 101 (1979).

16. J. Seto and T. Van Duzer, in Proc. 13th Int'l. Conf. on Low Temp. Phys., Plenum, New York, 1972, Vol. 3, pg. 328.

17. L.G. Aslamazov and M.V. Fistul, Zh. Eksp. Teor. Fiz. $\underline{81}$, 382 (1981) [Sov. Phys. JETP $\underline{54}$, 206 (1981)].

18. V.Z. Kresin Phys Rev. B, $\underline{34}$, 7587 (1986).

19. Y. Tanaka and Tsukada, Solid State Commun. $\underline{59}$, 683 (1986) and $\underline{61}$, 445 (1987).

20. R.B. van Dover, A. De Lozanne, and M.R. Beasley, J. Appl. Phys. $\underline{52}$, 7327 (1981).

21. A.W. Kleinsasser and T.N. Jackson, Jpn. J. Appl. Phys. $\underline{26}$, Suppl. 26-3, 1545 (1987).

22. H.H. Berger, Sol. State. Electron. $\underline{15}$, 145 (1972).

23. C.L. Huang and T. van Duzer, Appl. Phys. Lett. $\underline{25}$, 753 (1974).

24. M. Schyfter, J. Maah-Sango, N. Raley, R. Ruby, B.T. Ulrich, and T. van Duzer, IEEE Trans. Magnetics, $\underline{MAG\text{-}13}$, 862 (1977).

25. T. Nishino, E. Yamada, and U. Kawabe, Phys. Rev. $\underline{B33}$, 2042 (1986) and $\underline{B34}$, 4857 (1986).

26. T. Kawakami and H. Takayanagi, Appl. Phys. Lett. $\underline{46}$, 92 (1985).

27. A.W. Kleinsasser, T.N. Jackson, G.D. Pettit, H. Schmid, J.M. Woodall, and D.P. Kern, Appl. Phys. Lett. $\underline{49}$, 1741 (1986).

28. T.D. Clark, R.J. Prance, and A.D.C. Grassie, J. Appl. Phys. $\underline{51}$, 2736(1980).

29. H. Takayanagi and T. Kawakami, Phys. Rev. Lett. $\underline{54}$, 2449 (1985). and Int'l. Electron Dev. Mtg. Digest, IEEE Inc., Piscataway, N.J., 1985, page 98.

30. T. Nishino, M. Miyake, Y. Harada, and U. Kawabe, IEEE Electron Dev. Lett. $\underline{EDL\text{-}6}$, 297 (1985).

31. A.W. Kleinsasser, Phys. Rev. B $\underline{35}$, 8753 (1987).

32. T. Kawakami and H. Takayanagi, Jpn. J. Appl. Phys. $\underline{26}$, Suppl. 26-3, 2059 (1987).

33. Z. Ivanov and T. Claeson, Jpn. J. Appl. Phys. $\underline{26}$, Suppl. 26-3, 1617 (1987).

34. M. Hiraki and T. Sugano, Proc. 1987 Int'l. Superconductivity Electronics Conf., Tokyo, Japan and Trans. IECE Japan $\underline{E70}$, 389 (1987).

35. C.A. Mead, Proc. IRE $\underline{48}$, 359 (1960).

36. M. Heiblum, Sol. State Electron. $\underline{24}$, 343 (1981).

37. S. Luryi and A. Kastalsky, Physica $\underline{134B}$, 453 (1985).

38. M. Heiblum and M. Fischetti, in Physics of Quantum Electron Device, F. Capasso ed., Springer-Verlag, Berlin, 1987.

39. S.M. Sze and H.K. Gummel, Solid State Electron. $\underline{9}$, 751 (1966).

40. S. Luryi, IEEE Electron Dev. Lett. $\underline{EDL\text{-}6}$, 178 (1985).

41. A.F.J. Levy, Appl. Phys. Lett. $\underline{48}$, 1609 (1985).

42. M. Tonouchi, H. Sakai, and T. Kobayashi, Jpn. J. Appl. Phys. $\underline{25}$, 705 (1986).

43. C.R. Crowell and S. Sze, J. Appl. Phys. $\underline{37}$, 2683 (1966).

44. A.F.J. Levy and T.H. Chiu, Appl. Phys. Lett. $\underline{51}$, 984 (1987).

45. W.P. Dumke, in Low Temperature Electronics and High Temperature Superconductors, S.I. Raider, R. Kirschman, H. Hayakawa, and H. Ohta, Eds., Electrochem. Soc., Inc., Pennington, N.J., 1988, page 449.

46. H. Sakai, Y. Kurita, M. Tonouchi, and T. Kobayashi, Jpn. J. Appl. Phys. $\underline{25}$, 835 (1986).

47. T. Kobayashi, H. Sakai, Y. Kurita, M. Tonouchi, and M. Okada, Jpn. J. Appl. Phys. $\underline{25}$, 402 (1986).

48. J.C. Hensel, A.F.J. Levi, R.T. Tung, and J.M. Gibson, Appl. Phys. Lett. $\underline{47}$, 151 (1985). J.C. Hensel, Appl. Phys. Lett. $\underline{49}$, 522 (1986). E. Rosencher, P.A. Badoz, J.C. Pfister, F. Arnaud d'Avitaya, G. Vincent, and S. Delage, Appl. Phys. Lett. $\underline{49}$, 271 (1986).

49. D.J. Frank, M.J. Brady, and A. Davidson, IEEE Trans. Magnetics $\underline{MAG\text{-}21}$, 721 (1985).

50. A. Davidson, M.J. Brady, D.J. Frank, J.M. Woodall, and A.W. Kleinsasser, IEEE Trans. Magnetics $\underline{MAG\text{-}23}$, 727 (1987).

51. H. Tamura, S. Hasuo, and T. Yamaoka, Jpn. J. Appl. Phys. 24, L709 (1985).
52. A. Yoshida, H. Tamura, T. Fujii, and S. Hasuo, Proc. 1987 Int'l. Superconductivity Electronics Conf., Tokyo, Japan.
53. H. Tamura, N. Fujimaki, and S. Hasuo, J. Appl. Phys. 60, 711 (1986).

PROGRESS IN EPITAXIAL INSULATORS
AND METALS ON SILICON

Julia M. Phillips

AT&T Bell Laboratories
Murray Hill, NJ 07974 USA

1. INTRODUCTION

Research into the growth of epitaxial insulators and metals on silicon has been a fertile area of investigation for several years.[1] The various groups pursuing this line of research have given numerous motivations for this work, such as its possible applicability to 3-dimensional integration, new device structures, dielectric isolation and semiconductor passivation (in the case of epitaxial insulators), and superior device metallization (in the case of epitaxial metals). From a more fundamental standpoint, these materials also offer, by virtue of their high degree of structural perfection, the opportunity to study the relationships between the atomic structure and electrical properties of both the film itself and of its interface with the substrate. There has been progress in studying the applicability of new epitaxial materials to all of these areas. In this paper some specific examples of investigations in the areas listed above will be discussed.

CaF_2, $CoSi_2$, and Si form a natural set of epitaxial materials. All have cubic crystal structures. Si has the diamond structure, while CaF_2 and $CoSi_2$ have the fluorite structure. The lattice constants of all three materials are well matched, with room temperature values of 5.36Å, 5.43Å, and 5.46Å for $CoSi_2$, Si, and CaF_2, respectively. $CoSi_2$ has a resistivity of $\sim 15\mu\Omega$-cm at room temperature, making it a good metal. CaF_2 has a band gap of 12 eV and a room temperature resistivity of $10^{14}\Omega$-cm, making it a good insulator.

2. EXPERIMENTAL PROCEDURE

All of the films grown in our laboratory were produced in a molecular beam epitaxy (MBE) system having a base pressure of $\leq 10^{-10}$Torr. Si substrates were cleaned as described previously.[2] CaF_2 was evaporated from a graphite crucible in a Knudsen effusion cell operating at about 1200°C. Optimal epitaxial quality was obtained by maintaining the substrate temperature at ~ 600°C during growth on Si(100)[3] and at ~ 700°C during growth on Si(111).[4] $CoSi_2$ was formed by depositing Co onto Si(111) substrates held at ≤ 100°C ("near room temperature") during the metal deposition, followed by an anneal at 600°C in the MBE system.[5]

Y. I. Nissim and E. Rosencher (eds.), Heterostructures on Silicon: One Step Further with Silicon, 203–214.
© 1989 by Kluwer Academic Publishers.

Rapid thermal anneals (RTA) lasting about 30s were carried out in a commercially available flash lamp annealing furnace using an Ar ambient.[6] Similar results have also been obtained using a rapid thermal annealing furnace housed in the MBE system.[7] This development allows anneals to be carried out in a clean high vacuum environment, thus minimizing contamination and chemical reactions.

3. RESULTS
3.1. *Electrical Characterization of CaF$_2$/Si.*

One of the motivations for studying epitaxial insulators on semiconductors is the possibility that they may provide semiconductor passivation superior to the ordering obtainable using amorphous insulators. This might be expected by virtue of the improved atomic scale ordering at a crystalline insulator-semiconductor interface over that at an amorphous insulator-semiconductor interface. In order to test this possibility, the electrical characteristics of epitaxial fluoride films on Si have been investigated by several groups. Breakdown fields of about 5×10^5V/cm have been reported routinely for CaF$_2$ films. The best reported breakdown field for films of this material is 3×10^6V/cm at room temperature.[8] (This value is within a factor of two to three of the typical breakdown field in SiO$_2$ films on Si.)

Metal-insulator-semiconductor (MIS) capacitors have been fabricated in the CaF$_2$/Si system to study the electrical properties of the insulator-semiconductor interface. In the case of CaF$_2$/Si(100) the inversion capacitance is near the predicted value, and the overall behavior of the capacitance is correct, as shown in Figure 1.[9] The hysteresis is generally small in thin films (<100 meV), but it tends

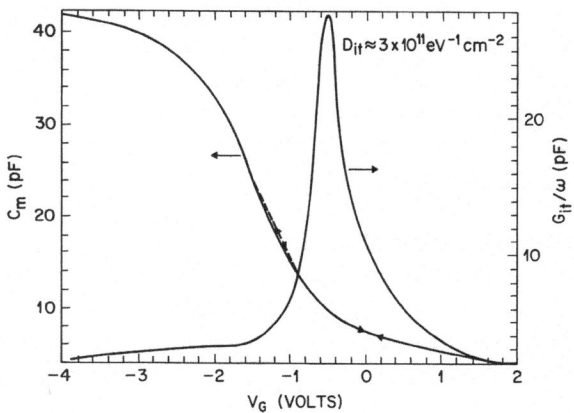

FIGURE 1. Measured capacitance and equivalent parallel conductance for an MIS diode of CaF$_2$ on $\sim$20Ω-cm p-Si(100) for f=1MHz and d$_{ins}$=730Å.

to increase dramatically in thicker films (>5000Å) or in thin films that have undergone mechanical stress. The lowest interface state density reported to date is $7 \times 10^{10} eV^{-1}$-cm^{-2} (nearly as low as that observed at the SiO_2/Si(100) interface). A more typical value for this system is $\sim 5 \times 10^{11} eV^{-1}$-$cm^{-2}$.

The C-V characteristics of MIS diodes fabricated in CaF_2/Si(111) are considerably different. In the case of films grown on p-type substrates, the interface state density is so large that no field-effect is observed.[10] (See Figure 6, which is discussed below.) Measurements on films grown on n-type substrates show a large region in the high-frequency (1 MHz) C-V curves, at an intermediate capacitance, in which the capacitance does not vary with the gate potential.[11]

3.2. *Field-Effect Transistors.*

Metal-epitaxial insulator-semiconductor field-effect transistors (MEISFET's) have been fabricated using CaF_2 as the gate insulator on Si(100). MEISFET's were fabricated on p-type Si substrates in the Corbino geometry.[8] The drain current (I_D) versus drain voltage (V_D) for a MEISFET is shown in Figure 2. The threshold voltage (V_{th}) is between -0.5 and +0.5V for most devices. Except for a shift of ~ 1V in V_{th} after prolonged application of a dc bias to the gate, the device characteristics are stable and reproducible. The room temperature mobility of the devices is about 400 cm^2/V-s. (This is within a factor of three of the room temperature mobility of the best field-effect transistors fabricated using SiO_2 as the gate insulator - about 1000 cm^2/V-s.) Transmission electron microscopic (TEM) analysis of epitaxial CaF_2 layers reveals that there are very few defects in the films exhibiting these device characteristics, though the films contain a significant number of (111) oriented crystallites. Dramatically worse device characteristics occur in devices fabricated using CaF_2 layers with large numbers of stacking faults and/or other defects. It is thought that defects promote the creation of interface traps by allowing impurities to diffuse to the interface and decorate misfit dislocations.

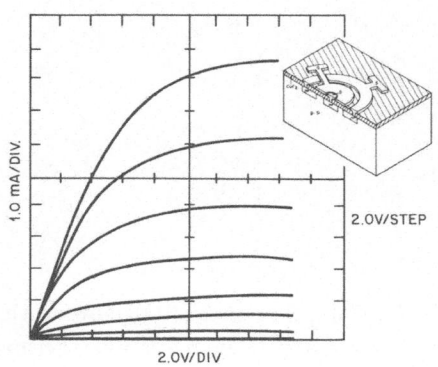

1.0 mA/DIV.

2.0V/STEP

2.0V/DIV

FIGURE 2. Drain current as a function of drain voltage for a MEISFET fabricated on p-Si(100) with a 6400Å CaF_2 film. The gate voltage is stepped between 0 and 14V. The inset shows the device geometry.

3.3. *Rapid Thermal Annealing.*

The epitaxial quality of MBE deposited CaF_2 films on Si(100) is a sharply peaked function of the substrate temperature during deposition. In addition, these films contain misoriented CaF_2(111) crystallites and have a very rough morphology, presumably due to the high energy of the CaF_2(100) surface. CaF_2 films grown on Si(111) have different, but equally serious problems in their tendency toward chemical instability and their poor electrical characteristics. An anneal after growth has proven useful in alleviating these difficulties. A time-temperature combination has been found which increases the molecular mobility of the CaF_2 molecules enough to improve the epitaxy and interfacial structure, but which does not activate chemical reactions.

We have used RTA to improve the epitaxial quality of CaF_2 films grown on Si(100).[12] Using RTA it is possible to create films whose crystallinity is superior to that of any unannealed film yet produced. Even films which are almost completely unoriented initially can be epitaxially recrystallized.[13] In addition, the morphology of the films improves so that the surface becomes featureless when examined by scanning electron microscopy, as shown in Figure 3. Misoriented (111) grains are removed, as well. The C-V characteristics of films grown on Si(100) are not affected by an RTA. There is dramatic improvement in the C-V characteristics of films grown on (111) substrates, however, which will be discussed below.

We believe that the driving force for the regrowth which leads to the improvement of these and other epitaxial layers (discussed below) is the minimization of the surface energy density in the films, both that associated with the substrate-overlayer interface and that associated with grain and microtwin boundaries.

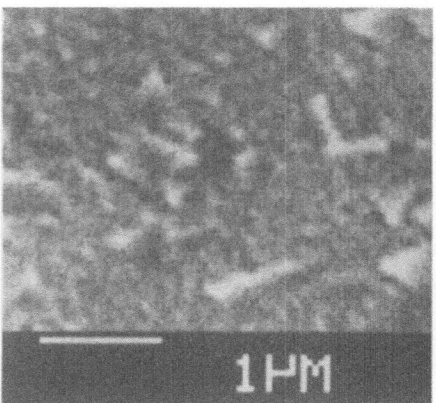

FIGURE 3. Scanning electron micrographs of a CaF_2 film on Si(100) showing the surface morphology of the film before (left) and after (right) a rapid thermal anneal.

FIGURE 4. Cross sectional TEM micrographs of a $CaF_2/CoSi_2/Si(111)$ heterostructure before and after RTA. Note the grain structure in the CaF_2 layer before RTA which merges into a single crystal upon annealing.

3.4. *Epitaxial Insulator-Metal-Semiconductor Structure.*

In order to realize 3-dimensional integration using epitaxial materials, it is necessary to demonstrate the compatibility of all three classes of electronic material, metal, insulator, and semiconductor, in a single heteroepitaxial structure. RTA has been indispensable in the growth of CaF_2, $CoSi_2$, and Si in such a heterostructure.[14] The dramatic improvement in the epitaxial quality of CaF_2 films grown on $CoSi_2/Si(111)$ substrates is illustrated in Figure 4. The as-grown CaF_2 film consists of grains which merge into a single crystal after RTA. In addition, the surface of the film becomes much smoother. In this figure, there is also a hint that the $CoSi_2$ layer has become more uniform as a result of the RTA. This observation has been corroborated in high resolution TEM micrographs of the as-grown and annealed heterostructure. The room temperature resistivity of the $CoSi_2$ layer is not affected by the RTA.

3.5. *Silicon on Sapphire (SOS).*

The dramatic improvements observed in the crystallinity of CaF_2 epitaxial layers after an RTA suggested that this technique should be applicable to improving epitaxial quality in general, provided that the melting temperature of the overlayer is less than or equal to that of the substrate.

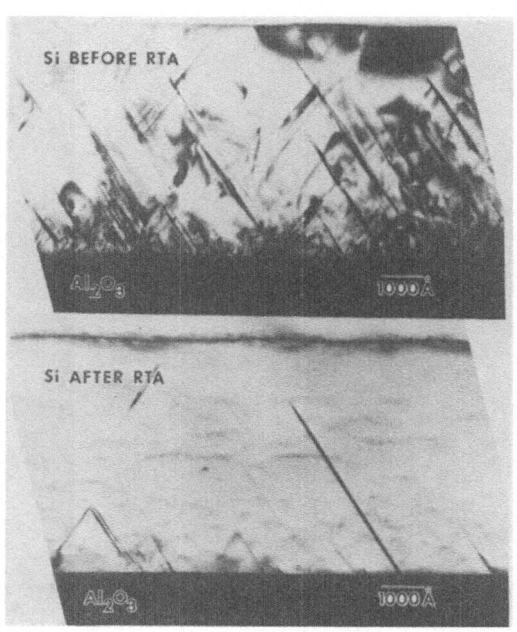

FIGURE 5. Cross sectional TEM <110> micrographs of an SOS film before and after RTA.

One of the oldest heteroepitaxial approaches to dielectric isolation is that of SOS. There have, however, been a number of problems with this system, many of them stemming from the high density of stacking faults and twins in the Si layer. As illustrated in Figure 5, an RTA for about 45s near, but below the Si melting temperature results in a dramatic reduction in the density of twin boundaries in the layer.[15] The Al concentration in the Si layer does not rise above $\sim 10^{16} cm^{-3}$ during this treatment, suggesting that improved device characteristics may be achievable in this material.

3.6. *Atomic Structure of the CaF$_2$/Si(111) Interface.*

The poor electrical qualities of the unannealed CaF$_2$/Si(111) interface have already been discussed. After RTA, however, Figure 6 shows that the 1 MHz C-V characteristics for films grown on p-type substrates are well-behaved, indicating that the interface state density has dropped by over two orders of magnitude, to $\sim 10^{11} cm^{-2} eV^{-1}$.[16] The pronounced effect of RTA on the electrical properties of this interface suggests that the detailed atomic structure of the interface may be affected by the anneal. This offers an excellent opportunity to take advantage of

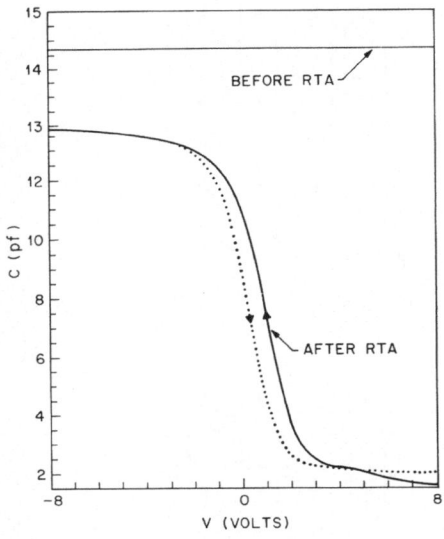

FIGURE 6. 1 MHz C-V characteristics of $CaF_2/Si(111)$ MIS capacitors before and after RTA.

the structural characterizability of epitaxial interfaces to investigate the relationship between this structure and the electrical properties of the interface.

Detailed analysis of high resolution cross section TEM (HREM) micrographs of the $CaF_2/Si(111)$ interface before and after an RTA has revealed that the atomic structure at the interface is indeed changed by an anneal.[17] We have considered the four possible bonding models for this interface shown in Figure 7, all of which assume that the interfacial Si always maintains its tetrahedral coordination at the interface. The models shown in Figure 7(a) and (b) have Ca-Si bonds at the interface. The interface Ca in Figure 7(a) is 8-fold coordinated, while it is 5-fold coordinated in Figure 7(b) due to the removal of a layer of F at the interface. The models shown in Figure 7(c) and (d) have F-Si bonds at the interface. The interface Ca is 7-fold coordinated in Figure 7(c) and 8-fold coordinated in Figure 7(d). Rigid shift measurements and multislice simulations of these models reveal that our HREM images are consistent with Ca-Si bonds at the interface, both before and after an RTA. There is, however, a subtle difference in the images in the two cases, in that there is an additional planar contraction at the interface after RTA which is not observed before the anneal. This can be explained by assuming that a layer of interfacial F which is present before the anneal is removed during the RTA.[17] This leads to strengthened Ca-Si bonds, and hence a shorter bond length.

These findings can be used to explain the change in the electrical properties of the $CaF_2/Si(111)$ interface as a function of RTA. The 8-fold coordinated interface has a layer of incompletely coordinated F. This F layer, because of its high electronegativity, will tend to form F^- by the accumulation of electrons. Since the F is not completely coordinated, formation of F^- is incompatible with a neutral

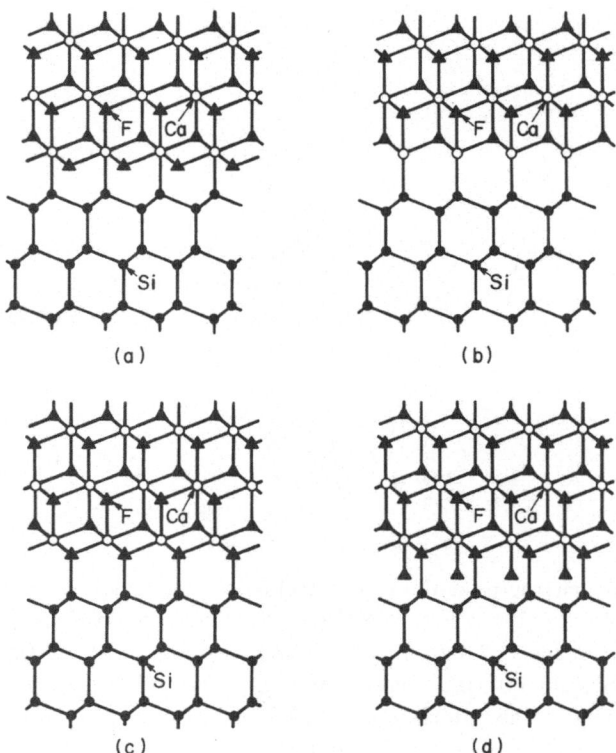

FIGURE 7. Model structures for the CaF_2/Si(111) interface: (a) Ca-Si interface bonds, 8-fold coordinated Ca; (b) Ca-Si interface bonds, 5-fold coordinated Ca; (c) F-Si interface bonds, 7-fold coordinated Ca; and (d) F-Si interface bonds, 8-fold coordinated Ca.

interface. This gives rise to a layer of net negative charge at the interface. In the case of p-type Si substrates, the layer of F^- will attract the majority holes, leading to an interface which is always in accumulation, as observed.[10] In the case of n-type Si, the F^- will repel the majority electrons and attract the minority holes. This leads to an interface which can be inverted easily, but which is difficult, if not impossible to accumulate. One can also hypothesize a plateau in the C-V curve when the number of holes is equal to the number of F^-, as observed.[11] After RTA, the layer of incompletely coordinated F is removed. This leads to an interface which tends to remain neutral and hence to have dramatically lower interface state density and reasonable C-V characteristics, as discussed above.

3.7. *Ultrathin Epitaxial CoSi$_2$ Layers on Si(111)*.

We have recently studied the structural and electrical properties of layers of epitaxial CoSi$_2$ on Si(111) having thicknesses between 10 and 100Å.[18,19] This system offers a unique opportunity to probe the physics of electrical transport in a structurally characterized metal layer of reduced dimensionality. One of the findings to emerge from this study is that electrically continuous single crystal films can be grown as thin as 10Å.[18] If the films are annealed at 600°C, the resistivity of a film depends very strongly on its thickness, as indicated in Figure 8. The transport is essentially metallic in all cases, corroborating the TEM observations that all four films are connected, albeit with pinholes in the case of films less than ∼30Å thick. The single most striking feature in these curves is the pronounced "size effect", the divergence of resistivity as the films become thinner. For the 12Å film the residual resistivity ρ_o is more than 40 times greater than the bulk residual resistivity, $\rho_{o\infty}$. It is difficult to account for resistivity increases of such magnitude in terms of bulk scattering. Scattering from grain boundaries, dislocations, etc. should be irrelevant since the samples are single crystal. The rise in resistivity is believed to be due to diffuse scattering at the free surface, consistent with cross-sectional HREM which indicates that the free surface profile is atomically rough and uncorrelated with a rms roughness amplitude perhaps as large as 5Å. Even with totally diffuse scattering, however, classical theory gives a

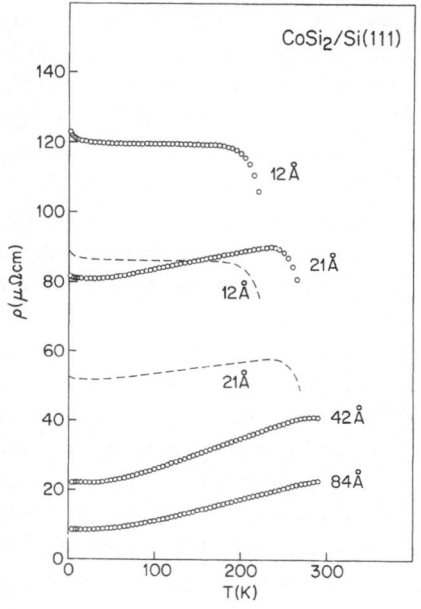

FIGURE 8. Resistivity of four CoSi$_2$ films as a function of temperature. The films are all epitaxial, as discussed in Ref. 18. Data corrected for pinholes are indicated by dashed curves.

ratio of $\rho_o/\rho_{o\infty}$ of only ~ 9. Classical theory should be inappropriate in this thickness regime, anyway, since the 12Å thickness is approaching the deBroglie wavelength, $\lambda_d \sim 7$Å, for $CoSi_2$. The observed strong divergence of resistivity appears to be more consistent with recently developed theories of quantum size effects.[20]

4. CONCLUSIONS

CaF_2, $CoSi_2$, and Si offer an excellent set of epitaxial materials for probing not only the potential applications of such heteroepitaxial systems but also for obtaining new understanding of the relationships between the structure and other properties of thin films and interfaces. The electrical properties of epitaxial CaF_2 films on Si are quite promising, even allowing the fabrication of new devices such as the MEISFET. The electrical properties of $CoSi_2$ films on Si have not been discussed here, but they are also promising, as detailed in reference 21, and these layers offer a number of possibilities for new devices, as well. Rapid thermal annealing has been found to be very useful for improving the heteroepitaxial quality of a growing number of systems, including CaF_2/Si, CaF_2/$CoSi_2$/Si, and Si on sapphire. The improvement of SOS crystalline quality by RTA is sufficiently impressive as to be of potential application in dielectric isolation in the fairly near future. If one wishes to use an epitaxial approach to 3-dimensional integration, it is necessary to ascertain the mutual compatibility of the metal, insulator, and semiconductor involved. In the case of CaF_2, $CoSi_2$, and Si, this has recently been demonstrated, again making use of RTA to epitaxially recrystallize the CaF_2. The finding that the electrical characteristics of the CaF_2/Si(111) interface can be improved dramatically by an RTA can now be understood in terms of changes in the atomic structure of the interface induced by the anneal. This finding points out one of the advantages of epitaxial interfaces, namely the possibility that they offer to correlate their detailed structure with their other properties. Ultrathin epitaxial $CoSi_2$ films again offer one the possibility to perform electrical characterization on structurally characterized films. These studies have revealed a divergence in the resistivity of the films with decreasing thickness, suggestive of a quantum size effect.

The investigation of epitaxial insulators and metals has entered a new phase. While early work was concerned to a large extent with the identification and straightforward characterization of epitaxial systems, more recent efforts have sought to obtain a more detailed understanding of fundamental aspects of epitaxial growth and of the interrelationships between various properties of epitaxial layers and their interfaces. This type of investigation has already led to some important insights. It is to be expected that a deeper understanding of these systems in particular and of heteroepitaxial systems in general will arise from these studies.

ACKNOWLEDGMENTS

It is a pleasure to acknowledge my colleagues who have collaborated on various aspects of this work: W. M. Augustyniak, J. L. Batstone, M. Cerullo, J. C. Hensel, D. C. Joy, L. N. Pfeiffer, and T. P. Smith, III.

REFERENCES

1. See Bean JC(ed): *Proceedings of the First International Symposium on Silicon Molecular Beam Epitaxy.* Pennington, NJ: Electrochemical Society, 1985. Also see Bean JC and Schowalter LJ(eds): *Proceedings of the Second International Symposium on Silicon Molecular Beam Epitaxy.* Pennington, NJ: Electrochemical Society, 1988.

2. Ishizaka A, Nakagawa K, and Shiraki Y: *Collected Papers of MBE-CST-2:* Tokyo: Jpn. Soc. Appl. Phys., 1982, p. 182.

3. Asano T and Ishiwara H: Thin Solid Films *93,* 143, 1982.

4. Asano T, Ishiwara H, and Kaifu N: Jpn. J. Appl. Phys. *22,* 1476, 1983.

5. Phillips JM, Batstone JL, Hensel JC, and Cerullo M: Appl. Phys. Lett. *51,* 1895, 1987.

6. AG Heatpulse 210, AG Associates, Inc., Palo Alto, CA 94303.

7. Cerullo M, Phillips JM, Anzlowar M, Pfeiffer L, Batstone JL, and Galiano M: Mat. Res. Soc. Symp. Proc., in press.

8. Smith TP, III, Phillips JM, Augustyniak WM, and Stiles PJ: Appl. Phys. Lett. *45,* 907, 1984.

9. People R, Smith TP, Phillips JM, Augustyniak WM, and Wecht KW: Mat. Res. Soc. Symp. Proc. *37,* 169, 1985.

10. Smith TP, III, Phillips JM, People R, Gibson JM, Pfeiffer L, and Stiles PJ: Mat. Res. Soc. Symp. Proc. *54,* 295, 1986.

11. Schowalter LJ, Fathauer RW, and Krusius JP: *Proc. Symp. on Silicon Molecular Beam Epitaxy:* Bean JC(ed): Pennington, NJ: The Electrochemical Society, 1985, p. 311.

12. Pfeiffer L, Phillips JM, Smith TP, III, Augustyniak WM, and West KW: Appl. Phys. Lett. *46,* 947, 1985.

13. Phillips JM, Pfeiffer L, Joy DC, Smith TP, Gibson JM, Augustyniak WM, and West KW: J. Electrochem. Soc. *133,* 224, 1986.

14. Phillips JM and Augustyniak WM: Appl. Phys. Lett. *48,* 463, 1986.

15. Pfeiffer L, Phillips JM, Luther KE, West KW, Batstone JL, Stevie FA, and Maurits, JEA: Appl. Phys. Lett. *50,* 466, 1987.

16. Phillips JM, Manger ML, Pfeiffer L, Joy DC, Smith TP, III, Augustyniak WM, and West KW: Mat. Res. Soc. Symp. Proc. *53,* 155, 1986.

17. Batstone JL, Phillips JM, and Hunke EC: Phys. Rev. Lett. *60,* 1394, 1988.

18. Phillips JM, Batstone JL, Hensel JC, and Cerullo M: Appl. Phys. Lett. *51,* 1895, 1987.

19. Phillips JM, Batstone JL, Hensel JC, Cerullo M, and Unterwald FC: submitted to J. Mat. Res.

20. Tesanović Z, Jarić MV, and Maekawa S: Phys. Rev. Lett. *57*, 2760, 1986; N. Trivedi and N. W. Ashcroft: unpublished and private communication.

21. Levi AFJ, Tung RT, Batstone JL, and Anzlowar M: Mat. Res. Soc. Symp. Proc., in press.

GROWTH OF CoSi$_2$ AND CoSi$_2$/Si SUPERLATTICES

J. Henz, M. Ospelt and H. von Känel

Laboratorium für Festkörperphysik, ETH-Zürich, 8093 Zürich
Switzerland

INTRODUCTION

The growth of epitaxial CoSi$_2$ films on Si(111) has been the subject of several research groups for quite a number of years (1-5). The motivation has mainly been spurred by the attractive electrical properties of the CoSi$_2$/Si system and by the wish to understand more about the fundamental mechanisms of eptiaxial growth. That is why it is of utmost importance to investigate the effects of different growth parameters on the quality of the silicide layers. As a matter of fact until now, most of the work has been done by solid phase epitaxy (SPE) (1-5). In this technique the deposition of a pure Co layer on the substrate held at room temperature is followed by an anneal to around 650 °C. Despite of the slight differences in the processing parameters from one group to the other, which are mainly due to the thickness of the deposited Co films, the annealing temperature or the heating rate (e.g. rapid thermal annealing), these layers have one common property, they have all a surface with a visible roughness. They further contain a high density of pinholes in the range of 10^5-10^7 cm^{-2}.

GROWTH OF CoSi$_2$:

In a previous publication (6,7) we have reported for the first time about a new approach to fabricate CoSi$_2$ with a smooth surface and no detectable pinholes. This was achieved by stoichiometric coevaporation of Si and Co at room temperature followed by an anneal up to 500 °C. The as grown layers are stable up to a least 600 °C, i.e. no pinholes can be detected by SEM and there is no roughness visible in Normarsky-microscopy. It has to be emphasized that the control of the different evaporation rates is the most important parameter. Even slight deviations from the ideal flux ratio of two to one lead to a different morphology of the layers. In the case of a Co-excess this can already be seen in situ by RHEED. In contrast to the optimal case, where a sharp 1x1 streak pattern is observed during almost the whole annealing process, i.e. from 100 °C up to 600 °C, for Co-excess the RHEED pattern changes form a 1x1 to a 2x2 in the region of 350 °C - 450 °C depending on the amount of Co-surplus. At this stage, the surface is still smooth and pinhole free. A further anneal to more than 500 °C makes the RHEED pattern change again, this time from the 2x2 to a 1x1. It is at this point that the surface gets rough and pinholes are formed. In Fig. 1a) we show a RHEED photograph of the <01$\bar{1}$> direction of a 35 Å thick CoSi$_2$ layer heated up to 350 °C and exhibiting a clear 2x2 reconstruction, indicating a Co-excess. In fig. 1b) is shown an SEM micrograph of the same sample after a further anneal to 550 °C. A large amount of pinholes can clearly be seen. To compare we have in Fig. 2a) and b) the RHEED pattern taken at 350 °C and the SEM image of a 35 Å thick sample having been heated up to

Y. I. Nissim and E. Rosencher (eds.), Heterostructures on Silicon: One Step Further with Silicon, 215–222.
© *1989 by Kluwer Academic Publishers.*

216

600 °C. The latter sample has been grown under optimal conditions, and thus no 2x2 reconstruction and no pinholes are visible.

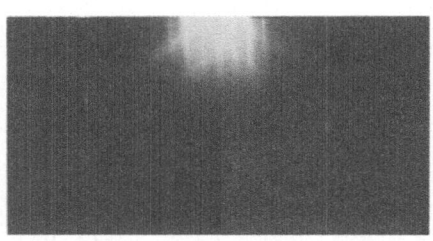

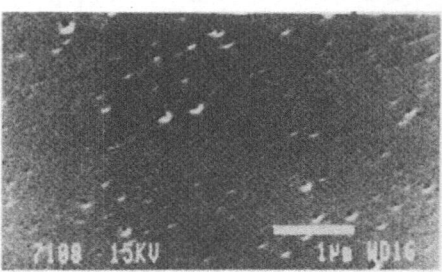

Fig. 1a) RHEED pattern taken from a 35 Å thick CoSi$_2$ layer at 350 °C. <01$\bar{1}$> azimuth. Slight Co-excess during co-evaporation.

Fig. 1b) SEM micrograph of the same sample after a further anneal to 550 °C.

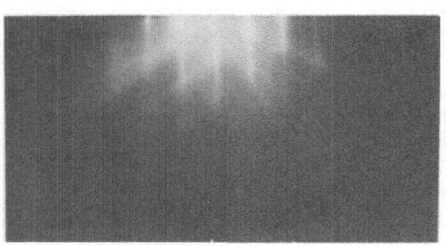

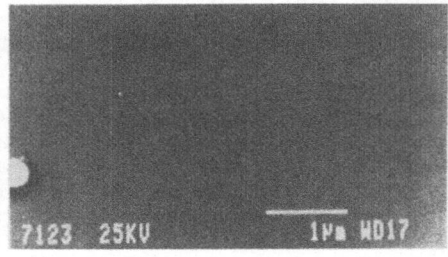

Fig. 2a) RHEED pattern taken from a 35 Å thick CoSi$_2$ layer at 350 °C. <01$\bar{1}$> azimuth. Co-evaporation exactly stoichiometric.

Fig. 2b) SEM micrograph of the same sample after a further anneal to 600 °C.

It should be mentioned that the orientation of these CoSi$_2$ layers is always B type, as can be seen directly in situ from the asymmetry of the Kikuchi-Bands in the <01$\bar{1}$> direction, or ex situ by RBS channeling experiments using appropriate thick films. A possible explanation for the different behaviour of the 2x2 and 1x1 reconstructed surfaces can be found by relating them to the C- and S-surfaces recently reported by Tung et.al. (8). According to these authors a CoSi$_2$ layer grown by SPE always has an S-surface when annealed up to 600 °C. An S-surface is one in which the silicide is terminated by two monolayers of silicon (9) on top of the CoSi$_2$ bulk. These two monolayers of Si are missing on a C-surface. C-CoSi$_2$ changes to S-CoSi$_2$ when it is heated up to 600 °C the required Si being provided by the opening up of pinholes. Thus we identify the 2x2 reconstructed CoSi$_2$ with a C-surface and the 1x1 with an S-surface. In this way, the behaviour of the two different surfaces can be explained and the decisive role of an accurate rate control becomes apparent. As little as a few percent of Co excess during the evaporation results in a C-surface leading to a rough, pinhole-rich layer when heated up to 600 °C. Evaporating a Si cap on the $CoSi_2$ before heating, as just recently proposed (10), renders the formation of a C-surface impossible and thus explains the absence of pinholes. If the Si cap is to thin, however, Si islands are formed on top of the $CoSi_2$ layer. We thus want to emphasize that thin (15 Å - 100 Å), perfectly smooth, pinhole free and single B type $CoSi_2$ layers can be grown at 500 - 600 °C by using stoichiometric evaporation at room temperature. A Si cap may help in case of small deviations from stoichiometry, but it can produce a slight roughness due to Si islands on top of the $CoSi_2$. Once a thin (30 - 50 Å) $CoSi_2$ layer is grown , one can use it as a template to grow much thicker $CoSi_2$ on top of it. Layers as thick as 1300 Å have been grown at 250 °C - 300 °C by coevaporation of Co and Si. Even at these low temperatures the crystalline quality is quite good, as a channeling yield of 4-7% shows, but it can be substantially improved down to 1.8% by annealing the samples up to 650 °C. As an example we show in fig. 3) the RBS spectra of a 1250 Å thick $CoSi_2$ layer annealed at 650 °C for 1 h. These thick layers are still smooth and perfect mirrors, as opposed to those fabricated by SPE, which show a milky surface when their thickness exceeds 200 Å. (2)

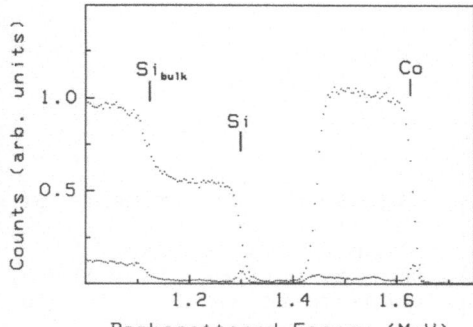

Fig. 3: RBS spectra of a 1250 Å thick $CoSi_2$ layer annealed at 650 °C. Random and (111) channeling direction are displayed, the yield in this direction is 1.8%.

Planview transmission electron microscopy experiments showed no dislocations in films with a thickness below 30 Å, i.e. they grow entirely pseudomorphic and pinhole-free (11). The angular dependence of the RBS yield around the [114] minimum shows still substantial tetragonal strain e_t in a 70 Å thick layer annealed at 500 °C, as is shown in fig. 4). From the displacement of the channeling minima of ~ 0.5° we calculate e_t ~ 1.8 %.

Another interesting question is the stability of the layers when they are exposed to air. XPS-measurements made immediately after the evaporation and annealing (without breaking the vacuum) were compared with those obtained after exposing the samples to air for some days. We found that after about one week a SiO_2 layer is formed on the surface with a thick-

ness of 6-7 Å. This layer is the same, no matter whether the
starting $CoSi_2$ surface was of S or C type. After reintroducing
these samples into UHV they are stable up to 650° C. This applies again to both types of original surfaces.

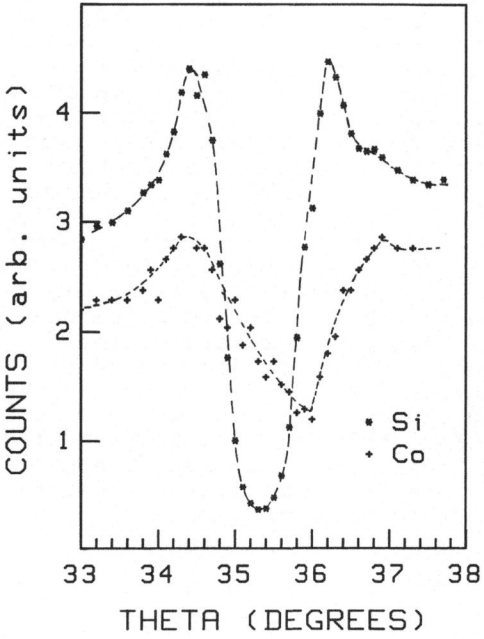

Fig. 4: Axial angular yield
profiles obtained from the
Co signal of a 70 Å thick
$CoSi_2$ layer on Si(111) and
the underlying bulk silicon.
([114]channeling minimum in
$CoSi_2$.)

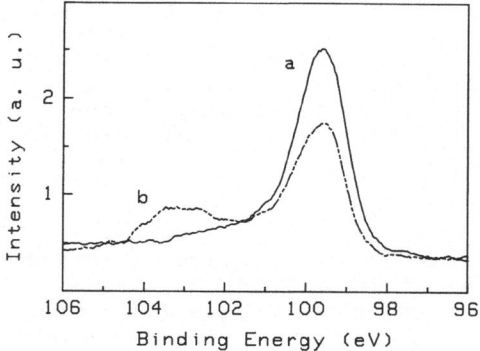

Fig. 5: XPS-measurements of
the Si 2p core level
a) pure $CoSi_2$ surface
b) same surface after the
exposure to air for 10 days.

$CoSi_2$/Si SUPERLATTICES

Having grown $CoSi_2$ according to the state of the art, there
remains the problem of silicon growth on top of it. Using
substrate temperatures as normally used in Si-MBE, an island
growth mode is observed (12,13). This can be suppressed by
lowering the temperature during evaporation and by growing the
Si-layer in several steps. Betweeen the evaporation steps, the
substrate is heated up to 650 °C. In this way, smooth pinhole-

free Si layers in the range of 20 Å to 4000 Å could be grown. The orientation of these layers with respect to the substrate is always type B, independent of the $CoSi_2$ thickness. Only when the $CoSi_2$ is of marginal quality (pinholes) does Si grow with type A orientation. We could not find a correlation between the strain in $CoSi_2$ and the orientation of the Si overlayer (14). Even on the thinnest (10 Å thick) $CoSi_2$ layers, which are certainly strained as shown before, Si grows with type B orientation (11). Another question is the orientation of a $CoSi_2$ layer on top of the overgrown Si. In this case we observe clear type B orientation of the $CoSi_2$ compared to the Si directly underneath, as long as the latter is thicker than ~ 30 Å. We thus have the following layer sequence ABBAABB... This can very nicely be seen in fig. 6), where is shown a high resolution TEM micrograph of a superstructure consisting of 32 Å $CoSi_2$ and 103 Å thick Si layers. This structure is made of seven $CoSi_2$ layers as can be seen in the low resolution TEM cross section of fig. 7).

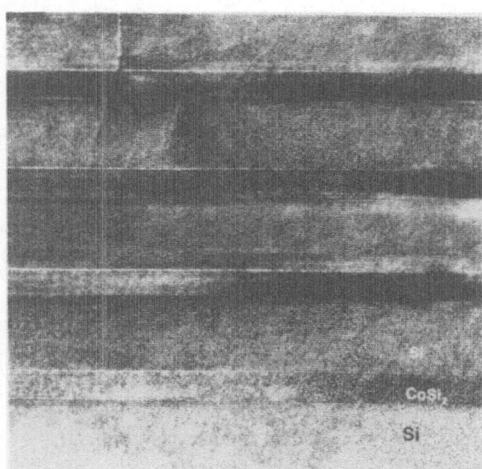

Fig. 6: High resolution TEM cross section of the super-lattice described in the text. Successive orientations can clearly be seen to be ABBAABB....

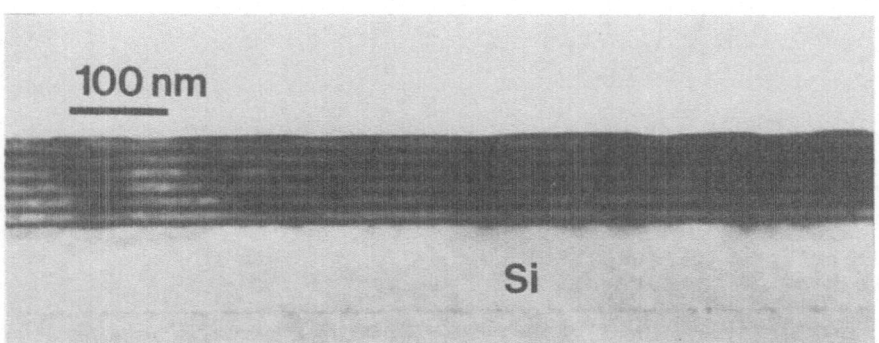

Fig. 7: Low resolution TEM cross section of the same sample as in fig. 6).

On this sample we have also measured rocking curves with X-ray diffractometry. We have used $CuK_{\alpha 1}$ radiation and measured near the (111) substrate reflection. We have been able to see reflexes due to the superlattice, which are of 0th, 1st and 2nd order as is shown in fig. 8). It is for the first time, that superlattice reflexes on an epitaxial $CoSi_2/Si$ system could be seen. From the angular spacing of the satellites the period of the superlattice can accurately be determined to be 136 Å, which is in reasonable agreement with the thickness measured with quartz balances during the evaporation. To measure the average thickness of the individual layers, random RBS measurements have been made. Here too, the layered structure can be directly seen as shown in fig. 9). For the $CoSi_2$ layers we have obtained an average thickness of 32.5 Å and thus for Si one of 103.5 Å. Using these values and the position of the 0th order superlattice X-ray peak we can calculate the perpendicular strain in the $CoSi_2$ layers. Assuming that the Si is not strained, we get a strain of 0.6% for the $CoSi_2$ layers, i.e. they are only partially strained. For pseudomorphic $CoSi_2$ layers on Si(111) the perpendicular strain would be expected to be 0.9 %. The fact, that they are not completely strained in our case, can be due to the total thickness of the superlattice, which probably exceeds the critical thickness as a whole. Nevertheless from fig. 6) and 8) it can clearly be seen that the silicon grows on $CoSi_2$ with the same orientation as the latter even though the silicide is strained. This is at variance with the results reported in ref. (14).

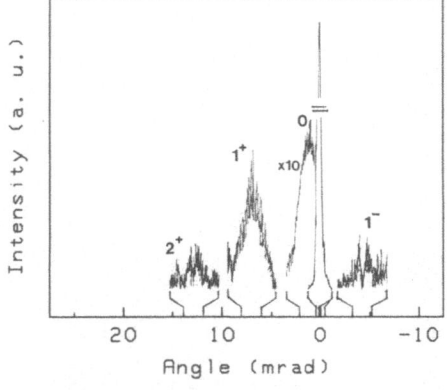

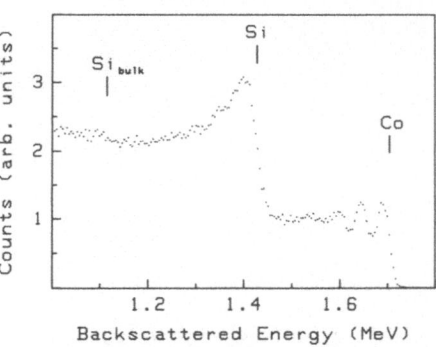

Fig. 8: X-ray rocking curves of the (111) substrate reflex and the satellites of the $CoSi_2/Si$ superlattices (the width of the peaks has been expanded).

Fig. 9: Random RBS spectra of the same sample as in fig. 8).

It should be mentioned, however, that the crystalline quality of the whole superlattice is far from being satisfactory as shown in RBS channeling measurements. In cross section TEM pictures an increasing amount of dislocations from the bottom to the top layer is also seen. It is not yet clear whether the decreasing quality is due to critical thickness effects or to the rather low annealing temperatures.

To show the uniformity of our samples over a large area we display in fig. 10) a Normarsky photograph of a structure with five $CoSi_2$ layers of about the same thicknesses as in the one mentioned before. To show the layer structure, we have etched layer by layer down to the first $CoSi_2$ film. As a result steps with a height of 135 Å appear, the terasses consisting of $CoSi_2$. As a side remark we might mention that since these samples have been heated up to 650 °C in situ, the electric properties of the $CoSi_2$ layers can be expected to be excellent (15).

Fig. 10: Nomarsky photograph of an etched superstructure made out of five $CoSi_2$ layers. The individual layers are clearly resolved.

To conclude, we have shown, that it is possible to fabricate pinhole-free, smooth $CoSi_2$ layers with thicknesses between 10 and 1500 Å by coevaporation of Si and Co at room temperature. The thicker layers are stable up to 650 °C. We have further shown, that Si can be grown on top of $CoSi_2$ with single B-type orientation. We are thus able to fabricate $CoSi_2$/Si super-structures with individual layer thicknesses covering a broad range.

ACKNOWLEDGEMENT
We would like to thank very much Mr. H.-J. Gübeli for his excellent technical support and Dr. R.E. Pixley from the Physics Department of the University of Zürich for the opportunity to use his RBS equipment and his help therewith. We further are very greatful to Mr. R. Wessicken for the TEM investigations and to Mr. P. Wägli for the SEM pictures. Financial support by the Swiss National Science Foundaton is also greatfully acknoweldged.

REFERENCES

1. R.T. Tung, J.C. Bean, J.M. Gibson, J.M. Poate and D.C. Jacobsen, Appl. Phys. Letters 40 (1982) 684
2. F. Arnaud D'Avitaya, S. Delage, E. Rosencher and S. Derrien, J. Vacuum Sci. Technol. B3 (1985) 770

3. B.D. Hunt, N. Lewis, E.L. Hall, L.G. Turner,
   L.J. Schowalter, M. Okamoto and S. Haskimoto, Mater. Res.
   Symp. Proc. S6 (1986) 151
4. S. Saitoh,H. Ishiwasa and S. Furukawa, Appl. Phys. Lett.
   37, (1980) 203
5. Y.C. Kao, M. Tejwani, Y.H. Xie, T.L. Lin and K.L. Wang, J.
   Vac. Sci. Technol. B3 (1985) 596
6. J. Henz, H. von Känel, M. Ospelt and P. Wachter, Surface
   Sci. 189/190 (1987) 1055
7. J. Henz, M. Ospelt and H. von Känel, Solid State Commun.
   63 (6) (1987) 445
8. R.T. Tung and F. Hellman, Mater. Res. Soc. Symp. Proc. 94
   (1987) 65
9. S.A. Chambers, S.B. Anderson, H.W. Chen and J.H. Weaver,
   Phys. Rev. B 34 (1986) 913
10. T.L. Lin, R.W. Fathauer P.J. Grunthauer and C.
    d'Anterroches, Appl. Phys. Lett 52 (10) (1988) 804
11. H. von Känel, J. Henz and M. Ospelt, Proc. of the Second
    Int. Symp. on Silicon MBE 88-8 (1987) 274
12. B.M Ditchek, J.P. Salerno and J.V. Gormley, Appl. Phys.
    Lett. 47 (11) (1985) 1200
13. F. Arnaud d'Avitaya, J.A. Chroboczek, G. Glastre, Y.
    Campidelli and E. Rosencher J. Cryst. Growth 81 (1987)
    463
14. R.T. Tung, J.L. Batstone and S.M. Yalisove, Proc. of the
    Second Int. Symp. on Silicon MBE 88-8 (1987) 247
15. S.Y. Duboz, P.A. Badoz, E. Rosencher, J. Henz, M. Ospelt,
    H. von Känel and A. Briggs, submitted to Appl. Phys. Lett.

FORMATION OF EPITAXIAL CoSi$_2$ FILMS ON Si(111) A LOW TEMPERATURE (≤400°C)

L.Haderbache, P.Wetzel, C.Pirri, J.C.Peruchetti D.Bolmont and G.Gewinner

Laboratoire de Physique et de Spectroscopie Electronique F.S.T. 4, rue des FréresLumière

68093 - Mulhouse Cédex, FRANCE

ABSTRACT

The growth of thin (≤100Å) epitaxial CoSi$_2$ layers on Si(111) is studied by angle resolved photoemission and low energy electron diffraction experiments. Various preparation methods at low temperature (≤400°C) are investigated. First it is demonstrated that layer by layer growth can be achieved by a series of sequential Co(1ML) and Si(2ML) evaporations and subsequent low temperature annealing. Ending the deposition series with either Co or Si results in a Co or Si rich CoSi$_2$ surface labeled CoSi$_2$(111)-Co and CoSi$_2$(111)-Si respectively with characteristic photoemission signature. According to photoemission, a minimum annealing temperature of 360°C is found to be necessary to produce epitaxial CoSi$_2$ films with a good crystallinity. CoSi$_2$ layers were also grown by coevaporation of Co and Si in their stoichiometric ratio onto Si(111) subjected to thermal processing at 360°C either during or after deposition. The CoSi$_2$, prepared in this way, invariably exhibits a bulk and surface excess of Si. In particular the film surface is systematically of the silicon rich CoSi$_2$(111)-Si type. In contrast a different preparation method which consists in deposition of Co only onto the Si(111) substrate held at ~360°C produces CoSi$_2$ films exposing a Co rich CoSi$_2$(111)-Co surface without any Si excess in bulk. It is concluded from these experiments that at 360°C diffusion of Si from substrate through the CoSi$_2$ layer is much easier than usually expected and quite sufficient to sustain further CoSi$_2$ growth without any extra Si supply.

1.INTRODUCTION

A strong interest in epitaxial CoSi$_2$ thin films on Si(111) has been raised up by the possibility of novel devices realisation based on the re-epitaxy of Si on CoSi$_2$ leading to Si/CoSi$_2$/Si heterostructures [1-3]. Numerous experiments were performed to study the nucleation and the growth of cobalt silicides on silicon by solid phase epitaxy (SPE) [4-8]. Good crystallinity CoSi$_2$ layers with interesting electrical properties are achieved by Co deposition onto a clean Si(111) surface and subsequent annealing at 550°C. Unfortunately pinholes are often observed in these layers, their density and size depending on annealing temperatures and CoSi$_2$ film thicknesses [9-10]. Films grown in such a way present two epitaxial orientations; type A silicide has the same orientation as the silicon substrate; type B silicide shares the surface

223

Y. I. Nissim and E. Rosencher (eds.), Heterostructures on Silicon: One Step Further with Silicon, 223–229.
© *1989 by Kluwer Academic Publishers.*

normal [1 1 1] axis with Si, but is rotated 180° about this axis with respect to the Si.

Recently, H.von Känel et al [11] have observed growth of ultrathin pinhole free $CoSi_2$ layers of type B orientation with a perfectly smooth surface by coevaporation at very low temperature (<400°C). However, silicide layers grown at low temperature have poor electrical conductivity probably due to crystallinity defects. Annealing above 600°C is required to improve the electrical properties to a reasonable level. In a previous paper, we have reported experiments on epitaxial $CoSi_2$ growth by SPE as a fonction of Co deposited thickness [7]. We have shown that very thin epitaxial layers can be achieved at low annealing temperatures, typically 360-420°C. In these cases, the Co thickness cannot exceed 6-7 monolayers (ML) ($1 ML=7.8 \times 10^{14}$ atoms/cm^2). It was also demonstrated that $CoSi_2$ films of any thickness can be terminated by either Co or Si rich (111) surfaces depending on the annealing conditions. These two distinct surface structures are labeled $CoSi_2(111)$-Co (Co enriched surface) and $CoSi_2(111)$-Si (Si enriched surface) [7]. The $CoSi_2(111)$-Si surface structure may be obtained by solid phase epitaxy (S.P.E.) and annealing above 500°C. The $CoSi_2(111)$-Co surface can be prepared from $CoSi_2(111)$-Si by a ~ 1ML Co deposit subsequently annealed at ~ 400°C [7]. Structural investigations [14,15] performed on cobalt disilicide have confirmed our finding that annealing above 500°C produces a Si rich surface. The Co (Si) enriched-surfaces are identified by means of their characteristic surface states [7, 14, 15]. For instance the $CoSi_2(111)$-Co terminated surface exhibits two surface states, located at -2.7 and -1.4eV respectively. On $CoSi_2(111)$-Si terminated surfaces only weak surface related features could be identified. R.T.Tung et al [12] propose that the driving force for the generation of pinholes in thin $CoSi_2$ layers in the temperature range ~550-650°C is closely related to change in the surface structure from $CoSi_2(111)$-Co, stable at low temperature, to the high temperature stable $CoSi_2(111)$-Si form. Thus, no pinholes were found after annealing at 600°C in $CoSi_2$ layers with an initial $CoSi_2(111)$-Si surface prepared by coevaporation at low temperature [12, 13].

The purpose of this study is to investigate the properties of $CoSi_2$ films obtained with various low temperature preparation techniques such as layer by layer growth or Co and Si coevaporation or a method that we call "reactive molecular beam epitaxy" (reactive MBE) consisting in slow evaporation of Co onto the Si substrate held at 360°C which results in continuous reaction of the Co species at the silicide film surface with substrate Si atoms having diffused through the film.

2. EXPERIMENTAL PROCEDURE

The experiments were performed in situ in an ultrahigh vacuum (UHV) chamber (base pressure 10^{-10} Torr) equipped with low energy electron diffraction (LEED) and ultraviolet (UPS) and x-ray (XPS) photoemission techniques. All measurements were done with He$_I$ (21.2eV) and MgK$_\alpha$(1253.6eV, unmonochromatized) photon excitation. The n-type Si(111) substrate was cleaned with standard ion etching and thermal annealing cycles to obtain the well known 7x7 Si(111) reconstructed surfaces. Co and Si layers of various thicknesses were then evaporated from home made Co and Si sources with a evaporation rate of ~ 0.4 and 0.7Å/min respectively and operating typically at a pressure of 5x10^{-10} Torr. The evaporation rate was calibrated by quartz thickness monitors and XPS core line intensity measurements.

3. RESULTS AND DISCUSSION

3.1 Layer by layer growth

Figure 1 presents valence band photoemission spectra for CoSi$_2$ films obtained by repeated cycles of sequential Co (1ML) and Si (2ML) evaporations subsequently annealed at ~ 360°C for 5 min. The procedure is started with a template layer prepared by a 3ML Co deposit subsequently annealed at 360°C for 5 min. As shown previously [7] the CoSi$_2$ surface formed at this stage can be identified with a CoSi$_2$(111)-Co surface caracterized by a surface state located at -2.7eV and a broader peak assigned to the contribution of a surface state at -1.4eV and a dominant feature at -1.8eV reflecting emission from bulk bands [16-17] (curve a) and a sharp (1x1) LEED pattern.

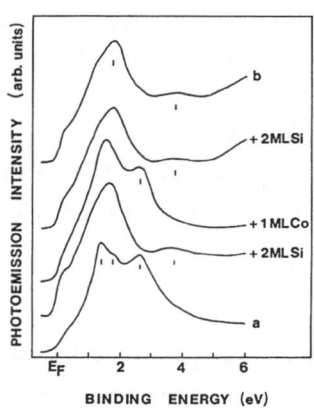

Fig .1 : Typical UPS spectra recorded with hν = 21.2 eV at normal emission on CoSi$_2$ films formed by repeated deposition of 1ML Co and 2ML Si subsequently annealed at ~360°C for 5 min. : (a) template layer prepared by 3ML Co deposited on the Si(111) substrate subsequently annealed at 360°C for 5 min., (b) UPS spectrum after 10 cycles ended by Co deposition.

After a 2 ML Si deposit on the CoSi$_2$(111)-Co surface annealed at 360°C a CoSi$_2$(111)-Si surface with two spectral features at -1.8 and -3.8eV is observed. The LEED pattern is still (1x1) but with distinct changes in the I(V) curves. The -1.8eV bulk emission feature has Co 3d non-bonding character [16, 17]. A further 1ML Co deposit on a CoSi$_2$(111)-Si surface results again in a CoSi$_2$(111)-Co surface (curve 3) if the surface is annealed at 360°C. It is important to note that if annealing is carried out at temperatures lower than 360°C, the surface does not exhibit the characteristic spectrum with well defined surface states. Since the ordering process is generally easier at surface than in bulk we believe that 360°C is a minimum annealing temperature in order to get a good crystallinity of the CoSi$_2$ film. Repeated application of this procedure makes possible the formation of new epitaxial $CoSi_2$ layers but after ~ 10 cycles ended by Co deposition the UPS spectrum is similar to the $CoSi_2(111)$-Si rather than $CoSi_2(111)$-Co surface spectrum. This suggests the presence of an excess of Si. Figure 2 presents the evolution of the $Co2p^{3/2}/Si2s$ core level intensity ratio at normal emission for different growth techniques. Again it appears that the $Co2p^{3/2}/Si2s$ core line intensity ratio is lower for layer by layer $CoSi_2$ growth than for S.P.E. $CoSi_2$ growth at 600°C indicating a Si enrichment of the $CoSi_2(111)$ film. Consequently the XPS core line intensity ratio data corroborate the UPS observations indicating an excess of Si in bulk and surface. Two possible reasons for this Si excess can be invoked. First, an uncertainty in the monolayer calibration of the order of 10% readily explains the UPS observations. However, the experiment was repeated several times and one invariably observes an excess of Si rather than Co. Possibly some Si supply from substrate must occur even at low temperature.

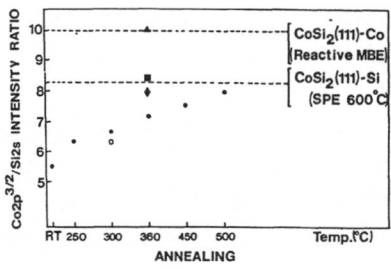

Fig.2 : Evolution of the $Co2p^{3/2}/Si2s$ core level intensity ratio for various epitaxial $CoSi_2$ growth techniques : (♦) layer by layer growth, (●) R.T. codeposition (10ML Co + 20ML Si) followed by annealing in the temperature range 250-500°C for 10 min., (○) codeposition (10ML Co + 20ML Si) onto the substrate held at 300°C, (▲) evaporation of 15ML Co onto the Si substrate maintained at 360°C, ■= (▲) + 2ML Si subsequently annealed at 360°C.

3.2 Coevaporation

3.2.1 Substrate at R.T.

This method consists of the codeposition of Co and Si at R.T. with evaporation rates that have to be exactly stoichiometric and subsequent stepwise heating in the temperature range 250-500°C for durations up to 10 min. Figure 3a shows the valence band spectra for ~10ML Co + 20ML Si codeposition at R.T. subsequently annealed at 250, 360 and 500°C. Curve I corresponds to the compound formed at R.T. At this stage the UPS spectrum is similar but not identical to that measured for bulk $CoSi_2$. One observes a broad peak near 1.8eV attributed to non bonding Co3d states in amorphous $CoSi_2$. The energy location of this feature agrees with that of the non bonding states in epitaxial crystalline $CoSi_2$ (upper spectrum) but the fine structure apparent in crystalline $CoSi_2$ is not yet visible. Furthermore the bonding states peak near 3.8eV characteristic of crystalline $CoSi_2$ cannot be distinguished in R.T. coevaporated films. Upon annealing at 250, 360 and 500°C that peak progressively appears becoming fully developped at 500°C where (1x1) LEED picture can be observed. At this stage the surface is of the

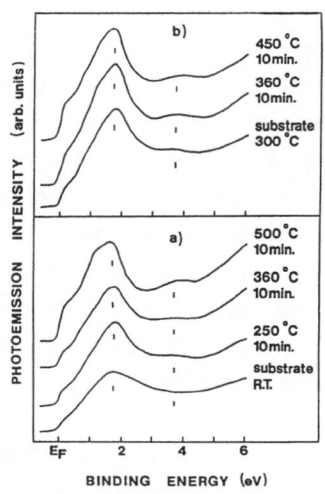

Fig.3 : Valence band photoemission spectra taken with $h\nu$ = 21.2 eV at normal emission for a coevaporated $CoSi_2$ layer (∼ 30 Å) deposited onto the Si(111) substrate held at : (a) R.T. and (b) 300°C, subsequently annealed at various temperatures.

$CoSi_2$(111)-Si type. As can be seen in Fig.2 the core level intensity ratio again indicates a large Si excess in the R.T. coevaporated $CoSi_2$ films. On annealing one observes a decrease in the Co2p/Si2s core level intensity ratio and in turn in the Si excess. We believe that this behaviour stems from Si/$CoSi_2$ phase separation in the coevaporated layer having a Si/Co ratio larger than 2 ($CoSi_2$ stoichiometry) and/or from the formation of Si islands on top of the silicide layer which exhibits initially a segregated Si overlayer.

3.2.2 Substrate at 300°C

Figure 3b displays the UPS spectra for a 10ML Co+20ML Si coevaporated layer on a substrate at 300°C subsequently annealed at 360 and 450°C. All spectra indicate a Si rich silicide surface of the $CoSi_2$(111)-Si type. The LEED diagrams are (1x1) at all temperatures. The $Co2p^{3/2}$/Si2s core level intensity ratio is again much lower than that measured on films prepared by the usual SPE technique indicating a substantial Si excess in the $CoSi_2$ bulk and/or a strong surface Si segregation.

Concluding this section we find that $CoSi_2$ films prepared by coevaporation show invariably a Si enrichment. Again one may wonder what the origin of the Si excess actually is. Besides uncertainties in the amount of deposited Si and Co we believe that a significant part of the Si excess is due to diffusion from Si substrate. In fact, it is well known that even at room temperature the first ∼ 4ML of deposited Co form a $CoSi_2$ like compound with Si species from substrate [1-8,18]. Consequently coevaporation of Co and Si in the $CoSi_2$ stoichiometric ratio results in a Si excess at least in the first $CoSi_2$ layers. The tendency of the deposited Si to segregate on top of the $CoSi_2$ film might also favor reaction of the underlaying Co rich phase with the substrate.

3.3 Reactive MBE

First a template layer is prepared by evaporating 3ML of Co onto the Si substrate maintained at R.T. and subsequently annealed at ~360°C for 5min. This results in an epitaxial $CoSi_2(111)$-Co template layer (curve 1 Fig.4). Subsequently, various Co thicknesses are deposited onto the $CoSi_2(111)$-Co template layer maintained at ~ 360°C. Figure 4 presents a series of energy distribution curves for various $CoSi_2(111)$ film thicknesses (9-54Å). Such silicide layers exhibit sharp (1x1) LEED patterns and the peaks at -1.4, - 1.8 and -2.7eV are well defined and easily identified in the ultra-thin films indicating a good bulk and surface crystallinity. Thus cobalt disilicide formed in this way exhibits a Co-rich surface ($CoSi_2(111)$-Co). It can be seen in Fig.2 that the $Co2p^{3/2}/Si2s$ intensity ratio is close to 10 for thick films i.e. within experimental error that measured on standard SPE films displaying a

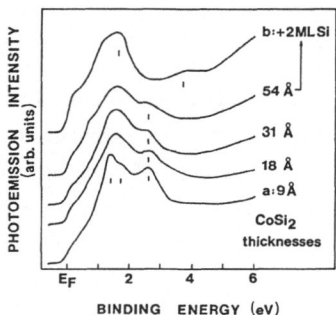

Fig. 4 : UPS spectra taken with $h\nu$ = 21.2 eV at normal emission for various film thicknesses obtained by first annealing at 360°C a 3ML R.T. Co deposit for 5 min. (template layer, curve a) followed by various Co deposits onto the template layer maintained at 360°C ("reactive MBE technique"). Curve b : 2ML Si deposited onto the final $CoSi_2(111)$-Co surface and subsequently annealed at 360°C.

$CoSi_2$-Co surface as prepared in Ref.7 A ~ 2ML Si deposition leads to a decrease in the Co $2p^{3/2}/Si2s$ intensity ratio to the value observed for a $CoSi_2(111)$-Si obtained by SPE at 600°C. The corresponding UPS spectrum is also identical to that measured on $CoSi_2(111)$-Si surfaces.

Thus, it is noteworthy that, first no Si enrichment can be detected in $CoSi_2$ films grown by the "reactive MBE" method and second that $CoSi_2$ growth readily occurs at the $CoSi_2$ film surface even for thicknesses as large as ~100Å. Since the films prepared at such low temperatures are probably continuous [10] diffusion of Si (or Co) through $CoSi_2$ must take place even at temperatures as low as 360°C. This is an unexpected result since it is usually believed that diffusion through $CoSi_2$ is difficult because of a low density of point defects.

4. CONCLUSION

In summary we have shown that various preparation methods at low temperature yield epitaxial $CoSi_2$ layers with distinct bulk and surface properties. Layer by layer deposition or evaporation of Si and Co in the nominal $CoSi_2$ ratio invariably favor a surface and bulk stoichiometry with an excess of Si apparently due to the fact that the Si substrate supplies part of the Si entering in the silicide formation. On the contrary, layer prepared by the reactive MBE method where only Co is deposited on the surface held at 360°C show stoichiometries close to the nominal value.

Thanks are due to Dr.F.Arnaud d'Avitaya and C.D'Anterroches from the Centre National d'Etudes des Telecommunications for many helpful discussions.

REFERENCES

1. E.Rosencher, S.Delage, Y.Campidelli and F.Arnaud d'Avitaya, Electron Lett.20, 762 (1984)
2. J.C.Hensel, A.F.J.Levi, R.T.Tung and J.M.Gibson, Appl.Phys.Lett.47, 151 (1984).
    R.T.Tung, J.C Bean, J.M Gibson, J.M Poate and D.C.Jacobson Appl.Phys.Lett.40, 684 (1982)
3. J.Derrien and F.Arnaud d'Avitaya, J.Vac.Sci.Technol.A5, 2111 (1987)
4. C.Pirri, J.C.Peruchetti, G.Gewinner and J.Derrien, Phys.Rev.B30, 6227(1984)
5. C.Pirri, J.C.Peruchetti and G.Gewinner, Surf.Sci 152/153, 1106 (1985)
6. C.Pirri, J.C.Peruchetti, G.Gewinner and J.Derrien, Phys.Rev.B29, 3391 (1984)
7. C.Pirri, J.C Peruchetti, D.Bolmont and G.Gewinner, Phys.Rev.B33, 4108 (1986)
8. C.Pirri, J.C.Peruchetti, G.Gewinner and D.Bolmont, Solid State Communications 57, 361 (1986)
9. F.Arnaud d'Avitaya, S.Delage, E.Rosencher and J.Derrien, J.Vac.Sci.Technol.B3 (2), 770 (1985)
10. R.T.Tung, AF.J.Levi and J.M Gibson, Appl.Phys.Lett.48, 635 (1986)
11. H.von Känel, J.Henz, M.Ospelt and P.Wachter, presented at 7th European Physical Society, Pisa, 1987
    J.Hunz, M.Ospelt and H.von Känel, Solid State Commun.63, 445 (1987)
12. R.T Tung and J.L.Batstone, Appl.Phys.Lett.52, 648 (1987)
13. T.L.Lin, R.W.Fathauer, P.J.Grunthaner and C.d'Anterroches,Appl.Phys.Lett.52, 804 (1988)
14. S.A Chambers, S.B Anderson, H.W Chen and J.H Weaver, Phys.Rev.B34(2), 913 (1986)
15. R.Leckey, J.D.Riley, R.L Johnson, L.Ley and B.Ditchek, J.Vac.Sci.Technol.A6, 63 (1988)
16. G.Gewinner, C.Pirri, J.C.Peruchetti, D.Bolmont, J.Derrien and P.Thiry, in press in Phys.Rev.B
17. C.Pirri, G.Gewinner, J.C.Peruchetti, D.Bolmont and J.Derrien, in press in Phys.Rev.B
18. J.Y Veuillen, J.Derrien, P.A Badoz, E.Rosencher and C.d'Anterroches, Appl.Phys.Lett.51, 1448 (1987)
    J.Derrien, M.De Crescenzi, E.Chainet, C.d'Anterroches, C.Pirri, G.Gewinner and J.C.Peruchetti, Phys.Rev.B36

RECENT DEVELOPMENTS IN THE EPITAXIAL GROWTH OF TRANSITION METAL SILICIDES
ON SILICON

L.J. CHEN, J.J. CHU, W. LUR, H.F. HSU, AND T.C. LEE

Department of Materials Science and Engineering, National Tsing Hua
University, Hsinchu, Taiwan, Republic of China

I. INTRODUCTION
    Recently a number of silicides were grown epitaxially on silicon. It
now appears that almost all transition metal silicides can be grown
epitaxially on silicon. [1-6] Previous studies on the epitaxial growth of
silicides on silicon were mostly on the growth of silicides on large area,
undoped samples. Furnace annealings were usually performed to grow
epitaxial silicides. [2-4] However, in device applications, silicides were
often grown on laterally confined, ion implanted silicon. [2,7] In order to
minimize dopant redistribution for shallow junction devices, rapid thermal
annealing (RTA) has been increasingly used to activate the dopant and form
the silicides. [8] In an effort to realize the practical applications of
epitaxial silicides in microelectronics devices, the effects of lateral
confinement, dopants, and rapid thermal annealing on the growth of
epitaxial silicides on silicon have been investigated.
    To study the effects of lateral confinement experimentally is more
demanding than it is to investigate the behavior of extended layers,
because good depth perception as well as good lateral resolution are now
required. [2] The presence of dopants was known to influence the silicide
formation. Silicide formation, on the other hand, also causes a dopant
redistribution which could affect the junction properties. In addition,
the presence of metal films on ion implanted silicon may alter the defect
structure commonly observed in post-implantation annealed samples.

II. EXPERIMENTAL PROCEDURES
    For ion-implanted samples, 15-25 Ω-cm, boron doped (001) and (111)
oriented silicon wafers were used for implantation. For through-oxide
implantation, following a standard cleaning procedure, wafers were
implanted with 110 KeV BF_2^+, 25 KeV B^+, 43 KeV F^+, 150 KeV As^+, or 65 KeV
P^+ to a dose of $5\times10^{15}/cm^2$ through a 40 nm-thick thermal oxide at room
temperature. The ion energies were chosen so that their projected ion
ranges were approximately equal. The wafers were oriented 7° off the
incident beam direction to minimize the channeling effect. The beam
current density was kept to be less than 5 µA/cm². Parts of the ion-
implanted samples were heat treated with rapid thermal annealing by a
resistively heated graphite heater (Varian IA-200) at 1050 °C for 15 s.
The RTA process was previously found to optimize p-n junction
characteristics of BF_2^+ implanted samples. [9] For some of the samples, ions
were also implanted directly on bare n-type (001)Si, then annealed at 1000
°C in O_2 ambient for 5 min followed by annealing in N_2 for an additional
25 min.
    As-implanted and annealed samples as well as bare silicon samples were
dipped in a diluted HF solution (HF:H_2O = 1:50) for 3 minutes immediately
before loading into an electron gun evaporation chamber. Metal thin films,

Y. I. Nissim and E. Rosencher (eds.), Heterostructures on Silicon: One Step Further with Silicon, 231–238.
© 1989 by Kluwer Academic Publishers.

5-60 nm in thickness, were then deposited onto the silicon substrates at room temperature in a vacuum better than 1×10^{-6} Torr. The deposition rate was about 0.1 nm/s. For sputter-deposition, the substrates were first etched by Ar backsputtering in situ to remove about 30 nm-thick Si surface layers followed by a sputter deposition of 30 nm in thickness of nickel films.

Annealing conditions of the samples will be specified in the respective sections. The preparation of patterned wafers and the characteristics of the heat treatment apparatus were described in the appropriate references and will not be repeated in the present paper.

III. GROWTH OF EPITAXIAL NiSi$_2$ AND CoSi$_2$ ON LATERALLY CONFINED SILICON
A. NiSi$_2$ Epitaxy on Silicon
1. Electron Gun Deposited Nickel Films on (111)Si
i. Two-Dimensional Windows

Epitaxial NiSi$_2$ of single orientation was grown on (111)Si inside miniature size oxide openings. Striking size effects on the growth of NiSi$_2$ epitaxy were observed. The formation temperature of NiSi$_2$ on (111)Si was found to be as low as 550 °C inside oxide openings 1.8 μm or smaller in size. Epitaxial NiSi$_2$ of single orientation which is identical to that of (111)Si substrate was formed inside oxide openings of or smaller than 1.8, 1, and 0.8 μm in size in samples annealed at 550-750, 800, and 850-900 °C, respectively. [10] The results obtained in samples annealed in diffusion furnace and by RTA were found to be similar. It is suggested that tensile stress exerted by the oxide may serve to compensate the intrinsic compressive stress so that the nucleation barrier of epitaxial NiSi$_2$ is lowered for the formation of epitaxial NiSi$_2$. The stress was also conceived to promote the growth of type A over type B epitaxy inside miniature size oxide openings.

ii. Linear Openings

100% type A epitaxy was obtained only in samples heat treated by RTA at 850-900 °C. The linear openings were found to be less effective than two-dimensional windows in promoting the growth of NiSi$_2$ epitaxy of single orientation on (111)Si.

2. Sputter-Deposited Nickel Films on (111) and (001)Si

100% type A NiSi$_2$ epitaxy on (111)Si was achieved in all sizes and shapes of oxide openings in samples RTA treated at 900 °C for 3 s, 850 °C for 10 s, and 800 °C for 30 s. In samples annealed at 785 °C for 30 s, only polycrystalline NiSi$_2$ grains were observed. However, 100% type A epitaxy was observed in all (001) samples treated under the aforementioned annealing conditions. Examples are shown in Fig. 1.

B. CoSi$_2$ Epitaxy on (111)Si
1. Electron Gun Deposited Cobalt Films on (111)Si Heat Treated by RTA
i. Two-Dimensional Windows

Epitaxial CoSi$_2$ of single orientation (type B) was found to grow on silicon inside oxide openings, 1-10 μm in size, in samples annealed at 870-1000 °C for 500 s and 1050 °C for 300 s. Both annealing time and temperature were found to be critical in obtaining 100% epitaxy. About 30% in areal fraction of CoSi$_2$ were found to be polycrystalline in samples annealed at 850 °C for 500 s and 900-1000 °C for 300 s.

Strong size effects on the morphology of epitaxial CoSi$_2$ on silicon were found. In 1000 °C annealed samples, CoSi$_2$ coverage was found to

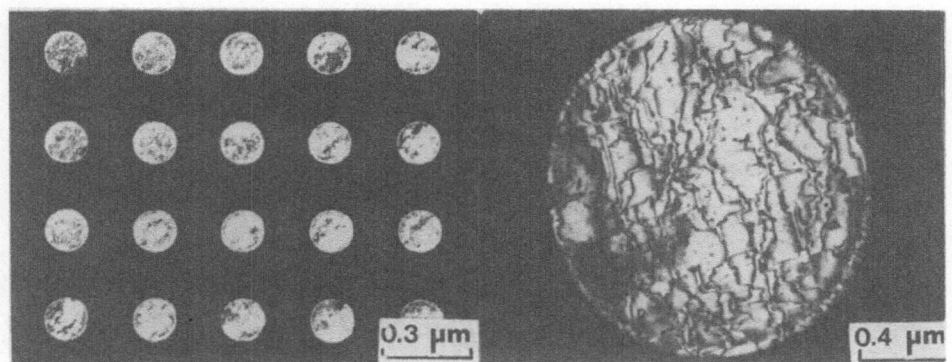

Fig. 1(a) (200)NiSi$_2$ dark field
    micrograph (DF), sputtered Ni,
    (001)Si, 900 ^{o}C, 3 s.

Fig. 1(b) Bright field micrograph
    (BF), same conditions as those
    in Fig. 1(b).

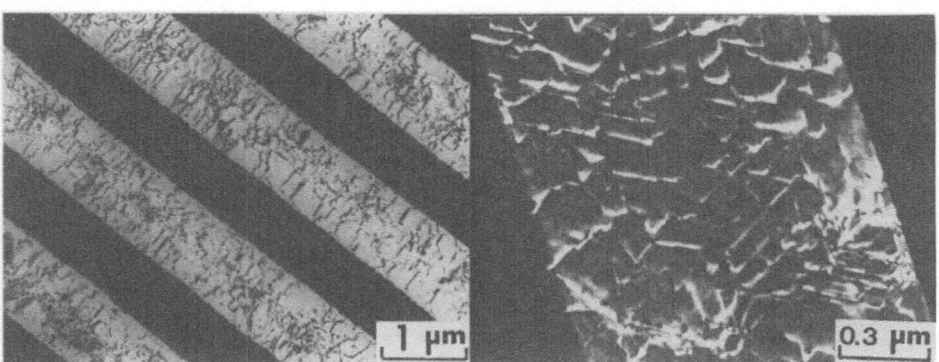

Fig. 1(c) BF, sputtered Ni,
    (001)Si, 880 ^{o}C, 3 s.

Fig. 1(d) Weak beam DF, sputtered
    Ni, (111)Si, 880 ^{o}C, 3 s.

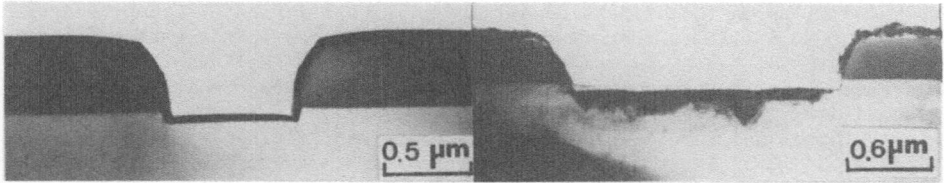

Fig. 1(e) BF, cross-sectional view
    (c-s), sputtered Ni, (111)Si.

Fig. 1(f) BF, c-s, sputtered Ni,
    900 ^{o}C, 3 s.

conform to the shape of the oxide openings 2.5 μm or larger in size .
However, regular faceted structure, with edges parallel to <1$\bar{1}$0>Si
directions, was found to form in oxide openings 1.4-2.0 μm in size. In
950 ^{o}C annealed samples, the size of oxide openings for which the
transition from conformal coverage to faceted morphology occurred was
scaled down to 1.4 μm. Conformal coverage was found inside all oxide
windows in 850 ^{o}C annealed samples. An example is shown in Fig. 2(a). The
morphology of epitaxial CoSi$_2$ coverage in windows 0.6 μm in size was

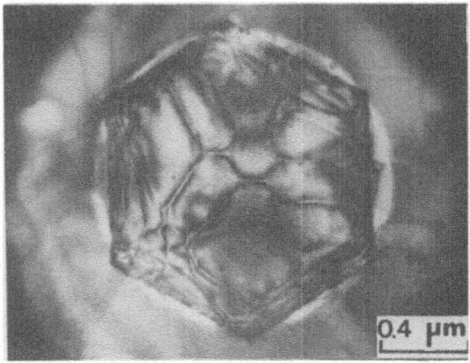 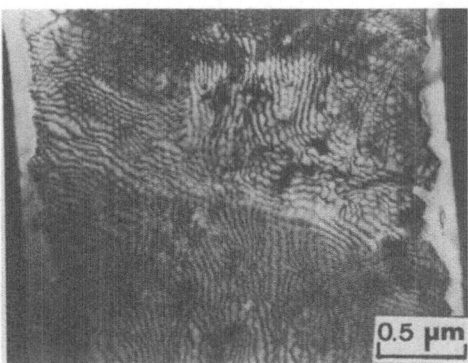

Fig. 2(a) BF, e-gun deposited   Fig. 2(b) BF, e-gun deposited
     Co, (111)Si, 900 °C, 500 s.   Co, (111)Si, 900 °C, 500 s.

generally irregular. It is suspected that the irregular appearance is due
mainly to the insufficient surface cleaning in the smallest size windows
rather than to the intrinsic size effect.

ii. Linear Openings

     No significant difference between the formation conditions of 100%
type B CoSi$_2$ epitaxy in linear openings and those in two-dimensional oxide
windows was found. Faceted structures were evident at the edges of
opening strips in samples annealed at 900 °C for 500 s. In samples
annealed at 1050 °C for 500 s, the morphology of the disilicide became
highly irregular. An example is shown in Fig. 2(b).
     The exact causes for the vast difference found in the behaviors of
the growth of epitaxy inside oxide openings between the metal films
deposited by electron gun evaporation and sputtering are not clear at this
time. XTEM observations showed that the sputtered films conform closely
with the outline of the oxide openings and the recess induced by Ar$^+$
backsputtering-cleaning. In contrast, the coverages of electron gun
deposited films on the oxide sidewall are less well defined. Examples are
shown in Fig. 3. It is conceived that the differences in the geometrical
configuration of the films and the resulting stress as well as the
substrate cleanliness may influence the growth of silicide epitaxy inside
oxide windows.
     Ishibashi and Furukawa reported the formation of pinhole free
epitaxial CoSi$_2$ on laterally confined (111)Si as the widths of patterned
strips are less than 3 μm. Furthermore, two-dimensional patterns were
found to be more effective in alleviating the island formation. It was

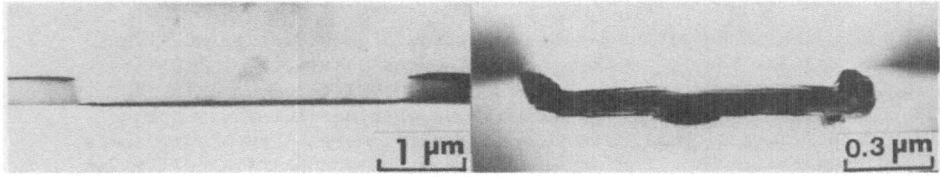

Fig. 3(a) BF, e-gun deposited Co,   Fig. 3(b) BF, c-s, e-gun deposited
     (111)Si.   Co, (111)Si, 950 °C, 500 s.

suggested that the films in the small pattern are more uniform since the energetics favors the configuration and the morphology of the epitaxial silicides can be understood by taking account of the surface energy involved. [11]

Formation of patterned silicide structure on silicon is an essential step for the fabrication of permeable base transistor and other high speed buried metal layer devices. The growth of silicon on metal gratings was reported by several groups. [11-14] In particular, characteristics of permeable base transistors were demonstrated for Si/CoSi$_2$/Si structures. [15,16]

IV. EFFECTS OF DOPANTS
A. NiSi$_2$ Epitaxy on Ion Implanted Silicon
1. Effects of Doping Species

Striking effects of B$^+$ and BF$_2^+$ implantation on the growth of epitaxial NiSi$_2$ on silicon were observed. As a result of ion implantation, epitaxial NiSi$_2$ was found to grow at 220-280 $^\circ$C instead of the usual formation temperature of about 800 $^\circ$C on blank silicon. [17] The results indicated that B atoms played an important role in inducing the epitaxial growth of NiSi$_2$ on silicon at low temperatures. Subsequently, it was found that F atoms also promote the epitaxial growth of NiSi$_2$ at low temperatures. [18] Good correlation was found between the atomic size factor and resulting stress and NiSi$_2$ epitaxy at low temperatures.

2. Effects of Through-Oxide Implantation

The formation of epitaxial NiSi$_2$ was seen to be a sensitive function of annealing temperature in through-oxide implantation samples, whereas

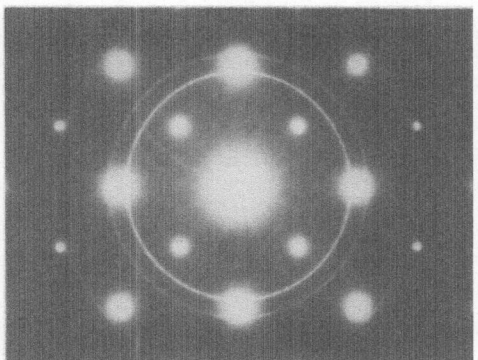

Fig. 4(a) Diffraction pattern, BF$_2^+$/600, 260 $^\circ$C.

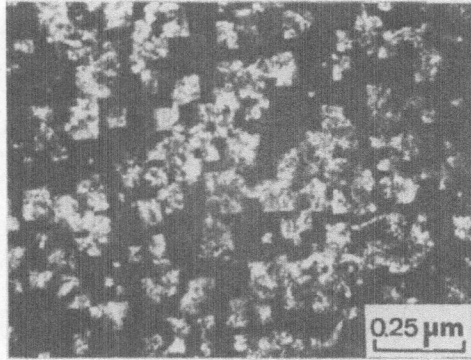

Fig. 4(b) (200)NiSi$_2$ DF, B$^+$/600, 260 $^\circ$C.

Fig. 4(c) (200)NiSi$_2$ DF, c-s, BF$_2^+$/800, 260 $^\circ$C.

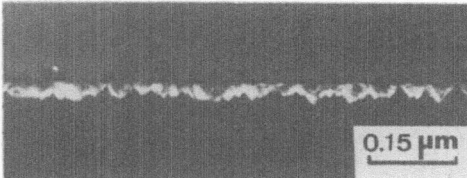

Fig. 4(d) (200)NiSi$_2$ DF, 5-nm thick Ni, BF$_2^+$/1000, 400 $^\circ$C.

prominent $NiSi_2$ epitaxy was observed in 220-280 ^{o}C annealed specimens with BF_2^+ directly implanted on bare (001)Si followed by annealing at 1000 ^{o}C for 1/2 h. The differences are likely due to variations in initial structure and/or impurity redistribution. [18,19]

3. Effects of Substrate Treatments

BF_2^+ implanted samples were first annealed at 600, 800, or 1000 ^{o}C for 1/2 h, which are designated to be $BF_2^+/600$, $BF_2^+/800$, and $BF_2^+/1000$ samples, respectively. No silicides were found to form in BF_2^+ samples annealed at 220-280 ^{o}C. For $BF_2^+/600$ and $BF_2^+/800$ samples annealed at 250-280 and 240-280 ^{o}C, respectively, epitaxial $NiSi_2$ was found to form at the Ni/Si interface with a layer thickness of about 30 nm. In $BF_2^+/1000$ ^{o}C samples, the volume fractions of epitaxial $NiSi_2$ formed were found to be much lower than those found in $BF_2^+/800$ and $BF_2^+/1000$ samples annealed in the same temperature range. Only epitaxial $NiSi_2$ was observed in Ni (5 nm)/$BF_2^+/1000$ samples annealed at 400 ^{o}C. Examples are shown in Fig. 4.

4. Dopant Redistribution

SIMS profiling showed that, in $B^+/800$ samples, B^+ profile was broader compared to B^+ samples. Some boron atoms appeared to be lost due to outdiffusion. Piling up of B at silicide/Si interface was evident.

Ni was found to react completely with silicon in BF_2^+ samples annealed at 260 ^{o}C. B was found to pile up at the surface and was depleted from the nickel silicide region.

In as-BF_2^+-implanted samples, $^{19}F^-$ profile was found to follow $^{11}B^-$ profile but slightly more concentrated at the surface. Dramatic changes were seen in $^{19}F^-$ profile in $BF_2^+/600$ samples annealed at 260 ^{o}C. The $^{19}F^-$ profile shows a maximum in the surface region and another maximum in the region where $NiSi_x$ stoichiometry begins to favor Si. The piling up of the B atoms at the silicide/Si interfaces shall exacerbate the influence of dopants on the silicide formation.

B. CoSi₂ Epitaxy on Ion Implanted Silicon

Single-crystalline $CoSi_2$ films were grown on B^+ implanted (111)Si annealed in vacuum at 1000 ^{o}C. TEM analysis indicated that it is of type B epitaxy. The films were found to be free from pin holes. The $CoSi_2$/Si interface was observed to be rather flat with undulation less than 10 nm in amplitude. Almost single-crystalline $NiSi_2$ films were grown on As^+ and P^+ implanted samples. However, the thickness variation of $CoSi_2$ films on As^+ implanted samples was of the order of the film thickness.

The areal fractions of surface coverage and percentage of type B epitaxy of epitaxial $CoSi_2$ were measured to be about (70%, 70%) and (80%, 80%) respectively, in blank and BF_2^+ implanted samples. Pinholes of $CoSi_2$ films, about 4 μm in average size, on blank silicon were found to be crystallographic in shape. Irregular pinholes, 3 μm in average size, on the other hand, were observed in BF_2^+ implanted samples. Examples are shown in Fig. 5.

V. RAPID THERMAL ANNEALING

The epitaxial growth of near noble silicides, including $CoSi_2$, $NiSi_2$, $FeSi_2$, Pd_2Si, and PtSi on (111)Si was achieved by RTA. [20,21] Single-crystalline $CoSi_2$ films have been grown in solid phase epitaxy regime on (111)Si by electron gun deposition of Co thin films followed by RTA in Ar ambient. The effect of gas ambient was found to be of critical importance in the growth of single-crystal $CoSi_2$ on (111)Si. The effect of heating

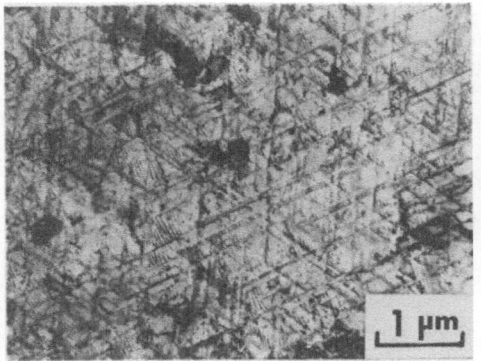

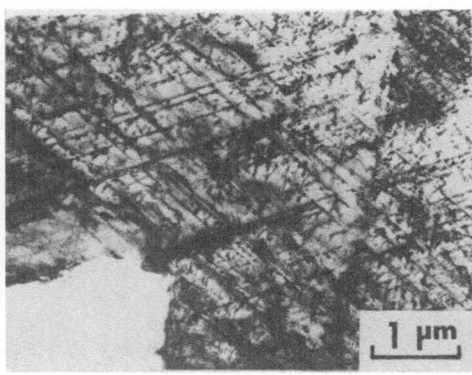

Fig. 5(a) CoSi$_2$ epitaxy, B$^+$, (111)Si.

Fig. 5(b) CoSi$_2$ epitaxy, BF$_2^+$, (111)Si.

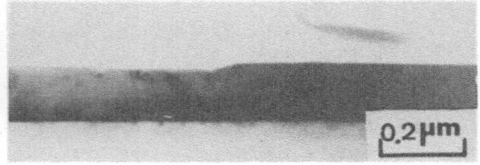

Fig. 5(c) CoSi$_2$ epitaxy, c-s, As$^+$, (111)Si.

Fig. 5(d) CoSi$_2$ epitaxy, c-s, B$^+$, (111)Si.

rate appears to be less direct than that of gas ambient. The best NiSi$_2$, FeSi$_2$, Pd$_2$Si, and PtSi epitaxy, in terms of average size, fraction of silicon surface coverage, and regularity of interfacial dislocations, were found to be of comparable quality to those grown by conventional furnace annealing.

Localized epitaxial TiSi$_2$ was grown on (111)Si by RTA in Ar ambient.[22] The best epitaxy was obtained in samples annealed at 1100 °C for 20 s. Almost full coverage of TiSi$_2$ (epitaxial and nonepitaxial) on silicon surface was found. The epitaxial regions about 20 μm in average size, were observed to cover 70% of the surface area. Some of the epitaxial regions were observed to be as large as 40 μm in size. Dominant modes and average size of TiSi$_2$ epitaxy in RTA samples were found to be different from those in vacuum furnace annealed specimens. Ambient gas induced silicide/Si interface energy changes are suggested to promote the growth of differently oriented grains. The main advantages of RTA in inducing TiSi$_2$ epitaxy appear to be better control of annealing ambient, temperature, and time for short-time anneals in the small RTA apparatus than in a furnace.

VI. CONCLUDING REMARKS

The knowledge derived from studies of laterally unconfined films on undoped silicon is basic to the understanding of silicide films as they are applied to microelectronics devices. That knowledge is now sufficient, however, because in applications, silicide films are always confined, the interactions of metal films with dopants in silicon are inevitable, and RTA is increasingly being used. These factors have been shown to exert strong influences on the epitaxial growth of silicides on silicon. For achieving high performance novel classes of devices, it is important to investigate the consequence of these influencing factors. Although it is much more demanding experimentally to gain insights on the non-ideal "real" systems, the potential benefits to the advancement of basic science and technology are tremendous. A great deal of more efforts are needed to meet the challenge in the future.

ACKNOWLEDGMENT
The research was supported in part by the Republic of China National Science Council.

REFERENCES

1. K.N. Tu and J.W. Mayer, in Thin Films Interdiffusions and Reactions, edited by J.M. Poate, K.N. Tu, and J.W. Mayer (Wiley, New York, 1978), p. 359.
2. M.A. Nicolet and S.S. Lau, in Materials and Process Characterization, edited by N.G. Einspruch andG.B. Larrabee (Academic, New York, 1983), p. 329.
3. L.J. Chen, H.C. Cheng, and W.T. Lin, Mater. Res. Soc. Symp. Proc. 54, 245 (1986).
4. J. Derrien and F.A. d'Avitaya, J. Vac. Sci. Technol. A5, 1412 (1987).
5. J.J. Chu, L.J. Chen, and K.N. Tu, J. Appl. Phys. 62, 461 (1987).
6. J.J. Chu, L.J. Chen, and K.N. Tu, J. Appl. Phys. 63, 1163 (1988).
7. S.P. Murarka, J. Vac. Sci. Technol. B4, 1325 (1986).
8. J.C.C. Fan, Mater. Res. Soc. Symp. Proc. 35, 39 (1985).
9. I.W. Wu, R.T. Fulks, and J.C. Mikkelsen, Jr., J. Appl. Phys. 60, 2422 (1986).
10. C.S. Chang, C.W. Nieh, and L.J. Chen, Appl. Phys. Lett. 50, 259 (1987).
11. K. Ishibashi and S. Furukawa, Jpn. J. Appl. Phys. 24, 912 (1985).
12. A. Ishizaka and Y. Shiraki, Jpn. J. Appl. Phys. 23, L499 (1984).
13. B.A. Vojak, D.D. Rathman, J.A. Burns, S.M. Cabral, and N.N. Efremow, Appl. Phys. Lett. 44, 223 (1984).
14. E. Rosencher, G. Glastre, G. Vincent, A. Vareille, and F.A. d'Avitaya, Electron. Lett. 22, 699 (1986).
15. K. Ishibashi and S. Furukawa, IEEE Trans. ED-33, 322 (1986).
16. G. Glastre, E. Rosencher, F.A. d'Avitaya, C. Puissant, M. Pons, G. Vincent, and J.C. Pfister, Appl. Phys. Lett. 52, 898 (1988).
17. S.W. Lu, C.W. Nieh, and L.J. Chen, Appl. Phys. Lett. 49, 1770 (1986).
18. L.J. Chen, C.M. Doland, I.W. Wu, J.J. Chu, and S.W. Lu, J. Appl. Phys. 62, 2789 (1987).
19. L.J. Chen, C.M. Doland, I.W. Wu, A. Chaing, C.C. Tsai, J.J. Chu, S.W. Lu, and C.W. Nieh, I. Electron. Mater. 17, 75 (1988).
20. H.C. Cheng, I.C. Wu, and L.J. Chen, Appl. Phys. Lett. 50, 174 (1987).
21. H.C. Cheng, I.C. Wu, and L.J. Chen, Mater. Res. Soc. Symp. Proc. 74, 654 (1987).
22. I.C. Wu, J.J. Chu, and L.J. Chen, J. Appl. Phys. 60, 3172 (1986).

FORMATION OF BURIED EPITAXIAL Co SILICIDES BY ION IMPLANTATION

K.Kohlhof, S.Mantl and B.Stritzker

Institut für Schichten- und Ionentechnik,
Kernforschungsanlage Jülich, P.O.Box 1913,
5170 Jülich, Fed. Rep. Germany

ABSTRACT

Implantation of 200 keV Co ions at 350°C into (111) Si and subsequent annealing led to the formation of buried Co silicides. Two step annealing at temperatures at about 600°C and 1000°C is required to recrystallize the Si layer on top and the substrate and to form an intermediate epitaxial, low ohmic $CoSi_2$ layer. From sheet resistivity measurements after rapid thermal processing we deduced an activation energy of 2.5 eV for the silicidation. Ion channeling experiments proved the $CoSi_2$ layer to be single crystalline (minimum yield about 6%) and aligned to the (111) Si without a rotation. Specific resistivity values of 15 $\mu\Omega$ cm were obtained. Conventional furnace annealing and rapid thermal processing led to similar results.

INTRODUCTION

In recent years buried metallic layers in Si have attracted great attention because of their potential applications in "metal base-" and "permeable base-transistors" and as buried interconnects in three-dimensional integrated circuits. For such applications low ohmic, epitaxially grown silicides like $CoSi_2$ with a small lattice mismatch with Si is required. The conventional way to form such $Si/CoSi_2/Si$-heterostructures is the silicidation of an evaporated Co layer capped with molecular beam (MBE) grown Si. We present a novel technique, firstly proposed by A.E.White [1], involving Co implantation into Si with subsequent two step annealing.

EXPERIMENTAL PROCEDURES

Co implantation was performed with a 200 keV medium-current ion accelerator (Eaton NV-3204). Doses up to $3*10^{17}$ Co cm^{-2} were implanted into (111)-Si substrates, heated to 350°C to minimize the radiation damage. The surface normal of the wafers was cut 4° off the <111> orientation. The subsequent two step annealing was performed alternatively in an evacuated tube furnace at 10^{-7} Torr or by rapid thermal processing (AET RV-1002) in a high purity Ar ambient. The silicide formation

239

Y. I. Nissim and E. Rosencher (eds.), Heterostructures on Silicon: One Step Further with Silicon, 239–245.
© 1989 by Kluwer Academic Publishers.

240

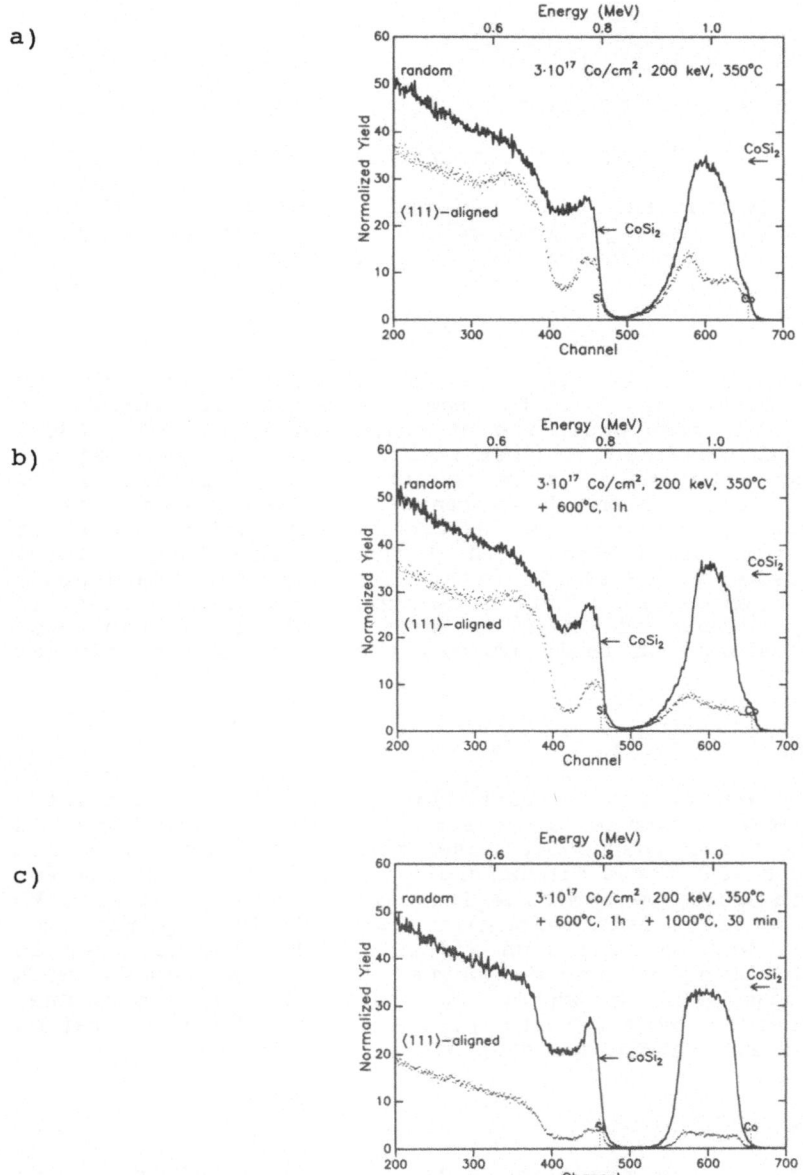

Fig. 1: RBS spectra after implantation of $3*10^{17}$ Co cm^{-2} into Si at 200 keV

required temperatures in the range between 600°C and 750°C before recrystallization and defect recovery was carried out around 1000°C. Silicide formation was controlled by four point sheet resistivity measurements. The morphology was analyzed by eletron microscopy, the crystalline quality and layer composition by 1.4 MeV He ion channeling.

RESULTS AND DISCUSSION

Fig. 1 shows random and aligned spectra of a (111)-Si substrate implanted with $3*10^{17}$ Co cm^{-2} at 200 keV. After implantation (Fig. 1a) Co is widely distributed up to the surface in a partly recrystallized Si matrix. The channeling spectrum shows a pronounced dechanneling peak caused by dislocation loops up to a depth of twice the ion range. Thus, significant disorder is observed also in the substrate. After a 600°C, 1 h anneal in vacuum (Fig. 1b) the Co distribution has sharpened and a CoSi$_2$ layer was formed with improved crystalline quality. However the defect peak in the channeling spectrum indicates defect dislocations, presumably at the second interface. After a 30 min anneal at 1000°C we obtained

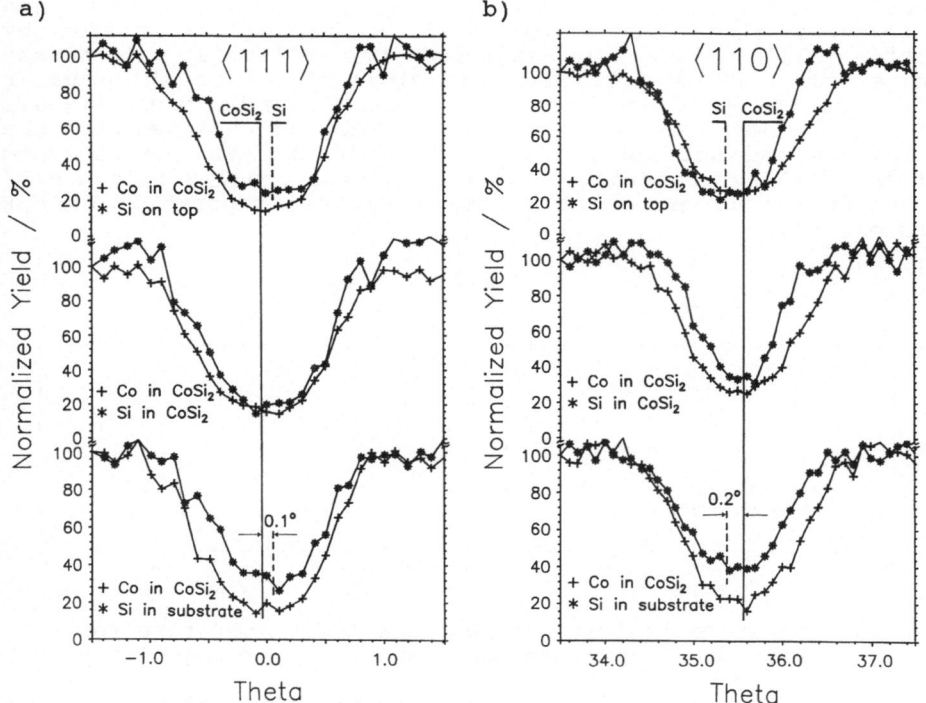

Fig. 2: Angular scans along {1$\bar{1}$0} plane, through the <111> and <110> directions

a 1300 Å thick CoSi$_2$ layer buried under 500 Å Si on top (Fig. 1c). The channeling spectrum reveals that all layers are of very good single crystalline quality (X_{min} about 6% in CoSi$_2$). The second high temperature step anneals out nearly all extended defects, also at the CoSi$_2$/Si-substrate interface. First we investigated the structure and in particular the strain in the heterostructure by ion channeling. Fig. 2a gives some examples of angular scans measured along a {1$\bar{1}$0} plane through the <111> surface normal. The angular dependence of the Si signals (*) from the cap, the CoSi$_2$ layer und the substrate are compared to the Co signal (+). Small shifts of the midpositions of about 0.1° (top and bottom curves in Fig 2a) are indications of small angle grain boundaries at both interfaces caused by the 4° misorientation of the wafers. Similar observations have been made in Ref. 2 and it seems that such a structure is energetically favoured compared to an ideal pseudomorphic structure. Fig. 2b shows corresponding angular yield scans recorded along a {1$\bar{1}$0} plane through an inclined <110> orientation. An angular shift of about 0.2° of the <110> direction of the CoSi$_2$ layer relative to the <110> orientation of the substrate and the Si cap demonstrates that the CoSi$_2$ lattice is tetragonally strained. However, the measured shifts are significantly smaller than the value of 0.6° estimated on the basis of the biaxial strain model for a pseudomorphic structure. This is not too surprising since the strain will be partly accomodated by misfit dislocations, as expected for silicide thicknesses larger than 300 Å [3]. From the similarities of the angular scans of the Si (cap), Co (CoSi$_2$) and Si (substrate) signals (Fig. 2b) we conclude that the CoSi$_2$ is aligned to the substrate and not rotated by 180°. This implies the presence of an A/A/A heterostructure in contrast to the A/B/A or A/B/B/ structures that are normaly formed by the evaporation or MBE process [4].

a)   b)

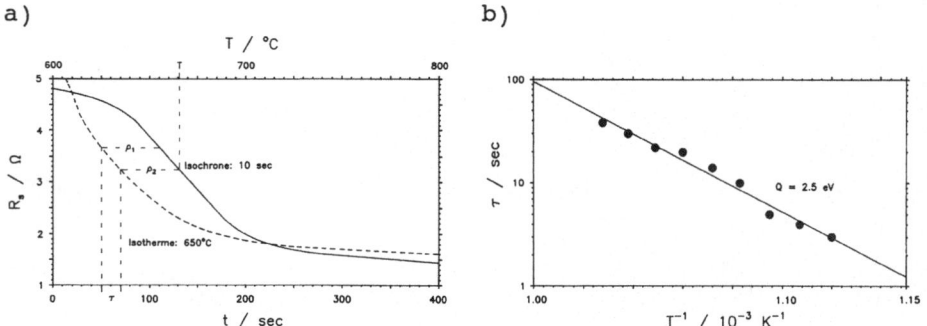

Fig. 3a: Sheet resistivity versus time and temperature for isothermal and isochronal annealing, respectively

Fig. 3b: Arrhenius plot for isothermal annealing time τ and isochronal annealing temperatur T

To study the silicidation kinetics we recorded the sheet resistivity after short rapid thermal processing cycles according to the Meechan-Brinkman procedure [5]. In Fig. 3a the sheet resistivity is plotted versus annealing time for a 650°C treatment (dashed line). This is compared to the sheet resistivity measured after 10 sec anneals at temperatures starting at 600°C and ranging to 800°C (solid line). From Fig. 3a we can deduce the time necessary to obtain a resistivity decrease from ρ_1 to ρ_2, e. g. at 650°C, or alternatively the temperature T needed to obtain the same resistivity change within a given time, e. g. 10 sec. From the Arrhenius plot of ln τ versus 1/T (Fig. 3b) an activation energy of 2.5 eV was derived corresponding very well with literature values ranging from 2.3 eV [6] to 2.6 eV [7] for the normal $CoSi_2$ formation by Co diffusion. Thus we conclude that in the first annealing step sharpening of the Co profile occurs dominantly by Co diffusion. Further experiments showed that the two step annealing procedure cannot be replaced by a single high temperature step. In spite of the fact that the silicide formation is nearly completed after the first 600°C anneal, interfacial defects are still present, leading to a higher resistivity value of about 19 $\mu\Omega$cm compared to the final value of 15 $\mu\Omega$ cm after the second anneal. On the other hand annealing at 1000°C only yields also to good single crystalline layers but with resistivities significantly higher than after the two step process. This may be explained by the faster nucleation of $CoSi_2$ in amorphous than in crystalline Si [6]. The different nucleation behaviour in disordered and recrystallized Si makes it also necessary to perform the implantation at elevated temperatures. High dose Co implantation at room temperature amorphizes Si and enhances the diffusion of Co to the surface. Therefore, only a silicide at the surface can be produced after room temperature implantation [8].

From the technical point of view thinner silicides of about 100 Å to 200 Å are of great interest. Smaller ion doses as used above yield to thinner $CoSi_2$ layers as long as the Co peak concentration after implantation exceeds about half of the stoichiometric value (i. e. 16.6 % Co) [1]. Fig. 4a shows RBS spectra of Si after implantation of $1*10^{17}$ Co cm^{-2} at 200 keV at 350°C and after the 2-step annealing procedure. The peak concentration was about 10 %. It is obvious that the initially broad Co distribution sharpens but does not reach stoichiometry although the good single crystalline quality is confirmed. Cross section TEM showed the Co-Si layer to consist of $CoSi_2$ islands with atomically sharp interfaces embedded in a single crystalline Si environment. During heating the widely spreaded Co atoms coalesce into clusters but no uniform layer is formed if the Co concentration is below a critical value. The distribution of Co atoms can be narrowed by reducing the implantation energy in order to decrease the energy straggling. For the sample from Fig. 4b the same implantation parameters were choosen as for the sample from Fig. 4a, except the implantation energy was reduced to 100 keV yielding to a

244

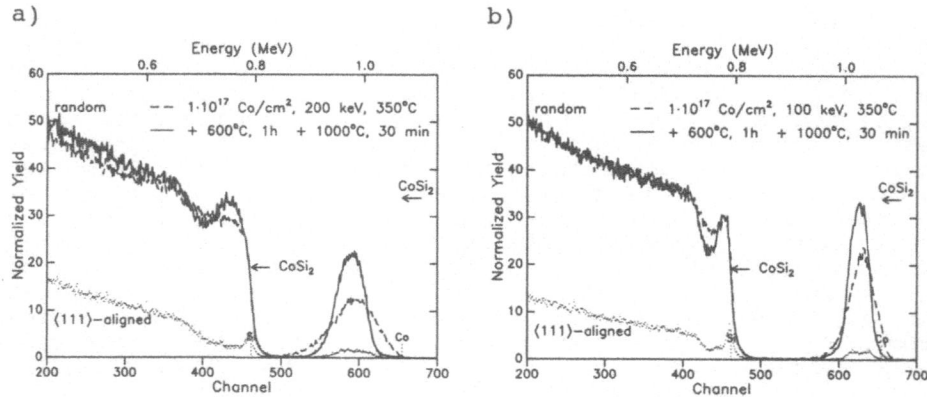

Fig. 4: RBS spectra after implantation of $1*10^{17}$ Co cm^{-2} into
Si a) at 200 keV b) at 100 keV

Co peak concentration of about 18 %. After the two step
annealing the narrower Co distribution leads to coalescing of
the Co atoms to a homogeneous, single crystalline 500 Å thick
silicide buried under 1300 Å Si.

SUMMARY

Epitaxial, low ohmic CoSi$_2$ layers embedded in (111) Si can be
formed by Co implantation followed by a two step annealing.
Depending on the desired layer thicknesses the implantation
energy and dose must be carefully choosen to obtain a
homogeneous layer. The first annealing step at about 600°C
nearly completes silicidation whereas the second step at
1000°C removes lattice damage, particularly at the interfaces
of the Si/CoSi$_2$/Si heterostructure. 200 keV Co implantation at
350°C with a dose of $3*10^{17}$ cm^{-2} led to a 1300 Å thick CoSi$_2$
layer with a resitivity of 15 $\mu\Omega$ cm buried under a 500 Å Si
surface layer. A 500 Å thick CoSi$_2$ layer capped with 1300 Å Si
was obtained after 100 keV Co implantation at 350°C with a
dose of $1*10^{17}$ cm^{-2} and subsequent annealing.

REFERENCES

1. A.E. White, K.T. Short, R.C. Dynes, J.P. Garno, J.M. Gibson,
    MRS Symp. Proc. $\underline{74}$ (1987) 481,
    Appl. Phys. Lett. $\underline{50}$ (1987) 95.

    A.E. White, K.T. Short, R.C. Dynes, J.M. Gibson, R. Hull,
    MRS Symp. Proc., Boston 1987, to be published.

2. G. Bai, D.N. Jamieson, M.A. Nicolet, T. Vreeland,
    MRS Symp. Proc., Boston 1987, to be published

3. Y.C. Kao, K.L. Wang, E. de Fresart, R. Hull, G. Bai,
    D. Jamieson, M.A. Nicolet,
    J. Vac. Sci. Technol. $\underline{B5}$ (1987) 745.

4. R.T. Tung, J.M. Gibson, MRS Symp. Proc. $\underline{67}$ (1986) 211.

5. Chapter 5 in "Theorie der Gitterfehlstellen",
    ed. A. Seeger.

6. C.D. Lien, M.A. Nicolet, S.S. Lau,
    Appl. Phys. $\underline{A34}$ (1984) 249.

7. F.M. d'Heurle, C.S. Petersson,
    Thin Solid Films $\underline{128}$ (1985) 283.

8. F.H. Sanchez, F. Namavar, J.I. Budnick, A. Fasihudin,
    H.C. Hayden, MRS Symp. Proc. $\underline{51}$ (1986) 439.

STRUCTURAL STUDY OF CoSi₂/Si (001) AND (111)

C.W.T. BULLE-LIEUWMA, A.H. VAN OMMEN and L.J. VAN IJZENDOORN

Philips Research Laboratories, P.O. Box 80.000, 5600 JA Eindhoven, The Netherlands

1.ABSTRACT

Nucleation and growth of $CoSi_2$ films by thermal reaction of vapour deposited Co on (001) and (111) Si have been studied by transmission electron microscopy (TEM).
On (001) Si, $CoSi_2$ occurs in a number of orientations including the aligned (001) orientation.
On (111) Si single crystalline layers are obtained, which are twin-oriented.
In addition $Si/CoSi_2/Si$ structures have been formed by high-dose implantation of Co into (001) and (111) Si and subsequent annealing. In this way single crystalline 'mesotaxial' $CoSi_2$ layers are obtained which are fully aligned with the Si-matrix.
Epitaxial growth of $CoSi_2$ on Si by conventional techniques (evaporation) and by high energy Co implantation is discussed.

2.INTRODUCTION

Silicide growth and silicide/silicon interfaces are of interest in solid state science and in micro-electronic technology (1,2). The advantages of $CoSi_2$ over several metal silicides is its low resistivity and the possibility of epitaxial growth on Si , due to the relatively small lattice mismatch of 1.2 % between $CoSi_2$ (a=0.5365 nm) and Si (a=0.5431 nm).
The thermal expansion coefficient of $CoSi_2$ is about 4 times larger than that of Si. This means that the lattice match improves at increasing temperature.
Transmission electron microscopy (TEM) has been used in various studies to characterize the microstructure of the $CoSi_2$ /Si system. The growth of mono crystalline films of $CoSi_2$ for (001) Si grown by conventional techniques has never been reported (3,4). On (111) Si the growth of single crystalline layers of $CoSi_2$ in which the twinned B-orientation is dominant over the aligned structure (A-orientation) has been reported at several instances (5,6).
In this paper we propose a nucleation and growth model of $CoSi_2$ on (001) and (111) Si on basis of our TEM observations and of models of the atomic structure of the interface.
    A new and promising technique is high-dose implantation of Co followed by subsequent annealing. The formed hetero-epitaxial $Si/CoSi_2/Si$ structures are of much interest due to their application as metal base and permeable base transistors (7). White et al (8) have demonstrated that single crystalline buried $CoSi_2$ layers can be formed by this technique in both (001) and (111) Si and named it 'mesotaxy'.
We present TEM results on buried $CoSi_2$ layers of as-implanted (001) and (111) Si wafers, and of subsequently annealed wafers.

3.EXPERIMENTAL

$CoSi_2$ films with thicknesses of 20 to 100 nm have been grown on (001) and (111) Si substrates. Co layers were deposited by evaporation at a background pressure of 10^{-8} Pa and at a temperature of 250 °C onto Si substrates. Subsequent annealing was performed for one hour at temperatures between 500 °C and 1000 °C in a N_2/H_2 ambient.
The $Si/CoSi_2/Si$ hetero-epitaxial structures were formed by implantation of 170 keV Co⁺ ions at a temperature of 400 °C and to doses of $1x10^{17}$, $2x10^{17}$ and $3x10^{17}$ Co⁺ ions /cm² in both (001) and (111) Si. The wafers were annealed for 30 minutes in a N_2 / H_2 ambient at a temperature of 1000 °C.

Y. I. Nissim and E. Rosencher (eds.), Heterostructures on Silicon: One Step Further with Silicon, 247–252.
© *1989 by Kluwer Academic Publishers.*

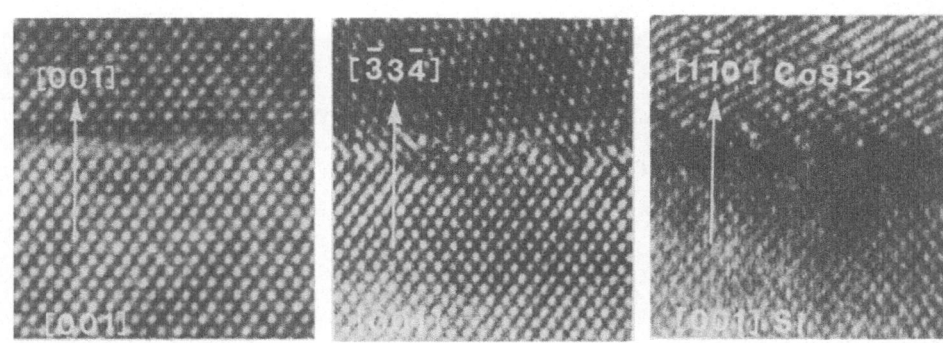

FIGURE 1. High resolution TEM images along the [110] direction of 100 nm CoSi$_2$ on (001) Si formed at 900 °C, showing CoSi$_2$ grains with different orientations with respect to the substrate.

For plan-view specimens, the wafers were cut ultrasonically into 3 mm discs and chemically thinned from the backside by jet-etching. Cross-sectional specimens were thinned mechanically, followed by Ar$^+$ ion-milling at 5 keV at an angle of incidence of 10°.
The wafers have been studied by conventional TEM (Philips EM 400T) and by high resolution TEM (Philips EM 430ST).

4.RESULTS

4.1. CoSi$_2$ on (001) Si

On (001) Si, X-ray diffraction (XRD) reveals that with increasing annealing temperature (500 °C - 1000 °C) the CoSi$_2$ films become increasingly (110)-textured. At temperatures of 500 °C the presence of both CoSi and CoSi$_2$ could be identified.
By TEM it is observed that the layer consists of CoSi and CoSi$_2$ grains (9,10). An interesting feature is that one CoSi$_2$ grain extends over various CoSi grains with different orientations. This suggests that it is likely that there is no orientational relationship between the CoSi and the CoSi$_2$ grains. At higher temperatures (600 °C - 1000 °C) the Co reacts completely with Si to form CoSi$_2$. The layer consists of CoSi$_2$ grains 0.5-1.0 μm in size, and are divided into subgrains by low-angle boundaries. From the electron diffraction patterns taken along the [001] Si zone axis it was observed that the silicide grains are preferentially oriented. Apart from an aligned orientation (001) (a)-orientation, the CoSi$_2$ grains occur in a number of orientations with an epitaxial relationship with Si. There is a strong preference towards the (110) (b)-orientation with the epitaxial relationship :

[110]CoSi$_2$ // [110]Si and [001]CoSi$_2$ // [1$\bar{1}$0]Si

Some typical orientations of CoSi$_2$ grains on (001) Si are shown in the HREM images in figure 1.

4.2. CoSi$_2$ on (111) Si

For (111) Si, single crystalline layers of CoSi$_2$ are obtained predominantly in the twinned (111) B-orientation, which is rotated 180° relative to the aligned A-type (111) orientation. The interfacial defect structure consists of misfit dislocations of edge-type with Burgers vector b=a/6<112>, running in <110> directions. These dislocations are associated with interfacial steps (5,10) as is shown from structure models of CoSi$_2$ / Si (111) projected on the (110) plane (see Fig. 3). For the atomic arrangement at the interface, the 5-fold coordination has been assumed (5). The steps are assumed to be perpendicular to the cross-section. There are opposite interfacial steps, corresponding to dislocations with b=a/6 <112> of respectively

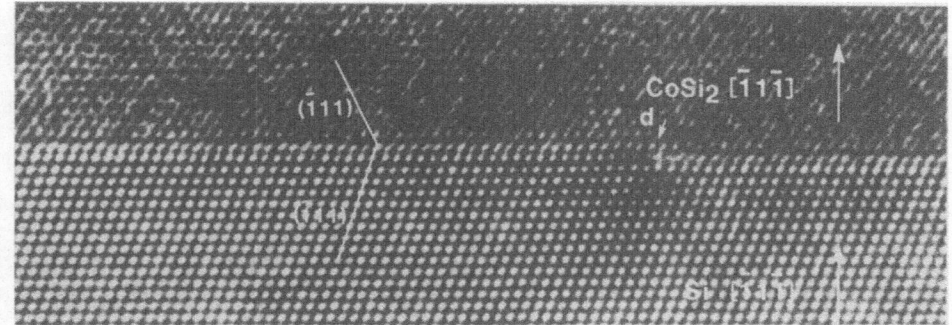

FIGURE 2. High resolution TEM image along [110] of B-oriented $CoSi_2$ on (111) Si, showing a step with height $d_{111} = a/3 \sqrt{3}$, corresponding to a dislocation with $b = a/6 <112>$ of edge-type.

30^0 type and 90^0 type. It appears that the dislocation of figure 3(1a) has a Burgers vector $b = a/6 [1\bar{1}2]Si$ and is of 90^0 type, i.e. edge-type. The dislocation in figure 3(Ib) has Burgers vector $b = a/6 [211]$ and is thus of 30^0 type. In the case of a step height of two monolayers we have the reversed situation. Now the partial dislocation in figure 3(IIa) is of 30^0 type and in figure 3(IIb) of 90^0 type. There are no dislocations for a step height of 3 monolayers. In figure 2, a HREM image along [110] Si of B-oriented $CoSi_2$ on (111) Si is shown, with a step height of one monolayer ($d_{111} = a/3 \sqrt{3}$), corresponding to a dislocation with $b = a/6 <112>$ of edge-type.

4.3. Si /CoSi₂/ Si (001) and (111)

After implantation, X-ray diffraction (XRD) reveals peaks due to $CoSi_2$ with the same orientation as the substrate (11). Investigation of the as-implanted Si (001) and (111) wafers by Rutherford backscattering (RBS) and TEM reveals that for the highest dosis of 3x

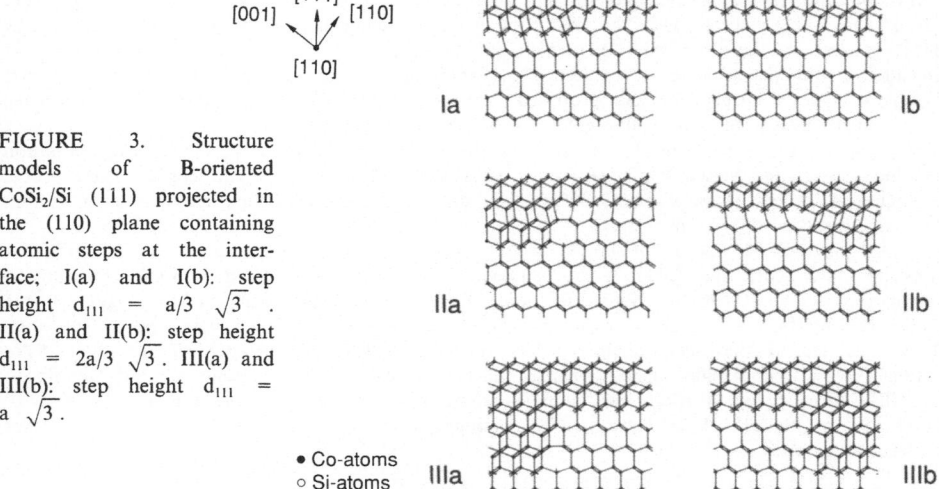

FIGURE 3. Structure models of B-oriented $CoSi_2$/Si (111) projected in the (110) plane containing atomic steps at the interface; I(a) and I(b): step height $d_{111} = a/3 \sqrt{3}$. II(a) and II(b): step height $d_{111} = 2a/3 \sqrt{3}$. III(a) and III(b): step height $d_{111} = a \sqrt{3}$.

• Co-atoms
○ Si-atoms

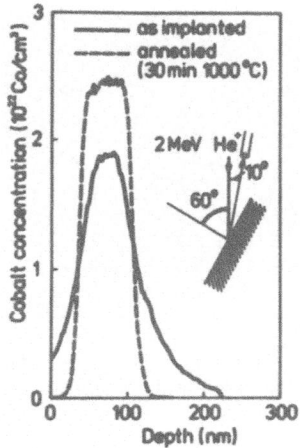

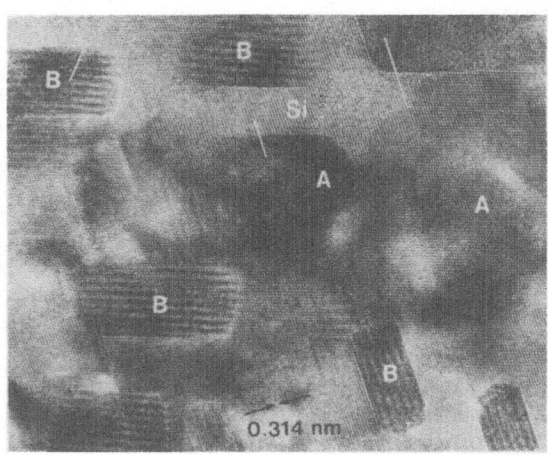

FIGURE 4. Cobalt concentration profiles after implantation and annealing, obtained from glancing angle RBS spectra.

FIGURE 5. High resolution TEM image of the as-implanted structure for (111) Si, showing aligned (A-type) and twin-oriented (B-type) CoSi₂ grains.

10^{17} Co⁺ ions/cm² a buried CoSi₂ layer aligned with the Si lattice has been formed. For the lower doses of 1×10^{17} Co⁺ ions/cm² and 2×10^{17} Co⁺ ions/cm², RBS (see Fig. 4) shows that after implantation the Co concentration does not reach the level of Co in CoSi₂ . Figure 4 shows the implantation profile of 2×10^{17} Co⁺ ions/cm² before and after annealing for (001) Si. This indicates that the concentration of Co before annealing is too low to form a continuous CoSi₂ layer.

Investigation of the implanted structure by high resolution TEM reveals that the implanted Co is present in the form of small CoSi₂ precipitates, which are coherent with the Si substrate. Both aligned (A-type) and twin-oriented (B-type) precipitates are formed as is shown in the high resolution TEM image of figure 5. It is particarly interesting to note the different forms of both types of precipitates. The aligned precipitates have a spheroidal shape with {100} and {111} facets. The twinned ones have an elongated shape and are bounded by {111} planes on the long sides on which twinning occurs.

In addition to the observed precipitates, implantation damage in the form of {113} defects is found also. Bourret (12) has proposed that these defects are precipitated Si interstitials, which have locally formed a hexagonal silicon phase.

Annealing of the wafers results in the dissolution of small precipitates and the formation of single crystalline buried CoSi₂ layers. This is also reflected in the Co concentration profile in figure 4, which changes from a Gaussian to a box-shaped distribution with a maximum Co concentration of Co in CoSi₂.

Cross-sectional TEM images of buried CoSi₂ layers formed by implantation of 2×10^{17} Co⁺ ions/cm² into (001) and (111) Si and subsequent annealing are shown in figure 6a and 6b respectively. The insets in figure 6a and 6b are high resolution TEM images of the CoSi₂/Si interfaces. It can be seen that the layers are fully aligned with the Si substrate. TEM analyses of the microstructure after annealing reveal the presence of threading dislocations below the silicide layers, whereas misfit dislocations are present at both interfaces. In (001) Si, the top Si film also contains some threading dislocations, but is otherwise of good quality. For (111) Si however, the micrograph in figure 6b shows that the top layer is heavily twinned.

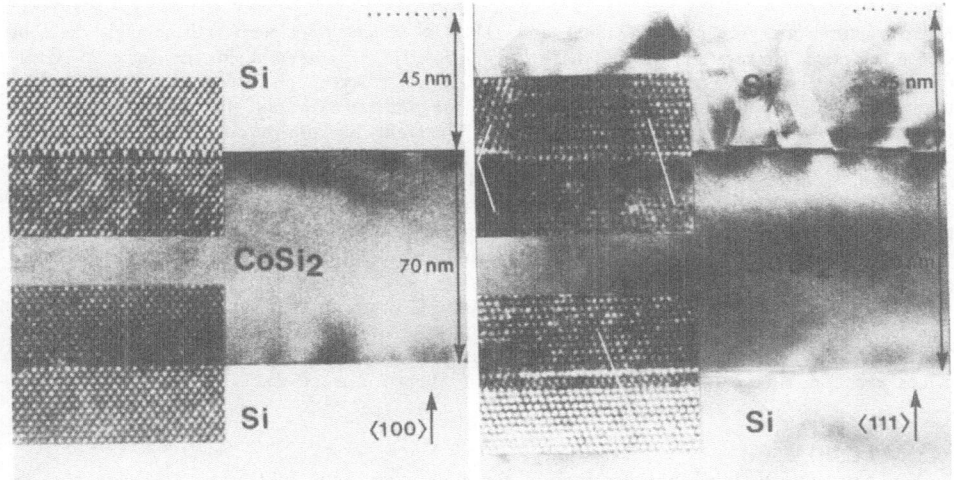

FIGURE 6a. Cross-sectional TEM micrograph of a buried CoSi₂ layer in (001) Si. Implantation dose: 2x 10¹⁷ Co⁺ ions/cm².

FIGURE 6b. Cross-sectional TEM image of a buried CoSi₂ layer in (111) Si. Implantation dose: 2x 10¹⁷ Co⁺ ions/cm².

5.DISCUSSION

The reduction of the interface energy is the driving force for the epitaxial growth of $CoSi_2$ on Si. In view of the small lattice mismatch between $CoSi_2$ and Si of 1.2 %, pseudomorphic growth for layers grown by conventional techniques is expected upto a critical thickness [13]. However, having an interface with well localized steps of monolayer height, the strain energy possibly leads to the formation of twin-oriented $CoSi_2$ and nucleates by arrangement of atoms around such a steps. When the interfacial steps are **unequal** to 3 monolayers or multiples of it, dislocations with Burgers vector $b = a/6 < 112 >$ are formed which relieve the misfit strain. In this way an extra plane of atoms is inserted in the $CoSi_2$ layer, which gives a better fit. We expected opposite interfacial steps corresponding to dislocations with $b = a/6 < 112 >$ of respectively 30⁰ type and 90⁰ type, whereas only 90⁰ type dislocations have been observed. Therefore we suggest that at the first nucleation stage, steps move along the interface or they are created in such a way, that only 90⁰ dislocations are formed. Dislocations of 90⁰ type are more efficient in relieving the misfit strain.

For (001) Si, nucleation of $CoSi_2$ can also be described by such a mechanism. One example of a twin-orientation for (001) Si is the (221) $CoSi_2$, which results from twinning on {111} planes. However, we only observed a small fraction of the grains with this orientation. As a possible explanation we propose that nucleation does occur at steps in the {221} orientation, but that the atoms undergo a small displacement, leading to the growth of {110} oriented $CoSi_2$. The driving force for the reorientation would then be a reduction of the long-range lattice mismatch. The (334) orientation of $CoSi_2$ on (001) Si shown in Figure 1 is also an example of twinning.

In addition to the experimental results, a computer program has been developed which calculates the matching between various orientations of $CoSi_2$ and Si. It appears that the {110} orientation has a quite good fit on (001) Si. When the same procedure is applied to (111) Si, we observed that only the {111} $CoSi_2$ orientation has a good match. The observations for $CoSi_2$ on both (001) and (111) Si can thus be explained by considering geometrical lattice match between $CoSi_2$ and Si.

Microstructural analyses of both (001) and (111) Si wafers implanted with a high dose of Co at elevated temperature, reveal that the implanted Co is present in the form of CoSi$_2$ precipitates. These precipitates occur in both aligned (A-type) and twinned (B-type) orientations. During implantation, CoSi$_2$ precipitates are nucleated within the Si lattice and are thus completely surrounded by the Si lattice. Therefore the nucleation of the aligned precipitates may be energetically more favourable. However, the 1.2 % lattice mismatch between CoSi$_2$ and Si causes strain energy, which possibly leads to the formation of the twin-oriented precipitates by twinning on {111} planes of Si. The growth of CoSi$_2$ along these {111} planes causes the elongated shape of these precipitates. The growth of the aligned precipitates occurs more uniformly in all directions, compared with the twinned ones, which results in preferential growth of the A-type precipitates.

Annealing of the wafers results in the formation of single crystalline layers of aligned orientation by a process in which larger precipitates grow at the expense of smaller ones. The preferential growth of aligned precipitates at the expense of the twin-oriented precipitates causes the formation of epitaxial layers with an aligned orientation.

6.CONCLUSIONS

We have shown that the observations for CoSi$_2$ on both (001) and (111) Si can be explained by considering geometrical lattice match between CoSi$_2$ and Si. The nucleation of the different grains for (001) Si and the twinned B-type orientation for (111) Si has been attributed to a nucleation mechanism at interfacial steps in combination with a relatively large mismatch.

Microstructural analyses of both (001) and (111) Si implanted with a high dose of Co, reveal that the implanted Co is present in the form of coherent aligned (A-type) and twin-oriented (B-type) CoSi$_2$ precipitates. Annealing of the wafers results in the formation of single crystalline layers of aligned orientation by a process in which larger precipitates grow at the expense of smaller ones. It is proposed that the preferential growth of aligned CoSi$_2$ precipitates at the top of the Co implantation profile causes the formation of buried epitaxial CoSi$_2$ layers with a fully aligned orientation.

REFERENCES

1. Murarka SP : Silicides for VLSI applications. Academic, New York, 1983.
2. Nicolet MA and Lau SS : VLSI electronics 6, edited by Einspruch N G. Academic, New York, 1983, p 330.
3. Chen LJ, Mayer JW, Tu KN and Sheng TT : Thin Solid Films 93, 1982, p 91.
4. Appelbaum A, Knoell RV and Murarka SP : J. Appl. Phys. 57, 1985, p 1880.
5. Gibson JM, Bean JC, Poate JM and Tung RT : Thin Solid Films 93, 1982, p 99.
6. d' Anterroches C : Surface Science 168, 1986, p 751.
7. Hensel JC, Levi AFJ, Tung RT and Gibson JM : Appl. Phys. Lett. 47, 1985, p 151.
8. White AE, Short KT, Dynes RC, Garno JP and Gibson JM : Appl. Phys. Lett. 50, 1987, p 95.
9. Bulle-Lieuwma CWT, van Ommen AH and Hornstra J : Proceedings Microscopy of Semiconducting Materials Conf., Oxford, Institute of Physics Conf. Series 87, 1987, p 541.
10.Bulle-Lieuwma CWT, van Ommen A H and Hornstra J : Materials Research Society Symposia Proceedings 102, 1988.
11.Van Ommen AH, Ottenheim JJM, Theunissen AML and Mouwen AG, Submitted to Appl. Phys. Lett.
12.Bourret A : Proceedings Microscopy of Semiconducting Materials Conf., Oxford, Institute of Physics Conf. Series 87, 1987, p 39.
13. Matthews JW : Epitaxial Growth B, edited by Matthew JW, Academic Press, New York, 1985.

GROWTH AND ELECTRONIC TRANSPORT IN THIN EPITAXIAL CoSi₂ - Si HETEROSTRUCTURES

F. Arnaud d'Avitaya, P.A. Badoz, A. Briggs*, C. d'Anterroches, J.Y. Duboz, G. Fishman**, G. Glastre, J.C. Pfister, C. Puissant, E. Rosencher and G. Vincent.

CNET-CNS - Chemin du Vieux Chêne - BP : 98 - 38243 Meylan Cedex France
* CNRS-CRTBT, BP 166 X 38042 Grenoble Cedex.
** Laboratoire de Spectromètrie Physique, BP 87, 38402 Saint Martin d'Hères Cédex.

ABSTRACT
Thin epitaxial CoSi₂ layers on Si have been grown by direct reaction in ultrahigh vacuum. The reaction path from Si + Co to CoSi₂ as a fonction of the initial Co thickness will be described, along with the evolution of interface roughness as seen by high resolution TEM. The generation of pinholes during high temperature annealing will be discussed as a function of geometry and stress relaxation.

Electrical measurements (resistivity, Hall effect and superconducting transition temperature) will be presented on thin layers made either by direct reaction or by codeposition, with particular emphasis on the effect of interfaces on electronic transport.

1. INTRODUCTION
Epitaxial silicon-silicide heterostructures are very attractive both for direct use in high speed devices like metal base and permeable base transistors, ballistic or tunnel devices and as models of Schottky junctions with enough crystalline perfection to expect intrinsic properties to dominate over defect characteristics. Among the epitaxial silicides, Cobalt disilicide CoSi₂ has been the most studied in recent years and a number of groups have shown that it can be grown in continuous thin films with good crystallinity in perfect epitaxial relationship with (111) Si substrates (1, 2, 3).

The paper is divided in two main parts. We will first describe the structural problems : formation of successive phases according to thickness and anneal cycle, interface roughness and pinhole formation. Part II will then be devoted to electrical transport properties and the role of interfaces in very thin layers, both for perpendicular and parallel transport. Both parts will be essentially concerned with CoSi₂ layers formed by room temperature evaporation of Co on (111) Si and subsequent annealing.

2. STRUCTURE AND EVOLUTION OF THE LAYERS
2.1. As-deposited layers
The chemical reaction between Co and Si is already observable directly after room temperature deposition, leading to a silicide layer of thickness ca 1 nm.

In the case of very thin Co films which react completely, this silicide layer is a mixture of CoSi and CoSi₂ (Fig. 1). When the film is thicker, only CoSi is detected, with the excess polycrystalline Cobalt and some CoSi and Co₂Si grains on top (Fig. 2). In both cases, atomic string contrast in HRTEM is clearly visible, showing that the silicide grains are epitaxial or at least highly textured with respect to the Si substrate (4). Although the nature of the migrating species is unknown, this epitaxial relationship

Y. I. Nissim and E. Rosencher (eds.), Heterostructures on Silicon: One Step Further with Silicon, 253–259.
© *1989 by Kluwer Academic Publishers.*

254

and the very small thickness suggest that Co penetrates over short distances into the Si at room temperature, leading to a coherent deformation of the lattice. The silicide phase formed at this stage will then act as a nucleus for the subsequent reaction wherever the deposited Co has not completely reacted.

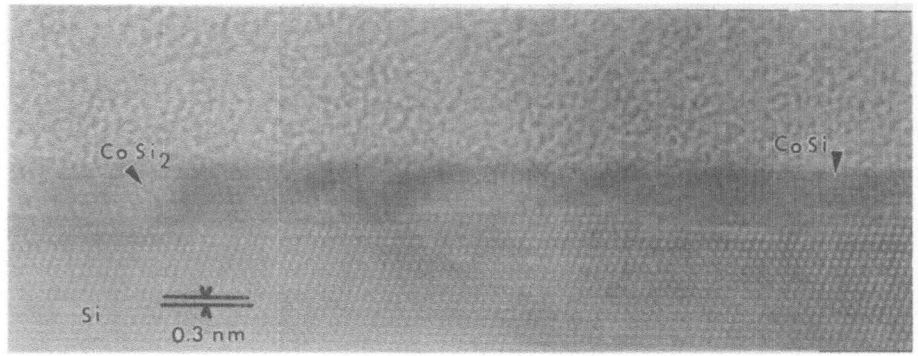

FIGURE 1.   High Resolution Transmission Electron Microscopy (HRTEM) observation of 0.6 nm thick cobalt film deposited at Room Remperature (RT) on top of a silicon substrate. The cobalt reacts during the deposition, which yields a mixture of CoSi and $CoSi_2$ grains.

FIGURE 2.   HRTEM observation of a RT deposited Co film (2.7 nm thick) the RT Co-Si reaction yields a CoSi intermediate layer 1.5 nm thick and a polycrystalline film composed by Co, Co_2Si and CoSi.

2.2. Low temperature anneals

Auger measurements during annealing between 200 and 600°C have been published showing a progressive change in the surface stoichiometry from Co to $CoSi_2$ with intermediate steps close to Co_2Si and CoSi (5). However TEM observations show that, as soon as the reaction starts, both Co_2Si and CoSi grains are formed. Residual Cobalt is observed, and confirmed by magnetic measurements, up to 400°C (6). $CoSi_2$ appears at higher temperatures of 600 to 650°C and rapidly makes up the entire layer. Thus the reaction appears to be largely dominated by nucleation processes in the first stages, with random nucleation events first leading to the coexistence of different silicide phases. Only when the atomic mobility is high enough does the most stable phase $CoSi_2$ finally nucleate since the corresponding critical nucleus is rather large. Growth is then very quick and the whole layer is transformed almost at once. This occurs around 600°C and results in a layer where all grains are epitaxial on (111) Si but with two in-plane orientations, called A and B, deduced from each other through a 180° rotation about the [111] axis.

It is also interesting to consider the roughness of the silicon-silicide interface during the same treatments. The initial interface has a small roughness of a few atomic steps corresponding to the residual polishing imperfections (Fig. 3). The first stages of the silicidation reaction, i.e. the formation of Co_2Si and CoSi phases, do not introduce large changes in the roughness, suggesting that either little Co diffusion occurs at this stage and Si is the dominant migrating species, or atomic transport across the interface is not a limiting step. The formation of the final $CoSi_2$ layer on the contrary is accompanied by the development of interface roughness comparable to the layer thickness with silicide protrusions into the silicon substrate mainly situated at the grain centers, not at the grain boundaries (Fig. 4). The outer surface remains essentially flat during the whole process. Two important indications result from this morphology :

i) the nonuniform final thickness is a clear proof that lateral Co migration occurs at some stage, presumably the $CoSi_2$ formation which creates the roughness.

ii) the excess penetration in grain centers strongly suggests a nucleation event at these points followed by lateral growth of $CoSi_2$ from the pre-existing Co-rich silicide.

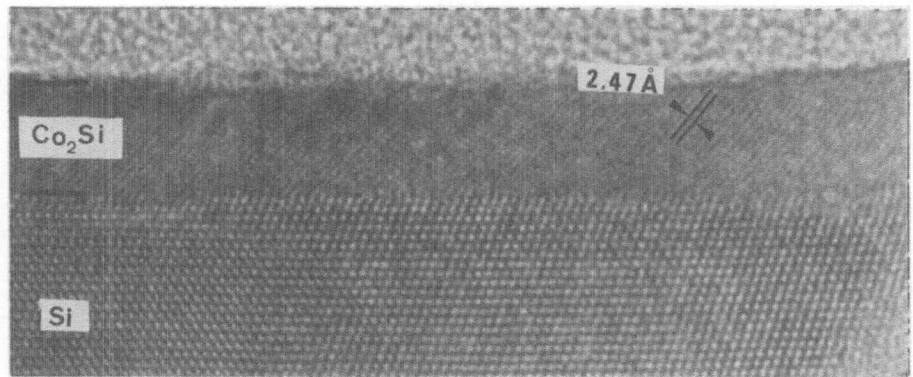

FIGURE 3.   HRTEM observation of a RT deposited Co Film, subsequently annealed at 350°C. A Co_2Si grain is clearly visible, while the metal/Si interface exhibits a very low roughness.

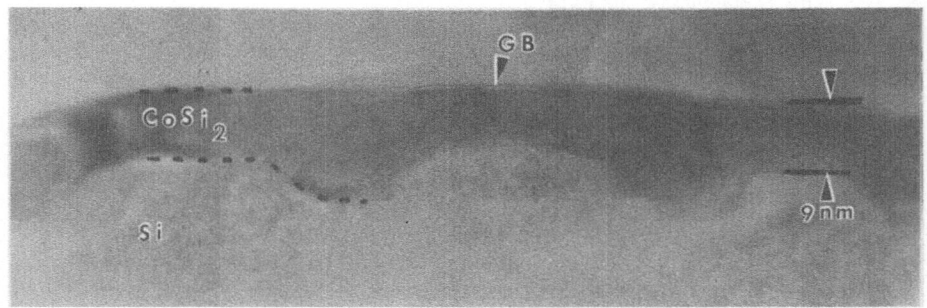

FIGURE 4.   HRTEM observation of a RT deposited Co film, subsequently annealed at 650°C during a few minutes. The interface roughness is characterized by protrusions into the Silicon substrate which are not associated with Grain Boundaries (GB).

2.3. High temperature anneals

For anneals at and above 650°C, the most conspicuous changes in the layer morphology clearly involve surface tension as a major driving force. Pinhole formation and interface planarization are the visible manifestations of atomic movements in this temperature range.

Pinholes appear mostly in association with either grain boundaries or even the presence of an extraneous phase e.g. pollution (7). Their nucleation is thus most probably induced by surface irregularities, defects and contamination in the early stages of layer formation. Their subsequent growth is however clearly a thermodynamic process, leading to faceting in the codeposited layers. There are also indications that a Si-rich surface greatly reduces the formation rate of pinholes, although the detailed mechanism is not well understood (8). The study of patterned layers also shows that no pinholes are formed near the layer edges (9), thus misfit strains, which are relieved close to the edges, are definitely important in the nucleation process.

Interface planarization shows a spectacular dependence on the surface composition of the layer : complete planarization is obtained in a few minutes at 650°C if Si is evaporated onto the surface (equivalent thickness ca 100 A), whereas even longer treatments at the same temperature without a Si flux produce no visible change in the interface (except of course the growth of some pinholes) (3). The same planarization effect is observed in plasma oxidation treatments although in this case the respective roles of silicon depletion to form an oxide and of bombardment induced defects are not yet ascertained (10).

3. ELECTRONIC TRANSPORT
3.1. Resistivity and Hall effect

Resistivity and Hall effet measurements as functions of layer thickness and temperature show that $CoSi_2$ formed by direct reaction is a metal with carrier concentration ca 3×10^{22} holes/cm^3, independent of thickness down to 1.4 nm and only slightly dependent on temperature. A comparison with published silicide band structure calculations shows that the main contribution to the density of states at the Fermi level is due to a band derived from Si 3p levels and analogous to the Si valence band. It is interesting to note however that $CoSi_2$ layers made by coevaporation and low temperature anneal (11,12) show consistently lower carrier densities although no difference in morphology is detected except for a much flatter Si-CoSi$_2$ interface. This indicates that stoichiometry may be an important parameter in determining the carrier concentration. The interface scattering however shows exactly the same behaviour for both kinds of layers and will be discussed in more detail in another section (2,13).

3.2. Perpendicular transport

Transport perpendicular to the Si/CoSi$_2$ interface has been studied in Semiconductor-Metal-Semiconductor (SMS) sandwich structures originally intended for the realization of metal-base transistors (3). The variation of the gain of such SMS transistors with applied voltages and design parameters has been exploited to yield parameters relevant for ballistic electrons, namely :

i) the mean free path of electrons injected over the Si/CoSi$_2$ Schottky barriers, i.e. energy about 0.63 eV above the Fermi level in CoSi$_2$, is equal to the mean free path deduced from conductivity measurements (85 A at 300 K, 350 A at 77 K) within the experimental accuracy of ± 10 %. This shows that inelastic processes are marginal compared to the elastic or quasi-elastic scattering mechanisms which dominate the conductivity (Fig. 5).

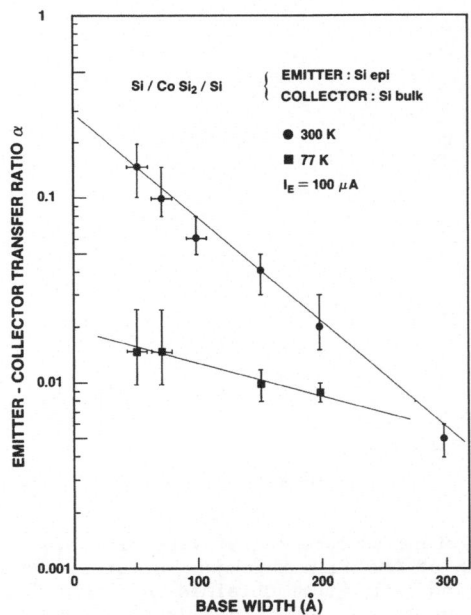

FIGURE 5.   Transfer ratio α vs CoSi$_2$ Film thickness for Si/CoSi$_2$/Si SMS at 300 and 77K. The emitter current is 100µA.

ii) the quantum reflection coefficient of electrons at the exit barrier is consistent with that expected from elementary quantum mechanics in 1 dimension considering the jump in k-vector from CoSi$_2$ with kinetic energy of a few eV (referred to the bottom of the conduction band) into Si with thermal energy. The limited accuracy in both band structure calculations and experiment precludes a really quantitative comparison.

iii) the scattering of electrons in the image potential well in Si is surprisingly strong, with a measured mean free path about 20 A compared to a value around 65 A deduced from hot electron experiments. No explanation has yet been proposed for this discrepancy.

3.3. Interface scattering

Interface scattering can be evaluated from the comparison of Si/CoSi$_2$ samples with different thicknesses. The resistivity versus temperature follows Matthiessen's rule rather nicely for all thicknesses, so that a meaningful study as a function of film thickness can be made on the low temperature residual resistivity. This study, as described in more detail in ref 14, shows that a simple approach using a partly specular scattering (Fuchs-Sondheimer model) at the interface is inadequate. The excess resistivity due to interface scattering shows an empirical thickness dependence $\rho_S \propto d^{-2.5}$ as shown in Fig. 6. This dependence, along with the right order of magnitude, has been recently derived theoretically from a model assuming surface short range roughness scattering and taking into account the 2-dimensional quantized subband structure in CoSi$_2$ in the effective mass approximation. The surface roughness necessary to fit the experimental results is characterized by its amplitude and correlation length, both of the order of atomic dimensions (15).

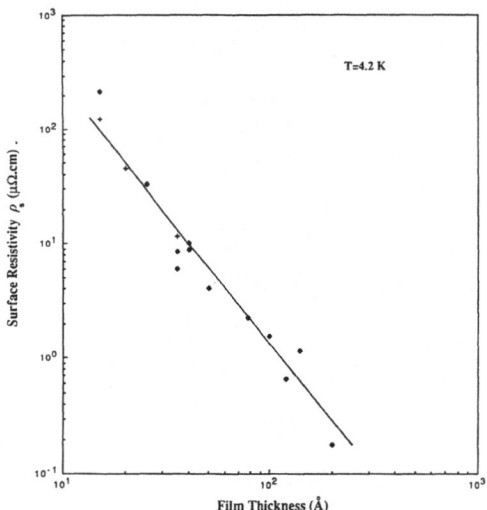

FIGURE 6.   Low-temperature surface resistivity $\rho_S = \rho_0 - \rho_0^{bulk}$ of CoSi$_2$ as a function of the film thickness d for two sets of samples. Diamonds indicate films obtained by Solid Phase Epitaxy at 650°C while crosses refer to films realized by codeposition of Co and Si followed by a thermal anneal above 500°C (11,12). The solid line is the best fit between data and the theory of Fishman and Calecki (15), based on intersubband transitions induced by a random potential in a highly degenerate electron gas quantized in a potential box.

The critical temperature T_c for superconductivity shows a behaviour qualitatively similar to the resistivity. T_c is slightly lower ($T_c \sim 1$ K) than the bulk value ($T_c = 1.2$ K) for films thicker than 100 A, then drops sharply to below 1.8 mK for thicknesses less than 35 A. Although no detailed theoretical explanation is available, it is interesting that surface effects on T_c set in for the same thicknesses and are as sharp as for resistivity.

The comparison with Si/CoSi$_2$/Si sandwiches where Si has been epitaxially deposited on top of the CoSi$_2$ film however destroys this encouraging picture. Although only a small number of samples have been measured, it is indeed quite clear that fundamental questions are raised about the qualitative value of the above discussion. The most disturbing experimental facts are :

i) the excess resistivity is higher (approximately twice) in a sandwiched CoSi$_2$ film than in a "free" film of equivalent thickness, although the interface roughness as measured from HRTEM is much lower, as described in section I.

ii) the superconducting temperature is almost equal to the bulk value in a sandwiched film of 100 A thickness (16).

These facts indicate that, although surface effects are significant both in resistivity and T_c measurements, the important parameters are different and the two interfaces Si/CoSi$_2$ and CoSi$_2$/vacuum are to be treated separately, with Si/CoSi$_2$ most effective on resistivity and CoSi$_2$/vacuum on superconducting properties. The sharp variation with thickness in both cases may be connected with the 2 dimensional subband structure which is basic in obtaining the proper $d^{-2,5}$ behaviour in the resistivity calculations, even if the assumptions of this theory, and particularly the physical meaning of the roughness parameters involved, may need reconsideration.

REFERENCES

1. Derrien J. and Arnaud d'Avitaya F.: J. Vac. Sci. Technol. A.5, 2111 (1987) and references there in.
2. Tung, R.T., in Silicon Moleculer Beam Epitaxy, C.R.C Press (in press).
3. Rosencher E., Arnaud d'Avitaya F., Badoz P.A., d'Anterroches C., Glastre G., Vincent G. and Pfister J.C.: Mat. Res. Soc. Symp. Proc. 91, 415 (1987) and references there in.
4. D'Anterroches C.: Surface Science 168, 751, (1986).
5. Arnaud d'Avitaya F., Delage S., Rosencher E. and Derrien J.: J. Vac. Sci. Technol. B.2, 770, (1985).
6. D'Anterroches C. and Madar R.: (unpublished).
7. D'Anterroches C., Yakupoglu H.N., Lin T.L., Fathaner R.W. and Grunthaner P.J.: Appl. Phys. Lett. 52, 434 (1988).
8. Tung, R.T. and Batstone J.L., Appl. Phys. Lett. 52, 1611, (1988).
9. Glastre G., Rosencher E., Arnaud d'Avitaya F., Puissant C., Pons M., Vincent G. and Pfister J.C.: Appl. Phys. Lett. 52, 899 (1988).
10. D'Anterroches C. and Straboni A., Private Communication.
11. Von Kanel H., Henz J. and Ospelt M., Physica Scripta T19, 158 (1987).
12. Henz J., Von Kanel H. Ospett M. and Wachter P., Surface Science 189/190, 1055 (1987).
13. Duboz J.Y., Badoz P.A., Rosencher E., Henz J., Ospelt M., Von Kanel H. and Briggs A., (unpublished).
14. Badoz P.A., Briggs A., Rosencher E., Arnaud d'Avitaya F. and d'Anterroches C., Appl. Phys. Lett. 51, 169 (1987).
15. Fishman G. and Calecki D. (unpublished).
16. Badoz P.A., Rosencher E, Briggs A. and Arnaud d'Avitaya F., Superlattices and Microstruct. 2, 425 (1986).

ORGANIC POLYMERS AND MOLECULAR MATERIALS ON SILICON

J.M. BUREAU

THOMSON-CSF, Laboratoire de Chimie pour l'Electronique
Domaine de Corbeville · BP 10 - 91401 - ORSAY - FRANCE

1. INTRODUCTION

Organic polymers are widely used today in electronics as potting materials, adhesives, laminated printed circuits boards, solder masks and connectors because of their insulating properties and stability together whith their unique processability and mechanical properties. But for these applications, organic materials are rather disconnected from the basic electronic device.

Closer to the chip, the use of organic resists makes possible the fabrication of VLSI circuits thanks to the combination of their coatability and their high sensitivity to radiation exposure. However, these resists are only used as tools and are removed from the chip.

During the past two decades, many attempts have been made to use organic polymers or ordered molecular materials within electronic devices, as it has become possible by new synthesis and new processing methods, to build and tailor new systems in order to obtain very specific properties, especially in the fields of optics and electronics. Some of these systems can now compete with inorganic materials and replace them advantageously in classical devices (heat-resistant polymers become familiar in the electronics industry as interlevel insulating layers) but many others offer new opportunities and can form the basis of novel devices.

Recently, the use of many organic materials has been considered as "active" components in applications such as xerography, batteries, videodisks, displays, transducers and microelectronics [1]. It is the purpose of this paper to focus on this last point.

Following are the major factors which contribute to the increasing interest in the research and the development of organic polymers and molecular materials for microelectronics.
- Deposition techniques are often easy and do not disturb semiconductor surface compared to conventional deposition techniques.
- Very thin films of precise order or composition can be deposited with excellent uniformity, purity, adhesion, chemical inertness and freedom of pinholes.
- These films are able to cover large surfaces while maintaining properties on a macroscopic and even a microscopic level.
- Organic materials offer a virtually limitless variety of molecules that can be designed for specific properties and applications.

In this paper, polymers and molecular materials are defined, deposition techniques relevant to microelectronic applications are shortly described and some applications, such as thin film dielectrics, integrated optics and conducting materials, are discussed.

2. ORGANIC POLYMERS AND MOLECULAR MATERIALS AS ELECTRONIC MATERIALS

2.1 Polymers
2.1.1 Definition.

Polymers are materials made of macromolecules formed by the repetition of thousands of units (monomers) linked by covalent bonds, and arranged in linear chains or in three dimensional networks. These units cannot be isolated without degrading the polymer.

Their solid state morphology depends on the chemical structure of the chains (geometry, flexibility, possible intermolecular bonds, ...) as well as on their processing technique. Three forms may be distinguished: amorphous (or glassy) polymers in which no long range order is

Y. I. Nissim and E. Rosencher (eds.), Heterostructures on Silicon: One Step Further with Silicon, 261–272.
© 1989 by Kluwer Academic Publishers.

found, giving interesting optical properties (e.g. PMMA), semicrystalline polymers in which small crystallites are dispersed in an amorphous matrix, and fully crystalline polymers only obtained in very specific cases.

Conjugated chain polymers (in which simple bonds alternate with double carbon to carbon bonds) are of great interest for electronic applications since delocalized electrons give thermal stability, conduction properties and optical nonlinear properties.

2.1.2 Methods of deposition.

Polymeric materials offer a wide range of processing techniques that can be used to build up high quality thin films on various substrates.

a) Solution techniques are useful to cover large surfaces with homogeneous films of various thicknesses. Spin or spray coating are very familiar techniques in the semiconductor industry for resist deposition and for planarization layer technology. Insoluble or intractable polymers can be deposited via a two stage route using a processible precursor prior the final conversion.

b) Vacuum techniques are also compatible with the wafer fabrication. Polymers can be deposited from glow discharge of a very wide range of volatile monomeric compounds, including those which do not polymerize from conventional methods. Plasma polymers are essentially free of pinholes, highly crosslinked and adhere well to various substrates. However the complex mechanisms underlying film formation are difficult to control and often produces unwanted byproducts or trapped ionic species. A great deal of work has been performed on plasma-polymerized fluorocarbon and organosilicon molecules.

c) Electrochemical polymerization of soluble monomers in an appropriate electrolyte is used to produce conductive polymer films of controlled thickness on conductive substrates [2].

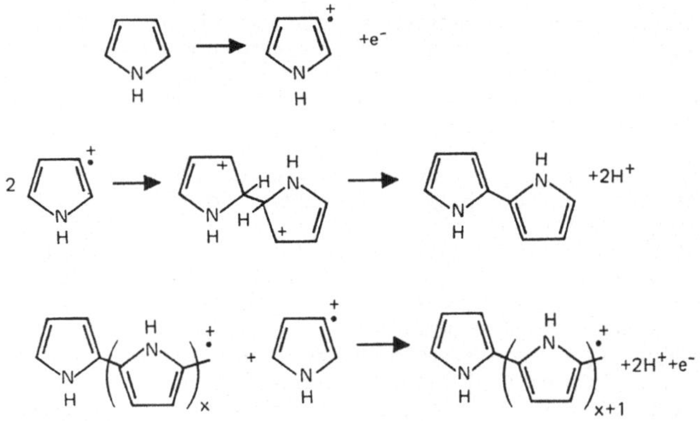

Figure 1.Electrochemical polymerization of pyrrole

Electropolymerization on metals or photoassisted electropolymerization on semiconductors is reported for electroactive polymers such as polypyrrole, polythiophene or polyaniline.

2.2 Molecular materials.

2.2.1 Definition.

Molecular materials are made of small molecular units which are isolately synthetized and which can be individually characterized, for example in solution or in the gas phase. These units are condensed and organized in the solid state to form materials in which interunit interactions are weaker than in inorganic solids. The structure of this assembly depends on individual properties of molecular units, such as geometry, polarity or ability to form intermolecular bonds but also on processing techniques. Order, orientation of units, packing symmetry and purity are of great importance for the bulk properties of the material.

Molecules exhibiting high hyperpolarizability and packed in a regular system give nonlinear optical properties. Molecular materials made of planar unsaturated units that can assemble in a cofacial columnar stack with π orbitals overlap exhibit large delocalization of charge carriers and hence electrical conductivity.

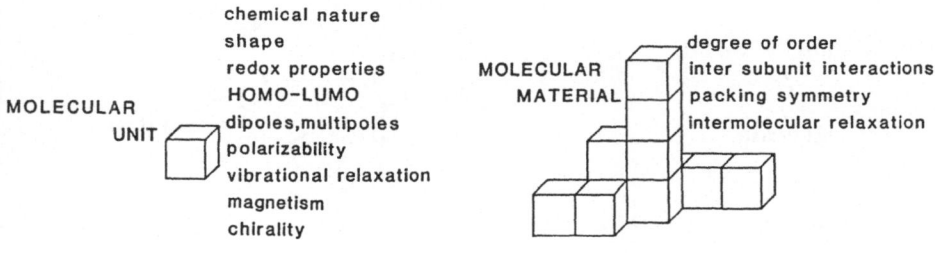

Figure 2. Concept of molecular material

2.2.2 Methods of deposition.

Vacuum sublimation is a method of choice when high purity thin films are required though it usually does not allow to control the packing arrangement of molecular units.

The technology of monomolecular Langmuir-Blodgett films has been developed some fifty years ago by Irving Langmuir and Katherine Blodgett but has received very much interest in the past twenty years for the electronic industry as it allows architectural control on the molecular level (as molecular beam epitaxy for III-V compounds).

Long chain amphiphilic (hydrophobic and hydrophylic end groups) organic molecules are placed, by means of a volatile solvent, on the surface of a liquid (usually purified water). These molecules, when carefully compressed, give a quasi-solid Langmuir monolayer with their chains vertically aligned. If a suitable substrate (e.g. a freshly etched silicon wafer) is now transported vertically through this compact monolayer, it is possible for the molecules to be transferred to the solid surface as a Langmuir-Blodgett monolayer (Figure 3). Typical layer thickness is about 3 nanometers.

During a dipping cycle (down and up) it is possible to deposit two layers whith molecules stacking head to head (Y-type), or only one layer in one direction (X- or Z-type). Whith great care it is now possible to build up high quality and high order multilayer films.

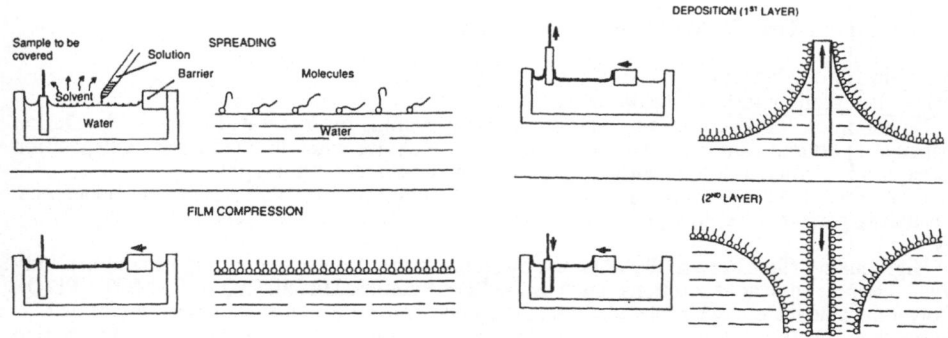

Figure 3. Technique of Langmuir-Blodgett films

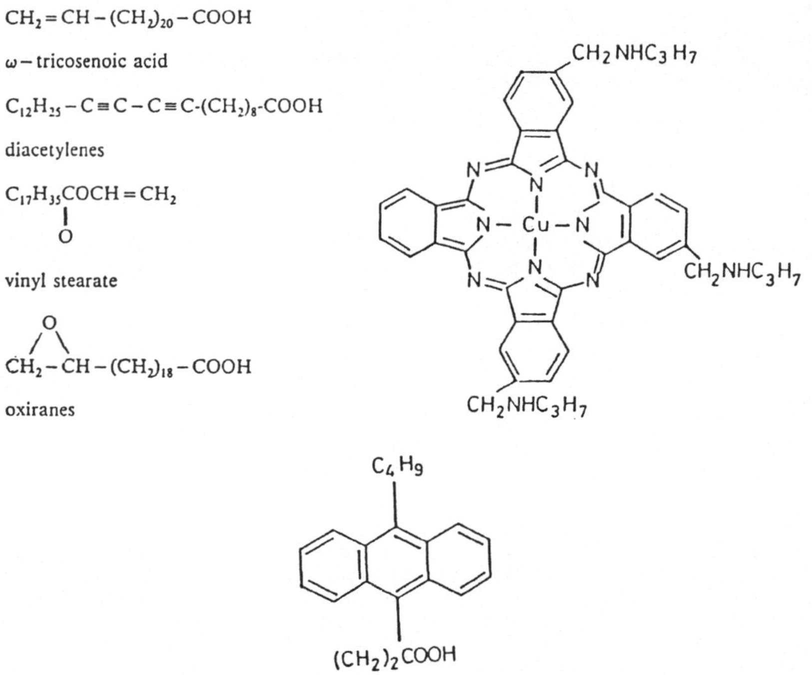

$$CH_2 = CH - (CH_2)_{20} - COOH$$

$\omega -$ tricosenoic acid

$$C_{12}H_{25} - C \equiv C - C \equiv C\text{-}(CH_2)_8\text{-}COOH$$

diacetylenes

$$C_{17}H_{35}COCH = CH_2$$
$$|$$
$$O$$

vinyl stearate

$$CH_2 - CH - (CH_2)_{18} - COOH$$

oxiranes

Figure 4. Molecules suitable for LB films deposition

Most work has been concentrated on LB deposition of long chain fatty acids but also of polymers and specific molecular units such as phtalocyanines [3]. A large variety of such assemblies of organic molecules or polymers have been reported and potential applications in lithography, microelectronic devices, optical systems and biotechnology have been demonstrated.

3. APPLICATIONS IN MICROELECTRONICS

3.1 Thin-film dielectrics
3.1.1 Planarizing interlevel insulation
Development of VLSI require multilevel metallizations on integrated circuits. A solution deposited polymer film is the intermetal insulator of choice when low temperature deposition is required or when a smooth surface is needed for subsequent lithography.

Heterocyclic heat-resistant polymers are advantageous as they combine adherence, insulating performance, low permittivity, high planarizing effect and technology compatibility [4].

High quality films are usually obtained by spin coating and thermal curing of commercially available polyamic acid solutions (such as Du Pont Pyralins, Hitachi PIQ or CEMOTA PI2Q). Polyamic acids are high molecular weight soluble precursors that can be transformed intopolyimide (insoluble) by the following non reversible reaction called cyclodehydratation or imidization:

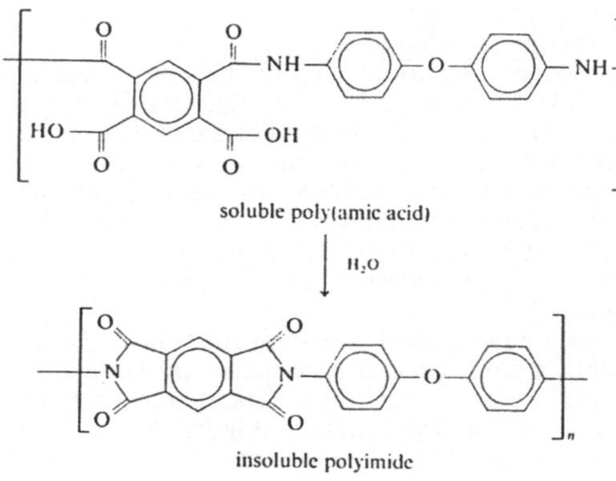

soluble poly(amic acid)

H₂O

insoluble polyimide

Figure 5.Imidization

These polyamic acid solutions, though widely used in the semiconductor industry, has several disadvantages :
- They are not very stable and their pot life is of few months if stored at low temperature and if moisture is avoided.
- Properties of imidized films are very dependant on the curing conditions (time, temperature and atmosphere).
- The imidization reaction is associated with water release (one molecule per ring) and is believed to be responsible of inhomogeneities and insulation defects within the film.

For these reasons, a great deal of work is being performed in order to synthetize fully imidized soluble polyimides or other soluble heterocyclic heat resistant polymers. Some of them are already commercially available (XU from Ciba-Geigy or PPQ from CEMOTA) but have only recently been reported for VLSI fabrication [5].

3.1.2 Ultrathin barrier layers

The metal-insulator-semiconductor (MIS) capacitor is a fundamental structure and an integral part of insulated gate field effect transistors (IGFETs), charge coupled devices and other field effect devices, such as ion or chemical sensors.

The lack of device quality inorganic dielectrics is motivating the work on ultrathin organic layers such as plasma-polymerized [6] or Langmuir-Blodgett films as gate insulators.

Conventional fatty acid LB films have been fabricated on silicion since 1972 [7].Organic insulating Langmuir-Blodgett films have the advantages of moderate temperature deposition, very fine thickness control and good uniformity. However they often lack thermal and mechanical stability.

Polymerized diacetylene LB films for MIS stuctures based on GaP and InP have been described with thermal stability up to 120°C [8]. In 1985 Fung et al. reported a ten layer polydiacetylene LB film used as the gate insulator for silicon IGFETs. The integrated IGFETs are fabricated by combining conventional IC technology and the LB technique and exhibit electrical parameters comparable with SiO2 IGFETs [9]. More recently, a modified LB technique (substrate parallel to the floating film) has been shown to allow the deposition of a prepolymerized heat resistant film of polybenzimidazole which remain fairly stable up to 400°C [10].

These results are encouraging for the use of organic films to replace inorganic dielectrics in conventional field effect devices and also makes envisageable the fabrication of integrated sensors. Since organic compounds often respond more positively than inorganic materials to external stimuli such as chemicals, pressure or temperature, it is possible to fabricate IGFETs which characteristics respond quickly and reversibly to small changes in ambient conditions. Ion sensitive FETs (ISFETs) with polymeric or other organic gate films have been used, for example for biomedical or clinical applications [11] as well as piezo and pyro FETs [12].

3.2 Organic materials for integrated optics
3.2.1 Passive light guiding for optical interconnections.

As the geometries of modern electronic circuits become smaller and denser and as computations speed and complexity increase, classical electrical interconnections (chip to chip or within a chip) have serious problems such as signal transmission capacity limitation due to RC time constants and signal interference due to stray capacitances.

Optical interconnects are a potential solution to these problems as they offer the combination of large bandwidth and large fanout. Two types of optical interconnections have been proposed: guided-wave type and free-space type. In the former, optical signals are transmitted from light sources to photodetectors via guided-wave circuits such as optical fibers or integrated waveguides [13].

Commercial systems already have optical fiber transmissions between boards but at lower levels (chip to chip or within a chip) several approaches are still in study. In one of these approaches the waveguiding medium may be a transparent material deposited on the semiconductor surface as a planar waveguide. Figure 6 shows a planar multimode configuration between silicon chips that incorporates crossovers, bends, branches and fanouts [14].

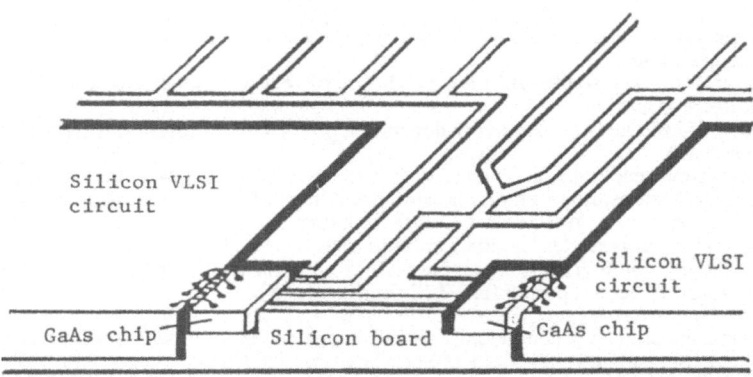

Figure 6.Planar optical interconnects

A large number of polymers can be processed as fully amorphous, show good transparency in the visible and in the near-IR regions and exhibit a wide range of refractive indices [15], but most of them have a restricted temperature performance.

As they are compatible with chip and mother board technologies, heat-resistant polymers are good candidates for integrated waveguides fabrication. They usually have high refractive indices and form amorphous low absorbing and low scattering layers. Such polymers are familiar in the electronic industry and integrated waveguides can be fabricated by conventional processes of integrated circuits (dip or spin coating, photolithography definition, wet or dry etching). Commercial polyimides have been studied as materials for low loss passive waveguides to interconnect several chips on a board or several active components arranged on a single chip [16]. Waveguides suitable for integrated optics must exhibit light losses of 1 dB/cm or less. Polyimide waveguides are reported to exhibit 0.3 dB/cm when dried at 90°C but after complete curing at higher temperature, losses tend to increase because of residual solvent or trapped water creating scattering centers.

Figure 7 shows an integrated waveguide achieved by Furuya et al between semiconductor integrated optics on the same chip using spin deposited polyimide to form the core of the waveguide [17].

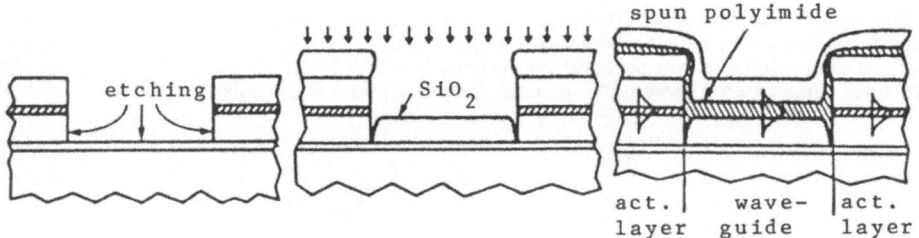

Figure 7.Monolithic integrated polyimide waveguide

3.2.2 Nonlinear materials for integrated optics

Nonlinear optical (NLO) effects in materials have attracted considerable interest for optical signal processing applications: laser modulation and deflection, electro-optical or all optical information control in integrated optical circuits.

Nonlinear effects in optically active materials are due to nonlinear polarization responses to electromagnetic fields. This polarization is given by:

$$P(E) = Po + \chi.E + \chi^2.E.E + \chi^3.E.E.E + ...$$

where χ is the linear susceptibility and the higher order χ's are responsible for nonlinear effects. The second order susceptibility corresponds to NLO effects (and hence possible applications) such as second harmonic generation (frequency doublers), frequency mixing (optical mixers), parametric amplification (optical parametric oscillators), Pockels effect (electro-optical modulators). The third order NLO effect leads to Kerr effect (high speed optical gates) and optical bistability (logic and memory operations) [18].

Such behaviours have been found both in organic and inorganic materials (lithium niobate is the current state-of-the-art waveguide NLO material for second harmonic generation) but recent studies have revealed the superiority of organic polymers and molecular crystals:
-Organic molecules are more versatile and able to be engineered to optimize their nonlinear responses.
-Since nonlinear optical processes in organic materials are mainly electronic in nature, their response time is shorter.
-Their laser damage threshold is higher.

Synthesis and processing techniques allow structural control for bulk response opimization (especially for second order effect which requires a noncentrosymmetric material) and make feasible integrated medium for optical interactions such as waveguides.

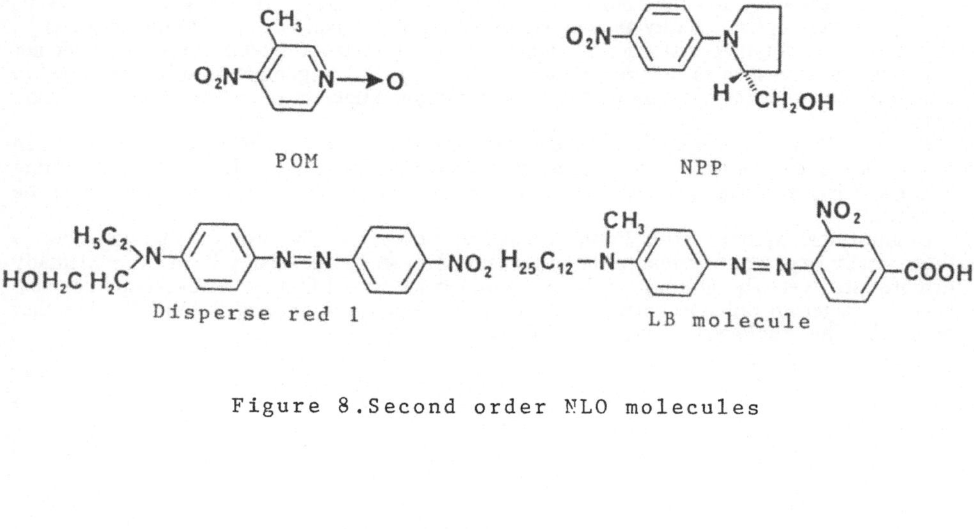

POM

NPP

Disperse red 1

LB molecule

Figure 8.Second order NLO molecules

The second order susceptibility in organic molecules depends on conjugation and intramolecular charge transfer. Molecules with a highly polarizable core (aromatic rings) and electron donor and electron acceptor end groups like POM or NPP are prototype molecular structures . Asymmetry, present in molecules is also required in the bulk material.

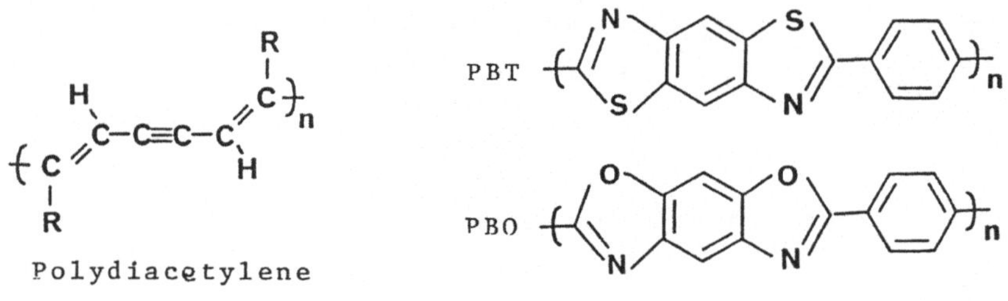

Polydiacetylene

Figure 9. Third order NLO polymers

The third order susceptibility is enhanced by one-dimension conjugation such as in polydiacetylenes, PBT or PBO (Figure 9). No asymmetry is required but parallelism of the conjugated chains is favourable.

In order to build up materials as waveguides with optimized NLO responses, that is to say with orientational order of NLO moieties, several approaches have been studied.

a)-crystallization. Though most of the work on NLO materials is dealing with bulk monocrystalline materials coupled to power lasers, monocrystals are not easily amenable to be integrated. Monocrystalline guides of several millimeters long and few tens of microns diameter have been fabricated by filling a groove etched in amorphous silica with a NLO organic material (NPP) and crystallized by a fusion zone technique.

It should be noted that only few NLO molecular organic compounds (less than 20%) crystallize in a non centrosymmetric system necessary for second order phenomena.

Among third order NLO materials, diacetylene exhibit the unique feature of allowing for solid state polymerization of crystalline monomer, giving polymeric units stacked parallel to a crystallographic direction.

b)-Langmuir-Blodgett films.Classical NLO molecules may be adapted so as to make them compatible with LB technique. Non centrosymmetric oriented structures have been built by deposition of X- or Z- type layers or by alternating two different layers in a Y- type configuration. Such films have been studied as NLO waveguides or as NLO claddings on a planar linear guide.

c)-guest-host systems with amorphous polymers. Fully amorphous polymers are materials possessing high optical quality as well as processing ability and where NLO moieties can be dispersed. As polymer coating techniques do not produce oriented films, a molecular alignment process such as electrical poling is necessary. This alignment must be performed in a high mobility phase (e.g. above the glass transition temperature) and be frozen in a rigid matrix.

Disperse Red 1 and other NLO molecules have been dissolved in PMMA and oriented above Tg [19,20]. However, the alignment is limited to 10-20% by the thermal randoming effect and has poor stability because of slow relaxation processes and dye diffusion in the polymer.

d)-guest-host systems with liquid crystalline polymers. The use of LC polymers is theoretically expected to enhance the alignment [21] but has not really been experimentally demonstrated yet. The grafting of the NLO molecules on a LC polymer backbone has also been envisaged in order to incorporate higher concentrations since their solubility is rather poor [22]. This would also reduce their diffusion in the matrix.

3.3 Electroactive materials

Since 1977 and the discovery by Shirakawa et al. [23] that the conductivity of polyacetylene could be raised to the metallic range, a great deal of work is devoted to organic materials exhibiting extended π electrons systems.

Electrical conductivity is found in linear conjugated chain polymers in which single bonds alternate with double bonds, and in molecular materials made of planar unsaturated units which can assemble in co-facial columnar stacks with π orbitals overlap.

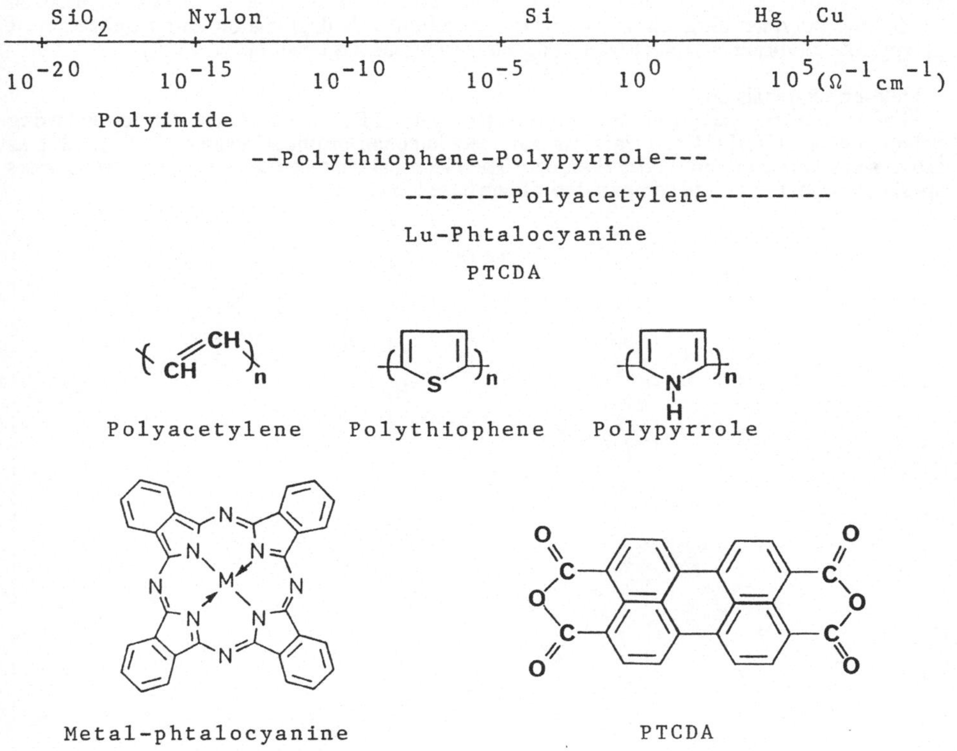

Figure 10.Conducting polymers and molecular materials

We do not discuss here wether the terms "organic semiconductors" or "synthetic metals" are adapted or not, but it should be noted that conducting organics are not substitutes of inorganic semiconductors or metals but specialty materials with a unique combination of properties.

Pure materials are usually insulators and can be p or n doped by with electron accepting or electron donating substances, giving the possibility of tailoring the conductivity in a wide range (Figure 10).

Conducting polymers can be synthetized chemically or electrochemically on various conducting substrates. They offer a wide range of potential applications especially in batteries, EMI shielding, and electrochromic displays [1]. Molecular materials are usually deposited by vacuum sublimation or by LB technique and applications such as chemical sensors or electrochromism have been studied [1].

Many attempts have been made in order to fabricate organic on inorganic electronic devices but convicing demonstrations have been few. Those which appear the most promising are presented here.

Stabilization of photoelectromical cells :

A photoelectrochemical cell for solar energy conversion or storage consists of a semiconductor electrode and a counter electrode immersed in a suitable electrolyte. The major problem in the development of such devices is the susceptibility of small band gap semiconductors (Si or GaAs) to photodegradation. Thin transparent electropolymerized conducting polymer films have been shown to inhibit this degradation and stabilization of polypyrrole modified n-Si electrodes in aqueous solutions has been reported [24].

Field-effect transistors :

Similar organic materials based insulated gate FETs have been fabricated using polythiophene (PT) [25] and vacuum sublimed lutecium-bisphtalocyanine (LuPc2) [26] as active semiconductors on oxidized silicon. The electrical characteristics of these two devices are shown in figure 11 and compared to silicon based devices.

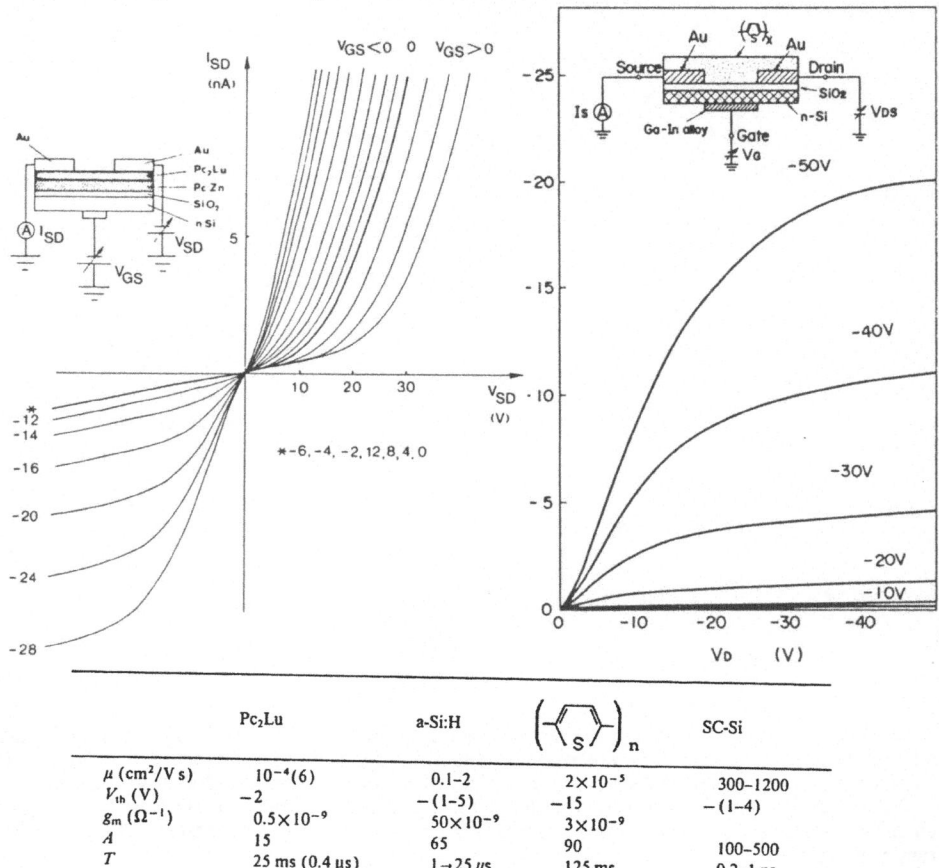

	Pc$_2$Lu	a-Si:H	$\left(\!\!\underset{S}{\langle\!\!\!\!\rangle}\!\!\right)_n$	SC-Si
μ (cm^2/V s)	$10^{-4}(6)$	0.1–2	2×10^{-5}	300–1200
V_{th} (V)	-2	$-(1-5)$	-15	$-(1-4)$
g_m (Ω^{-1})	0.5×10^{-9}	50×10^{-9}	3×10^{-9}	
A	15	65	90	100–500
T	25 ms (0.4 μs)	$1\to25\,\mu$s	125 ms	0.2–1 ns

Figure 11. Insulated gate FETs characteristics

These devices, fabricated with very simple techniques show fairly good characteristics. In addition, the conductivies of these materials could be very sensitive to the presence of certain gases (NO_2 and Cl_2 for phtalocyanines). This phenomenon could be exploited to fabricate microsensors.

Organic-on-inorganic diodes :
High quality organic on inorganic barrier diodes have been reported on silicon and III-V compounds with 3,4,9,10-perylenetetracarboxylic dianhydride (PTCDA), an organic dye which can be vacuum deposited [27]. PTCDA and related compounds form very high energy barriers with semiconductor surfaces and very high avalanche breakdown (of up to 230V for Si) is found. High reverse biases across the wafer and hence deep depletion of carriers makes possible the measurement of several semiconductor properties such as free carrier concentration, deep level density and energy and epitaxial thicknesses.

The ease of fabrication of organic-on-inorganic diodes and the ease of removal of PTCDA from the wafer by immersion in basic solutions provide a useful nondestructive means for determining the properties of the wafers. Low surface disturbing nature of the deposition allow to study intrinsic surface properties and the characteristics of the diodes allow diagnostics deeper into the substrate that can be done using conventional methods.

These organic-on-inorganic diodes may also be envisaged for device applications.

4. CONCLUSIONS

The purpose of this review was to consider the state of the art and the potentialities in the use of organic polymers and molecular materials in microelectronic devices.

Very often, the easy and non damaging deposition processes of high quality films on semiconductor substrates are the most attractive feature of these materials, but in addition, in many applications organic materials may exhibit superior intrinsic properties compared to inorganic materials and the possibilities of improving these characteristics by molecular engineering are still in progress.

Modifications of the optical and electrical behavior by particle beam irradiation has also been demonstrated and suggest the possibility of patterning.

The area of organics as electronic materials is still in its infancy, especially that dealing with "active" materials such as nonlinear optical and conducting ones. In order to improve the knowledge and to develop new applications, interdisciplinary research including chemical and electrical engineers is needed.

REFERENCES

1-R.S.Potember, R.C.Hoffman, H.S.Hu, J.E.Cocciardo, C.A.Viands, R.A.Murphy and T.O.Poehler, Polymer, Vol 28 No 4 (1987)
2-A.F.Diaz and J.Bargon, Handbook of Conducting Polymers p.81 Ed. T.A.Skotheim (1986)
3-M.C.Petty, Polymer Surfaces and Interfaces, p 163, ed. W.J.Feast and H.S.Munro (1987) Wiley and Sons
4-A.M.Wilson, "Use of Polyimides in VLSI fabrication", Polyimides synthesis,characterization and applications, Ed. K.L.Mittal, Vol.2, 715
5-Fall Meeting of the Electrochem.Soc., Hawai 1987, Ext. Abstr. n458, 470 and 473
6-M.Aktik, Y.Segui and Bui Hai, J.Appl.Phys., 51, 9 (1980), p.5055
7-J.Tanguy, Thin Solid Films 13 (1972) 33-39
8-A.S.Dewa, C.D.Fung, E.P.Dipoto and S.E.Rickert, Thin Solid Films 132 (1985) 27
9-C.D.Fung and G.L.Larkins, Thin Solid Films 132 (1985) 33
10-M.T.Fowler, M.Suzuki, A.K.Engel, K.Asano and T.Itoh, J.Appl.Phys., 62, 8 (1987)
11-A.Sibbald, Journal of Molecular Electronics, Vol.2, n2 (1986) 51
12-G.G.Roberts, Sensors and Actuators, 4, 131 (1983)
13-see for example: Optical Egineering, October 1986, Vol 25 No 20
14-L.D.Hutcheson, P.Hangen and A.Husain IEEE Spectrum, March 1987, p 30
15-R.M.Glen, Chemtronics, 1986, Vol 1, 98-106
16-H.Franke, J.D.Crow SPIE Vol.651 Integrated Optical Circuit Engineering III (1986)
17-K.Furaya, B.I.Miller, L.A.Coldren and R.E.Howard, Electron. Letters 18, (1982), 204
18-J.Zyss, Journal of Molecular Electronics, Vol. 1,25-45 (1985)
19-K.D.Singer, S.J.Lalama and J.E.Sohn, SPIE Vol. 578 Integrated Optical Circuit

Engineering II (1985)
20-P.Le Barny, S.Esselin, D.Broussoux, J.Raffy and J.P.Pocholle, to appear in SPIE Proceedings, Cannes 1987
21-G.R.Meredith, Macromolecules, Vol. 15, 1385-1389
22-C.Noel, C.Friedrich, V.Leonard, P.Le Barny, G.Ravaux and J.C.Dubois, to be published in Makromolecular Chemie
23-H.Shirakawa, E.J.Louis, A.G.MacDiarmid, C.K.Chiang and A.J.Heeger, J.Chem.Soc. Chem.Commun. (1977) 578
24-O.Ingans, I.Lundström and T.A.Skotheim, Handbook of Conducting Polymers p.525, Ed. T.A.Skotheim (1986)
25-H.Koezuka, A.Tsumura and T.Ando, Synthetic Metals, 18 (1987) 699
26-R.Madru, G.Guillaud, M.Al Sadoum, M.Maitrot, J-J. André, J.Simon and R.Even, Chem.Phys.Lett. 145, 4 (1988) 343
27-S.R.Forrest, M.L.Kaplan and P.H.Schmidt, J.Appl.Phys 55, 6 (1984) 1492

ORGANIC POLYMER FILMS FOR SOLID STATE SENSOR APPLICATIONS

JIRI JANATA[1]

Center for Sensor Technology, University of Utah, Salt Lake City, Utah 84112, U.S.A.

MIRA JOSOWICZ

Institut für Physik,Universität der Bundeswehr München, D-8014 Neubiberg, FRG

1. INTRODUCTION

Organic polymers from monolayers (~50Å) to thin films (~200 mm) are used in solid state chemical sensors for the purpose of introducing chemical selectivity to these devices. When a chemical species interacts with such thin organic layers by **adsorption** or by **absorption** it changes at least one of its physical properties. Such change is usually very weak and must be amplified by the physical part of the sensor (Table I). The type of this amplifier usually gives the group name to the sensor, i.e. electrochemical, optical, mass or thermal. The method of application of the organic layer and the mode of its coupling to the physical part of the sensor is critical and in many cases it profoundly affects the performance characteristics of the sensor.

There are too many examples of application of organic materials in sensors (Table I) to be covered in one short paper. Changes in mass or the thermal effects are the general attributes of any chemical interaction and therefore the most important issues in the sensors based on those properties are the design of selectivity and sensitivity of the layer. On the other hand the chemical interactions do not necessarily lead directly to a quantifiable change of some electrochemical or optical property and therefore in development of electrochemical and optical sensors attention has to be paid also to the choice of the primary interaction between the species of interest and the chemically sensitive layer.

In this paper the discussion will be limited to the application of organic layers to potentiometric sensors. There are some general facets of this group of sensors which need to be pointed out. First of all the basic requirement of an electrochemical measurement is that the electrical circuit is closed and that all its components function flawlessly in order to obtain information from it. This may look like a trivial statement, however, most of the failures of practical applications of electrochemical sensors are due to some

---

[1] This paper was originally presented at the Solid State Ionics Conference, September 6-11, 1987, Garmisch-Partenkirchen, and will be published in Solid State Ionics (1988).

Y. I. Nissim and E. Rosencher (eds.), Heterostructures on Silicon: One Step Further with Silicon, 273–280.
© *1989 by Kluwer Academic Publishers.*

malfunction of some component in this electrical circuit, not necessarily of the electrochemical sensor itself. The second equally trivial statement relates to the fact that most interactions within electrochemical sensors are ionic in nature but the information is always displayed and processed in the electronic form. This ionic/electronic coupling is an important function of any electrochemical sensor. Thirdly, the signal in amperometric or conductimetric sensors is area dependent (it is generally governed by the electrode of a smallest area) whereas in zero current potentiometric sensors it is not. This fact makes the location of the error in measurement with potentiometric sensors relatively difficult.

Table I
Organic Layers in Chemical Sensors

Type of Sensor	Type of Organic Layer (Thickness)	Example
Electrochemical	polymer + ionophore (100μ)	ions
	gel + enzyme (100μ)	substrates
	semiconductors(0.1μ)	gases,substrates
	LB film (50 - 2000Å)	gases,substrates
	gel (1μ)	humidity
Mass	general polymers (0.1μ)	gases
	LB films (50 - 2000Å)	gases, vapours
Optical	polymer + dye (1μ)	ions, gases
	gel + enzyme (100μ) substrates	
	gel + antibody (100μ)	antigens
	LB film (50Å - 0.1μ)	gases
Thermal	gel+ enzyme (100μ) substrates	

    With the exception of the bilayer conductimetric sensors the organic layer is usually a polymer which is used either as a convenient matrix which accommodates the selective molecule or generates the primary signal itself. Depending on the type and the size of the sensor they are applied by variety of techniques including co-polymerization, electropolymerization, solvent casting, or photolithography in order to form layers of different thicknesses (Table I).
    The above discussion relates to the organic layers as the active elements of the sensor. There is another, equally important function which organic polymers have in chemical sensors, as protective and insulating layers. This means that there are at least three types of interfaces involved in electrochemical sensors: two between the active polymer or the insulating polymer and the physical part of the sensor and one between the active and the insulating polymer. It is known that long term stability and life-time are related to the problems which originate at these interfaces. Generally, the same

processing techniques are used for the application of the insulators and the thickness of these layers usually matches the thickness of the chemically sensitive layer.

2. POTENTIOMETRIC ION-SELECTIVE SENSORS

The Eisenmann-Nikolskij equation (Eq.1) which relates the activity of the ion in solution to the output (potential) of the sensor does not depend the area of the membrane.

$$E_{out} = RT/z_jF \ln (a_j + \Sigma_i K_{j,i} a_j^{z_j/z_i}) \quad (1)$$

This greatly simplifies the fabrication of these sensors. In this equation $K_{j,i}$ is the selectivity coefficient and z's are the valencies. Other terms have their usual meaning. In macroscopic ion-selective electrodes the membrane is prepared by solvent casting (200 - 500 μm thick) separately from the sensor. The segments of such membrane are then mounted in the electrode body [1,2] . The basic ingredients of the membrane are the polymer (e.g. ~30% PVC) the plasticizer (~ 70%) and the ion exchanger or the ionophore (<1%). Thus, the membrane is basically an organic gel which is dissolved in a suitable solvent (e.g. cyclohexanone) for solvent casting. The potential difference which develops across the membrane/solution interface depends, in principle, only on the difference of the free energy of transfer of the ion of interest from the bulk of the solution to the bulk of the membrane. The Debye length in these membranes is < 1μm which means that at the thickness used in the sensors the bulk of the membrane is electrically neutral. There are no hard requirements on the conductivity of the membranes because no net current flows through the membrane during the measurement.

Scaling down of the ion-selective electrodes to the level of ion- selective field-effect transistors [3] leads to several problems. First of all the requirements on the integrity of the encapsulation are higher because the on-chip integrated electronics must be rigorously protected from the effects of electrolytes. Secondly, the ion-selective membranes must be applied over very small and closely spaced areas with the aim of building a multi-ion sensing device. It is therefore desirable to use photolithography for definition of these individual ion-selective areas on the chip. However, attempts to use this approach have been largely unsuccessful because the minimum thickness required for a functional membranes of this type is approximately 100 μm which is beyond the practical limits of photolithography. Furthermore, the interface between the active membrane and of the encapsulant represents an electrical shunt which is also mechanically weak. Because the area of the membrane decreases faster than its circumference with the diameter of the membrane the interfacial shunt becomes more and more serious as the microelectrode becomes smaller. Both the fabrication difficulties of multichannel ISFET and the issue of the mechanical stability and of the electrical integrity of these small membranes has been solved by doping the continuous blank membrane with the electrochemically active ingredients directly over the gate areas (Fig.1) [4]. In this way the formation of the different ion-selective membranes has been simplified and the troublesome membrane/encapsulant interface has been eliminated.

3.POTENTIOMETRIC ENZYME SENSORS

By incorporating enzymes into the organic layer a very high selectivity towards broad range of electrically neutral substrates can be obtained. In that case the organic polymer plays the role of supporting matrix which otherwise does not directly participate in the sensing mechanism. The principal role of the enzyme is to catalyze selectively the chemical reaction. The concentration of the product(s) and of the reagent(s) involved in this reaction within the enzyme-containing layer give rise to the analytical signal. Enzyme field-effect transistor [5-7] is an example of such non-equilibrium sensor in which the response is governed by the steady-state concentration of the electroactive species within the enzyme layer (Fig. 2). This concentration is affected both by the mass transport within the gel layer and by the enzyme mechanism. There is no unique expression which relates the concentration of the substrate in the sample to the output of these devices mainly because different reaction kinetics apply to different reactions under different conditions. Generally, the problem can be formulated as the set of simultaneous second-order partial differential equations of the type

$$\pm \frac{d[A]}{dt} = D_A \frac{d[A]^2}{dx^2} \pm \frac{V_m [A]}{F(H^+) (K_m + [A])} \qquad (2)$$

where [A] is the concentration of a general substrate, D_A is its diffusion coefficient, V_m is the maximum reaction velocity, K_m is the Michaelis -Menten constant and $F(H^+)$ is the pH function.

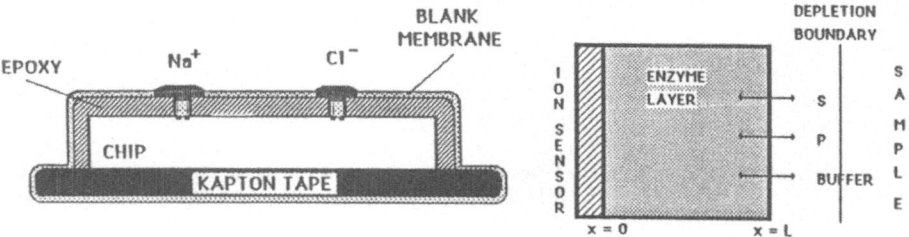

Fig.1   Crossection of a dual-gate ISFET chip prepared by doping two different electroactive compounds to the blank membrane. The diagram is drawn (almost) to the scale.

Fig.2   Schematic diagram of a potentiometric enzyme sensor. S is the substrate and P is the product.

The complexity of the solution depends on the initial and boundary conditions which, in turn, depend on the physical assumptions of the operation of these sensors. The usual assumptions in the order of seriousness are:

1. Linear concentration gradients within the gel layer
2. No pH dependence of the reaction rate
3. No effect of partitioning of species between the gel and the solution
4. No formation of the depletion layer at the gel/solution interface
5. No effect of the capacity of the mobile buffer
6. No formation of the Donnan potential at the gel/solution interface
7. No effect of the gel buffer capacity

The mathematical complexity increases with the elimination of these approximations in the above order. Although they lead to mathematically simple solutions the first two assumptions are clearly unrealistic. The third assumption affects the detection limit but does not seriously affect the general shape of the response curve. Assumption four is acceptable for well stirred solutions. The fifth assumption leads to serious error for sensors operating in the media of low buffer capacity. Assumptions six and seven can be accepted for rationally designed gel matrices. The solution of the resulting set of differential equation yields not only the response curve of this type of sensor but also the information about the optimum thickness of the gel layer and about the minimum amount of enzyme which needs to be incorporated in order to obtain the best response.

4. WORK FUNCTION SENSORS

Chemical modulation of electron work function is a relatively new principle applied to chemical sensing. It relies on the fact that the two principal components of the electron work function, the Fermi level and the surface potential of an organic layer, change when a chemical species is adsorbed onto or absorbed in such layer. In order to make use of this effect the organic layer has to be coupled to the amplifier capacitively [3]. For this purpose we have developed a suspended gate field-effect transistor (SGFET) into which the organic polymer is introduced by electrochemical deposition (Fig. 3). An advantage of this type of sensor is that a wide variety of organic semiconductors with different selectivity can be prepared and applied in a well controlled manner which facilitates the fabrication of multisensors [8,9]. Furthermore, because this is a potentiometric (equilibrium) sensor there is no special requirement on the conductivity of the organic layer. The only pre-requisite is that it makes an electronic contact with the suspended metal gate, the condition which is easily satisfied for electrodeposited layers.

The origin of the chemical signal can be expressed in thermodynamic terms. At equilibrium the number of moles n of all species and their chemical potentials μ in a phase (e.g. in a chemically selective layer) are related through the Gibbs-

Duhem equation

$$\Sigma_i \, n_i d\mu_i = 0 \qquad (3)$$

in which the chemical potential μ_i of species i is defined as the molar change of the free energy ∂G

$$(\partial G/\partial n_i)_{T,P} = \mu_i \qquad (4)$$

Thus, if a new species enters the organic layer the chemical potentials of all species in that layer must change. These include the change of the electrochemical potential of electron - the Fermi level and therefore the electron work function.

The graphical representation of the situation in the gate of SGFET is shown in Fig. 4 where the energy band diagram of the chemically selective layer is shown. In this figure the material is considered to be a p-type semiconductor. The position of the energy level for the dopant and thus the position of the Fermi level in the whole phase depends on the position of the intrinsic Fermi level of the pure material, E_{FI}, on the electron donor / acceptor properties of the dopant and, if the dopant is charged, on the occupational density of the donor states E_D. Therefore, for a n-type material the dopant energy level (donor) would be located close to the conduction band edge, E_c, and the Fermi level would be closer to that edge accordingly. In Fig. 4 the acceptor molecules (i.e. p-type semiconductor) are considered to be charged and therefore their distribution depends on their occupancy. This fact is shown by the symbol for a distribution (   ) in Fig. 4.

The electrochemical potential of electron can be expressed as the sum of electrostatic energy and of the chemical potential

$$\mu = \mu - e\phi_G \qquad (5)$$

where ϕ_G is the bulk (Galvani) energy of the chemically selective layer. Because this energy is referenced to vacuum level it consists of the energy contributions resulting from the bulk potential $\Psi = (\phi_B + Eg/2)$ and from the surface (dipole) potential χ. Thus,

$$\mu' = \mu' + e\chi + e\Psi \qquad (6)$$

The work function of the chemically sensitive layer ϕ_L is then

$$-\phi_L = \mu_L - e\chi_L \qquad (7)$$

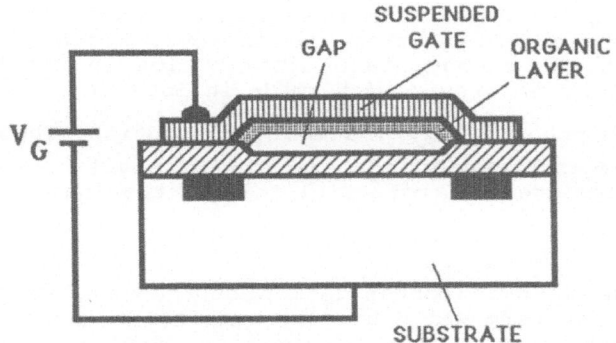

Fig.3   Diagram of the crossection of the suspended gate
field effect transistor (not to scale). The gap thickness is
typically 1000 - 2000Å and the thickness of the organic polymer
is between 500 and 1000Å.

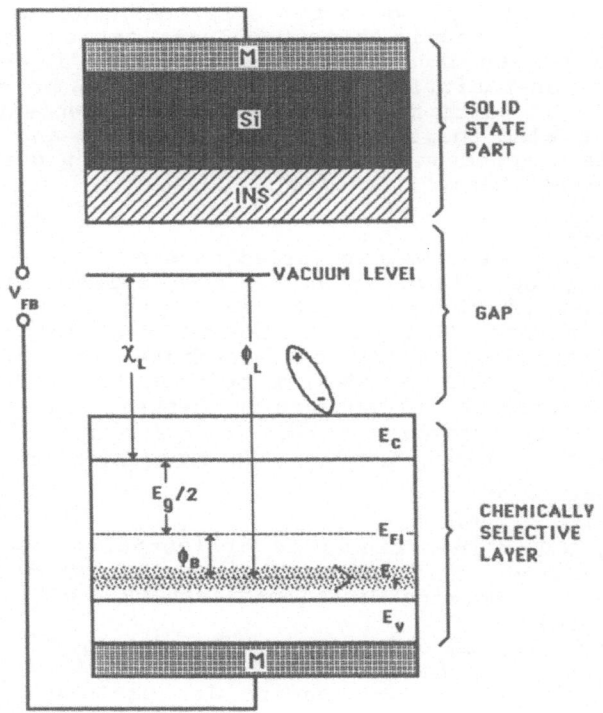

Fig.4   Chemical modulation of the work function ϕ_L of the
chemically selective layer. E_F is the real and E_{FI} is the
intrinsic Fermi level, E_C and E_V are the conduction and valence
bands, respectively and χ_L is the surface potential.

From Eq. 7 we see that the chemical modulation of the work function can originate from two effects: action of the guest molecule on the energy state distribution in the bulk of the phase, i.e. by absorption (term μ_L in Eq.7 or by modulation of the surface potential χ_L , i.e. by adsorption. These two terms have different dependence on the activity of the guest molecule. The chemical potential follows the logarithmic law

$$\mu_L = \mu_L^o + RT \ln a_i^L \qquad (8)$$

whereas the surface concentration depends on the type of the adsorption isotherm which usually has form of a power law. This may, in fact, create some problems in resolving the two types of contributions to the overall change of the work function because the relative degree of their contribution is not known a priory.

An example of organic layer which changes its work function due to chemical modulation is polypyrrole. It can be conveniently deposited on the suspended gate of a field-effect transistor by anodic electropolymerization. Because of its strong hydrogen bonding properties such layers bind water, alcohols, acetonitrile etc. [8,9]. It can be deposited under variety of conditions such as different deposition potential, different electrolyte, different solvents and additives. The result is a spectrum of materials which show different affinity to different chemical species.

5. SUMMARY

In this paper we have tried to show that the choice of the chemically selective organic layer depends on the specific conditions of operation of the type of the potentiometric chemical sensor. Thus the thickness, the electrical properties and the method of deposition is different for different layers. When crossing the boundaries between the potentiometric and other electrochemical sensors or between the different groups of sensors the requirements on the performance and application parameters of the chemically selective layer are even greater.

REFERENCES

1. Morf, W.E. "The Principles of Ion-Selective Electrodes and of Membrane   Transport", Elsevier, Amsterdam, 1981
2. Camman, K."Working with Ion-Selective Electrodes" Springer, Berlin, 1979.
3. Janata, J., in J.Janata and R.J.Huber, Eds.,   Solid State Chemical   Sensors, Academic Press 1985
4. Bezegh K., Bezeg A., Janata J., Oesch U., Aiping Xu, and Simon W. Anal. Chem., 59 (1987) 2846
5. Caras, S.D., Janata, J.,Saupe D.,and Schmidt,K., Anal. Chem. 57 (1985) 1917
6. Caras,S.D.,Petelenz,D.,and Janata,J.,Anal.Chem.57(1985)1920
7. Caras, S.D. and Janata, J., Anal. Chem. 57 (1985) 1924
8. Josowicz, M., and Janata, J., Anal.Chem. 58 (1986) 514
9. Josowicz, M., Janata, J., Ashley, K. and Pons, S., Anal.Chem. 59 (1987) 253

ELECTROCHEMICAL ENCAPSULATION OF SOLID STATE DEVICES

KARIN POTJE-KAMLOTH AND MIRA JOSOWICZ

Universität der Bundeswehr München, Werner Heisenberg Weg 39, D-8014 Neubiberg (F.R.G.)

1. INTRODUCTION

The need for encapsulation of solid state devices, particularly of those which are chronically exposed to chemical environment, has been long recognized as one of the outstanding problems in their use. These devices range from solar cells to chemical and physical sensors and the nature of the chemical environment varies from atmospheric exposure to biological fluids. Additional corrosion problems can occur due to presence of high electrical fields at the device surfaces and in dielectrics which may be as high as 100 kV/cm. The primary purpose of the encapsulation is to prevent the degradation of the materials from which the device is constructed which might result in the premature device failure. Equally important is the aspect of the electrical safety of the devices which are used for medical purposes or in the chemical environment in which the live electrical components could present operational problems such as contamination of the environment with the electrolytic products, explosion hazards etc..

There is no general encapsulation procedure available for all these applications mainly due to the fact that the size of these devices, and therefore the area which has to be protected, ranges from square meters for solar cells to a few square microns for integrated chemical sensors. The materials which have to be encapsulated also range from noble metals to semiconductors. Moreover the device geometry can be quite complex involving corners, cavities, steps etc. sometimes in very small dimensions. Combination of all these factors presents a difficult practical task. Other aspects which have to be considered are the economy, reproducibility and reliability of the encapsulation procedure because they are reflected in the price of the final product.

Encapsulation of chemical and physical sensors is one of the most important and difficult steps in their whole fabrication sequence. The design of sensors which are intended for use in liquids or gases require encapsulation only of the selected parts of the device such as bonding pads, bonding wires and of the header while the active part of the sensor has to be left exposed to the environment which it senses. The need for such "geometrically selective" encapsulation creates a difficult practical problem. Partially successful attempts to use photolithography [1] glass bonding {2] and other semi-automated techniques [3] have been made. In spite of this most encapsulations are still done by hand. Such procedures are non-uniform, time-consuming and therefore not economical. The common encapsulation material used for the hand encapsulation are high-

281

Y. I. Nissim and E. Rosencher (eds.), Heterostructures on Silicon: One Step Further with Silicon, 281–288.
© 1989 by Kluwer Academic Publishers.

grade epoxy resins which require for the curing process two temperature steps; one at the room temperature for about 12 h followed by the 12 h curing at higher temperature.

In this paper we have investigated a generic encapsulation procedure using electrochemically generated insulating precursor which is subsequently thermally cured to form insulating film. Such procedure should be applicable to all electronically conducting surfaces regardless of the dimensions and complexity of the device geometry. The materials which have been investigated were Pt, Au, W/Ti alloy, Al, Ta, Mo and Si because

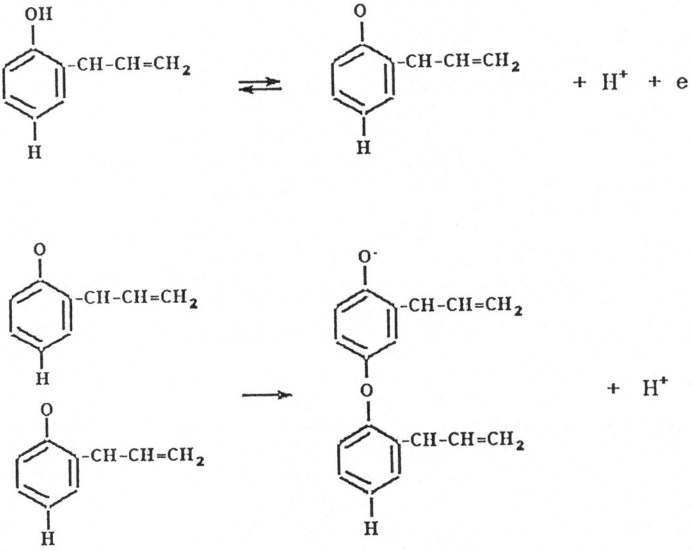

Scheme I

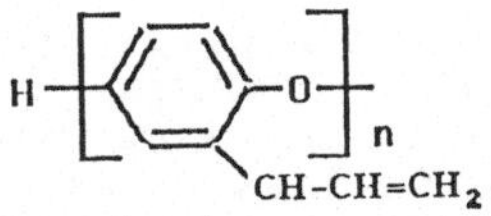

they are found in most solid state devices such as integrated chemical sensors.

The reactions on which this process is based have been described by Mengoli et. al. [4]. They involve electrochemical oxidation of 2-allylphenol in water/methanol/butylcellosolve mixture to yield poly(oxyphenylene) (Scheme I). The electrooxidation is done in the presence of allylamine in order to minimize the competing passivation of the substrate. The further role of allylamine is postulated to crosslink the linear polymer. Typically 2-5% of N is found in the cured polymer [5,6].

The efficacy of the electrochemical encapsulation procedure has been tested on the individual materials listed above and on chemically sensitive field-effect transistors (CHEMFET), in which these materials can be found in a complex geometrical combination.

2. EXPERIMENTAL

2.1.Chemicals

The substrates used for the studies were the cut off gold plated pins (1μm Au) of TO4 headers, Mo, Ta, W wires, p-type silicon chips and 1μm thin layers of Pt sputtered on microscope glass slides pre-coated with an intermediate layer (500Å) of Ti/W for adhesion. Two kinds of suspended gate field effect transistors (SFGET) obtained from the University of Utah [7] have been used; One with all bonding contacts made from Au and the other in which only the suspended gate was from Au and all the other metallization was from Al. All solution experiments were done at room temperature without further thermostatting.

2.2.Procedures

Before deposition the substrates were cleaned chemically in a solution containing 5.0 ml of EDTA (0.1M), 1.0 ml of H_2O_2 (30%) and 0.5 ml of NH_4OH (32%) for 5 min [8]. The poly(oxyphenylene) was electrodeposited from freshly prepared solution containing 0.23 Mol/l of 2-allylphenol (Merck), 0.4 Mol/l of allylamine (Merck), 0.2 Mol/l of butylcellosolve (ethyleneglycolmonobutylether, Merck) in water/methanol mixture (1:1 by volume). The electrodeposition was carried out in a one-compartment cell at room temperature by applying 4 V from a constant voltage power-supply (Zentro-Elektrik, Type LA 15/156 B) between the cathode (a platinum coil of 1.5 cm^2 area) and the substrate (anode). The current was monitored with a Keithley 177 microvoltmeter and recorded with a Metrawatt, Model SE 780 recorder. The electrodeposited films of poly(oxyphenylene) were rinsed with distilled water and cured at 150^0 C for 30 min.. The thickness of the films was measured with a profilometer (Sloan Dektak II). Their resistance was calculated from the slope of the current-voltage relationship measured at zero current crossing. These plots were obtained by applying a slow triangular voltage sweep (150 mV s^{-1}) while monitoring the current (same as in an ordinary cyclic voltammetric experiment) using either Wenking (Model VSG 72) or a EG&G PAR Potentiostat Model 273 in the solution of 2.5 mM ruthenium hexamine chloride $Ru(NH_3)_6Cl_3$ (RuHex), (Aldrich) in 1 M KNO_3. Silver/silver chloride (3M KCl) reference electrode with 1M KCl bridge was used as a reference electrode. The auxiliary electrode was a Pt coil of 1.5 cm^2 area.

3. RESULTS

There are at least two competing processes taking place during the electrooxidation: The passivation of the substrate and the electropolymerization of the allylphenol. Clearly, the fraction of the charge contributing to the one or to the other process depends on the substrate used and so does the time variation of the oxidation current. For noble metal substrates (Au and Pt) the formation of poly(oxyphenylene) dominates (Fig. 1a) while in the case of the other materials the formation of the oxide is the principal reaction. In the latter case no formation of the polymeric film has been observed despite the fact that the current continued to flow. However, in the case of Mo and W substrates a dark brown color has been observed to form close to the surface of the electrode but not adhering to it. At Pt substrate the initial rate of decay of the current depended on whether the electrode was initially etched or not. In order to ensure a reproducible deposition conditions the etching step has been included in the procedure at the beginning.

The thickness of the deposited/cured film has been found to depend on the deposition time (Fig. 2 and 3). It is important to realize that the profilometric trace is obtained by slowly dragging the stylus of the profilometer across the measured surface. In that respect the variance of the profilometric trace provides also some information about the roughness of the film. We believe that the higher roughness of the film obtained after the deposition for two hours corresponds to the more rapidly growing less dense film (Fig. 3). This conclusion is supported by the occurence of the maximum on the curve of the specific resistivity (Fig.4) measured on gold plated pins. Visual microscopic test reveals the presence of crevasses which may be responsible for the fluctuations of the profilometric trace (Fig. 2d) and for lower specific resistance. On the other hand films deposited between 1 and 1.5 h were smooth and virtually crack-free both by the microscopy and by the profilometry (Fig.2b,c). They also show the highest specific resistivity (Fig. 4).

Of course the real test of the insulation integrity is the measurement of the leakage current. It was done in two ways: first in the NaCl solution immediately after the curing of a 2 μm (1.5 h) film and then in various intervals up to four weeks of continuous immersion in 0.1M NaCl. The initial value was 4.5 x 10^{12} Ω cm and after the period of two weeks it decayed to 7.3x10^{11} Ω cm. The second test involved application of the

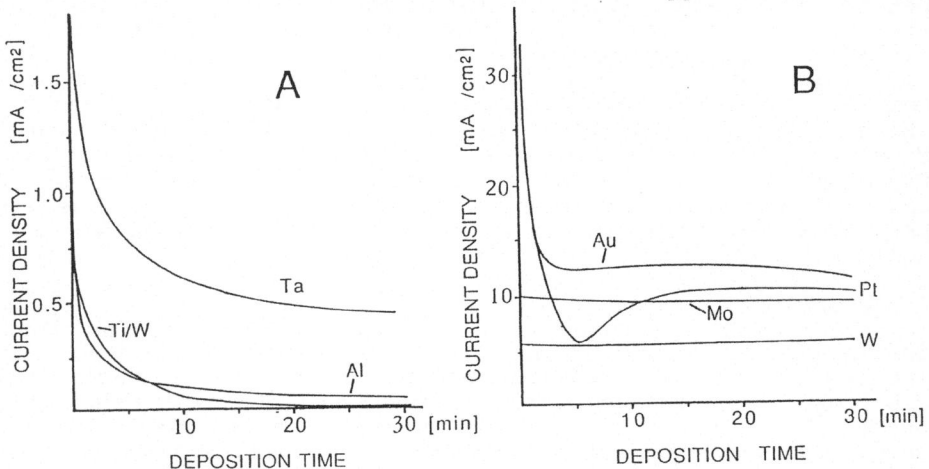

1. Current variation during electrodeposition of poly(oxyphenylene) (A): On Ta and Al wires and on thin films of Ti/W deposited on glass substrate. On (B): Mo, W and Au wires and thin Pt film deposited on a glass substrate precoated with Ti/W for adhesion. The deposition conditions are given in the text.

increasing voltage until the dielectric breakdown has occured.
In this case the etching of the substrate has been found to play
the most important role: Etched samples did not break down
within ± 10 V while the un-etched samples broke down at ± 1.2V
(Pt) and ± 2V (Au), respectively (Fig. 5). This results seems
to indicate the non-uniformity of the deposited film on un-
etched samples. In order to eliminate the possibility of the
high charge transfer resistance near zero current density the
leakage tests were also carried out in the presence of fast
redox couple (ruthenium hexamine). The current voltage curves in
the same solution were recorded both for the encapsulated and
bare Au pin (Fig. 6). Finally, the test of the "dry" insulation
was done by immersing the encapsulated Au pin in the 75.5%
Ga/24.5% In eutectic (Ventron-Alfa Products) and applying the
test voltage. No breakdown was obtained within the range of the
power supply (300V) for a 2 μm film.

Following the studies of the individual materials
encapsulation of the SGFET chips was attempted. As expected a
good encapsulation has been obtained on Au and Pt surfaces while
the Al and Si/SiO2 areas were not covered. This is in fact a
highly satisfactory result considering the fact that it is
possible to choose the metallization in such a way that the Al
is avoided in the construction of the device but used as a
convenient mask should a photographically patternable
encapsulation of the noble metal be desired. The lack of
encapsulation of Si (SiO2) is annoying but not serious because
the electrically live edges of the diced chips are usually
sufficiently far from the active areas of the chip and can be
encapsulated by some other technique. Nevertheless, work is in
progress to develop electrochemical encapsulation procedure also
for oxidized surfaces, particularly SiO2.

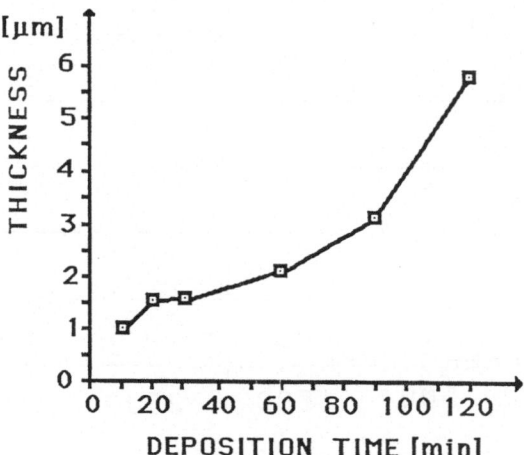

2. Dependence of the poly(oxyphenylene) film thickness on the
   deposition time. The film was deposited on chemically etched
   Pt substrate and cured at 150° C for 1h.

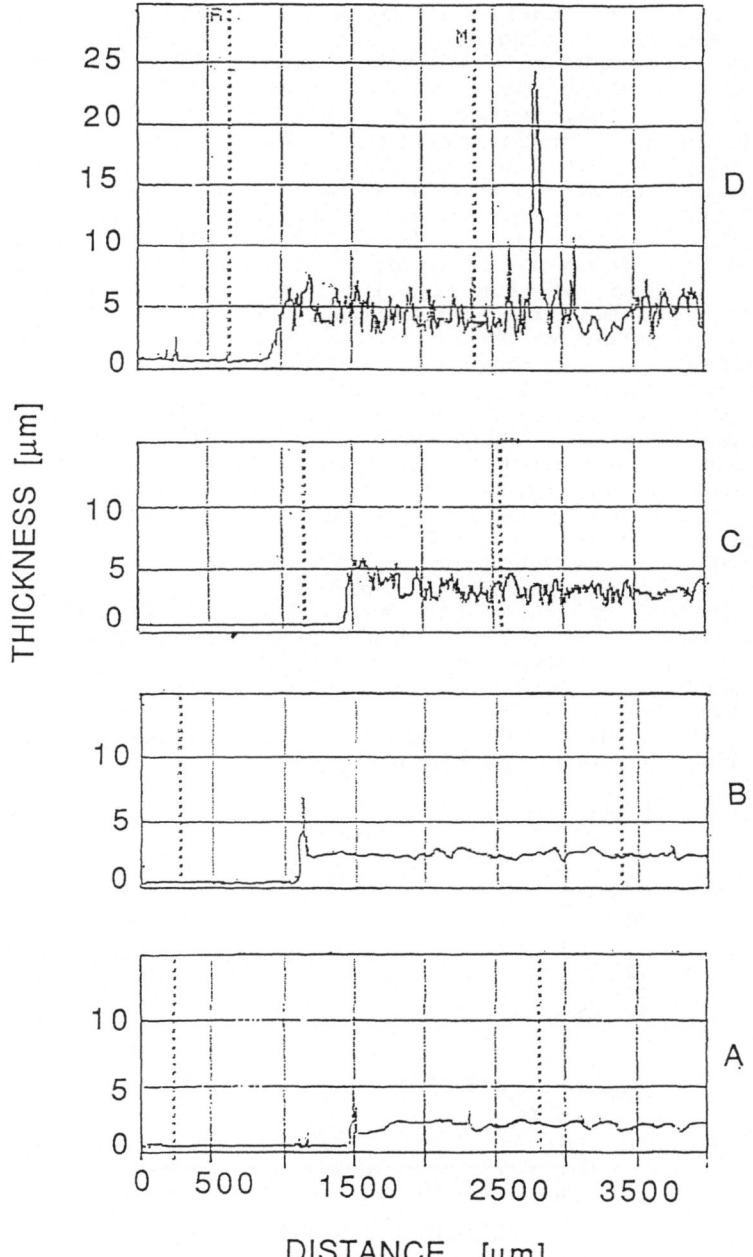

3. Profilometer traces for films of poly(oxyphenylene) deposited for (A) 30 min; (B) 1h; (C) 1.5h and (D) 2h under the conditions of Fig. 2

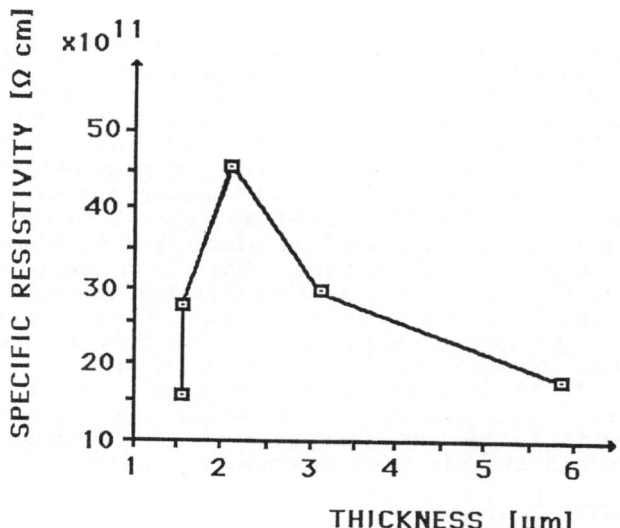

4. Specific resistivity of the poly(oxyphenylene) film as a function of film thickness. The film was grown and cured on an Au pin.

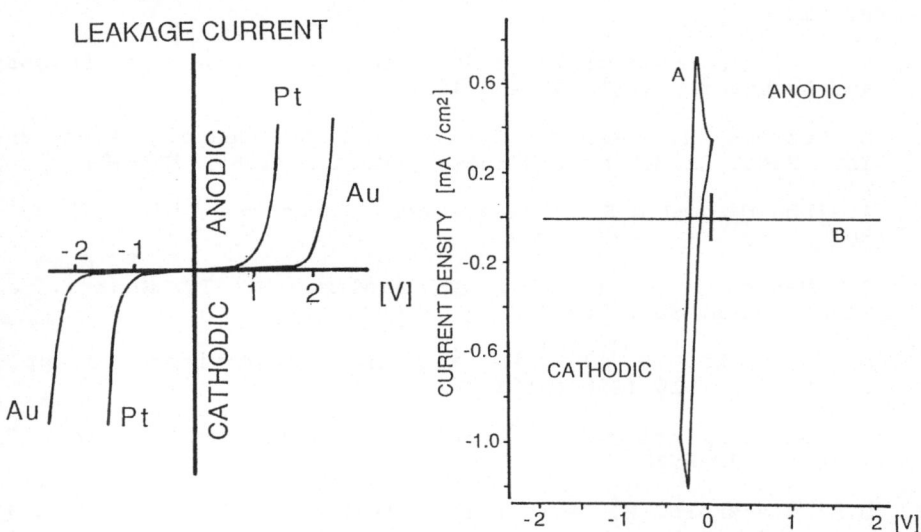

5. Breakdown leakage characteristics of the 2 μm film deposited on un-etched Pt and Au wires. The breakdown of films on etched substrates did not occur within ± 10V.

6. The blockage of the electrochemistry of ruthenium hexamine by the 2μm poly(oxyphenylene) film. The solution was 2.5 mM Ruhex in 1M KCl. (a) bare Au electrode and (b) same electrode encapsulated with 2μm film of poly(oxyphenylene).

4. DISCUSSION
It has been shown that partial, electrochemically generated encapsulation of solid state devices is possible. It works on the surfaces at which the electropolymerization reaction occurs preferentially as compared to the passivation of the substrate. It is possible to grow films up to 5 μm thick however, the specific resistivity decreases beyond approximately first 2 μm (Fig.4). Even though it seems to be advantageous to deposit thicker films because of the added mechanical protection which these films provide to the bonding wires and to other delicate features at the device surface. The long term continuous exposure to aqueous electrolytes degrades the overall resistance somewhat but the final value is still sufficiently high for most applications. The new encapsulation procedure is particularly useful for the devices which may be difficult to encapsulate otherwise due to their complicated geometry. The dry insulation is particularly outstanding.. The major application of this process appears to be for encapsulation of devices operating both in gases and liquids such as chemical sensors.

ACKNOWLEDGEMENT
The assistance in this project from J. Janata is acknowledged. This work was supported by the BMFT Grant Number AS 00585/88.

REFERENCES

1. N.J.Ho, J.Kratochvil, G.F.Blackburn and J.Janata, Sensors and Actuators, 4 (1983) 413-421

2. M. Decroux, H.H. Van den Vlekkert and N.F.DeRooij, Proc. 2nd Int. Meet. Chem.Sens., Bordeaux, France (1986) 403-404

3. A.Sibbald, and J.E.A. Shaw, Sens. Actuators 12 (1987) 297-301

4. G. Mengoli, P. Bianco, S. Daolio, M. T. Munari, J. Electrochem. Soc. 128 (1981) 2276-81

5. G. Mengoli, S. Daolio and M. M. Musiani, J.Appl. Electrochem. 10 (1980) 459-471

6. G. Mengoli, and M. M. Musiani, J. Electrochem. Soc. 134 (1987) 647C-652C

7. G.F. Blackburn, M. Levy, and J. Janata, Appl. Phys. Lett. 43 (1983) 700-703

8. M.Josowicz, J.Janata and M.Levy, J.Electrochem.Soc. 135 (1988) 112-115

SILICON ON INSULATOR

D. BENSAHEL

Centre National d'Etudes des Télécommunications, BP 98, 38243,
Meylan – Cedex, France.

1. HISTORY

Bulk – Si technology is the technology which has obtained the most
impressive results in the last 30 years. It is a mature and well established
technology. However, as the demand for high speed, high density circuits
increases, the technological processes are becoming more and more sophisticated.
New solutions must be found to solve the problems resulting from reduction in
size and the physical effects of small geometry devices. Silicon On Insulator
(SOI) technology is an old dream which has existed since the beginning of bulk
technology as this structure can naturally offer some clever answers to the
critical problems encountered in bulk – Si technologies. What is SOI ? It is the
concept of a thin monocrystalline layer of silicon on an electrically insulating
substrate. This give rise to two questions: (1) How do we obtain the thin
silicon layer ?, (2) what is the insulating layer ?. Unfortunately, these
questions are linked and, up to now, it has not been possible to grow or deposit
monocrystalline silicon layers on an insulating amorphous substrate.

However, several attempts have been made, i.e., by direct epitaxy of
silicon on a crystalline substrate whose lattice parameter is that of or close
to silicon. From a material point of view, CaF2 or Al2O3 (Saphire) are good
candidates (<1% and 4% lattice mismatch with silicon, respectively). But they
have severe drawbacks:

– Although monocrystalline CaF2 is obtained with a reasonable good quality in
large diameters, it is an expensive substrate. So far, its use is still
restricted to Ultra High Vacuum systems. Moreover, it cannot be treated in
hydrogen ambient for CVD deposition as it reacts, giving gaseous HF.

– On the other hand, Saphire does not suffer from the preceding problems and
has been developed through the so – called SOS (Silicon On Saphire) technology, in
which silicon is epitaxially deposited on the substrate. However, this
technique, which is still at the industrial stage, has not yet replaced the
bulk – Si technology as it is an expensive technology with limited process
developments. For example, the wafer size is limited to 3", and thermal
processes are limited in order to avoid Al diffusion from the substrate to
silicon. In addition, many crystallographic defects remain in the layer and at
the Si/Saphire interface.

As stated above, the insulating layer must fit the trends, and Quartz and
SiO2 are thus the best candidates. Since quartz is an expensive substrate and
has a thermal expansion coefficient very different from that of Si, its
applications have been limited to low temperature processes where moreover, the

Y. I. Nissim and E. Rosencher (eds.), Heterostructures on Silicon: One Step Further with Silicon, 289–301.
© 1989 by Kluwer Academic Publishers.

quality of deposited Si is not of prime concern (flat panel applications). SiO2 either deposited or thermally formed has been, still is and will always be THE insulating material of Si technology and its derivatives. Among its different advantages, we may note its mechanical and thermal stability, its flow behaviour when doped, its easy etching properties, the large amount of knowledge accumulated on the Si/SiO2 interface... The SOI structure thus developed is obtained by a silicon layer on a deposited or thermal oxide on bulk monocrystalline silicon.

– How can the upper silicon layer be formed ?. Since the underlying oxide is amorphous, we can no longer think of using CVD processes as seeds are not present. However, by patterning the oxide, we can create an artificial grating which will influence crystallization. This technique, named Grapho – epitaxy /1/, has given some encouraging results but the density of the remaining defects such as stacking faults and sub grain boundaries is too high for practical applications.

SOI studies therefore slowed down until the development of the cw laser annealing experiments initiated by Gat et al. /2/: in his experiment, a polysilicon film deposited on an oxidized Si substrate is locally melted and recrystallized by the focused spot of an Ar+ laser. At the trailing edge of the spot, the film recrystallizes. This technique is known as Zone Melting Recrystallization (ZMR) and its consequences have stimulated all SOI activity.

Due to the new interest in SOI structures, several other techniques have also merged after several years of basic research. Two of them are now in competition. The first is the so – called SIMOX method and the second the Si – porous approach. They both aim to obtain an oxide buried layer in monocrystalline silicon. SIMOX means Separation by IMplantation of OXygen, i.e., a large oxygen dose is implanted in bulk – Si and further annealed. This technique is conceptually the easiest and the most attractive for the industrial community. However, the oxygen dose is in the E18 At/cm2 range which can only be obtained in a huge implanter.

The porous – Si approach is a derivative of the electrochemical etching of Si. In an electrochemical cell filled by HF and water, under certain current and voltage conditions, bulk – Si becomes porous. Since this structure develops a high surface/volume ratio, it can be seen as an apparent high speed oxidation technique. SOI obtained by this technique thus necessitates the formation and oxidation of a buried porous – Si layer below a monocrystalline upper Si layer.

All these techniques are described in more detail below, after which, we will discuss the different advantages, drawbacks, and trends of SOI technology.

2. ZONE MELTING RECRYSTALLIZATION (ZMR)

In ZMR, a molten zone is scanned across a polycrystalline film to produce large grained or single crystals. Molten zones can be created by several methods: for the treatment of a whole wafer in one scan, the heating systems used are either graphite strip heaters or line – focused sources /3/. For a localized treatment, a laser spot from a cw Ar+ laser is now commonly used. In these systems, heating is mostly radiative and the physics involved is expected to be the same. Moreover, for large molten zones, the reflectivity increase in

Si at melting renders the radiative process very stable and therefore easy to control. SOI technology requires thin films (<1 um) and the recrystallization of such thin films is more difficult than "classical" bulk growth. In the following we describe several phenomena which have been studied or which still remain questionable.

2.1 Melting front

In Si – bulk growth in the liquid phase, any morphological and crystalline memory is erased and the quality of the film is independent of the intrinsic melting process and initial characteristics of the material. In the thin film case however, it is not the same as instabilities occurring in the melting front can be detrimental to the recrystallized films. Indeed, in the melting front, "Explosive Melting" /4/ occurs: the solid/liquid interface does not advance continuously at the same speed as scanning, but occurs in a succession of jumps or bursts. Each burst involves a partial melting of the film.

Two assumptions are made to explain the explosive behaviour in thin films: (i) the spread in melting temperature of the various grains, (ii) the superheating in polycrystalline silicon. It can therefore be shown that explosive melting depends on the film thickness and the Si/SiO2 interface properties. Schematically, explosive melting has been explained as follows: when the molten zone is scanned, liquid Si reaches non – transformed Si and the grains with a low melting temperature suddenly melt, producing a burst. Resolidification then occurs, seeded by grains with a high melting temperature. This zone stabilizes the melting front for a moment at a high temperature, after which a new burst is generated. This behaviour has dramatic consequences in the case of SOI by ZMR. Since it is a partial melting, each new burst creates a free volume (silicon contraction upon melting), and a certain amount of Si remains solid and acts as a pile which supports the cap. Thus, everything depends upon the cap behaviour. A thick cap does not flow nor lose its shape when the liquid reaches the burst zone. Consequently, bubbles are created. Depending on the extent and precise nature of the bursts, and on the wetting conditions between Si and SiO2, these bubbles will have different futures. Under poor wetting conditions, the bubbles will extend into the whole molten zone, resulting in complete dewetting of the film. It is this effect which has slowed down the studies on ZMR by SOI for one or two years. If the wetting conditions are less severe, dewetting can be stopped by the unmelted grains, resulting in the classical small "voids" encountered in ZMR by graphite, lamp heaters e – beams or cw lasers. These voids are a few micrometers in lateral dimensions and are surrounded by an area in which the Si film is thinned. For example, voids have been reported in laser – ZMR under certain scanning conditions. It can be demonstrated that in this case, voids are also due to the behaviour of the expance of molten zone as it crosses the SOI and seeded zones, i.e, zones were the liquid is in contact with the seed regions (in solid Si).

Fortunately, several ways have been developed to control and avoid the "void" phenomenon: they include (i) good control of the lateral gradients, (ii) control of the polysilicon quality, and (iii) control of the flowing conditions of the cap layer depending on the time and space scales.

2.2 Melting zone

Since liquid silicon has an increased reflectivity coefficient, in – situ observations and monitoring of the melting area can be easily carried out. However, since several effects are present in the melting and solidification fronts, a certain width of molten zone must be defined. On the other hand, too wide a molten zone will lower the power window as melting of the underlying bulk – Si must be avoided. Moreover, the width of the molten zone must be adjusted while keeping in mind the thermal gradients generated at the two fronts.

2.3 Solidification front

In unseeded Si ZMR, the films have a <100> texture since (100) planes correspond to a minimum of interfacial energies between Si and SiO2. However, in the plane, all orientations are possible which result in the existence of Grain and SubGrain Boundaries (GBs and SGBs). This is a general phenomenon whatever the scanning and thermal gradient conditions. In this latter case, it has been shown that the crystallographic nature of the GBs and SGBs are very dependent on the thermal gradient occurring along the molten zone. No clear behaviour has been established and according to different authors, the spacing of the defects can either decrease or increase with an increasing thermal gradient. However, all the authors agree that low scanning (0.2 mm/sec.) and low thermal gradients (<4°K/mm) (i) cause the defects to disappear (especially GBs) and/or (ii) decrease the misorientation between grains resulting at best in isolated threading dislocations whose density can be in the E+5/cm2 range. However, one fundamental point still remains unanswered: where do the defects come from ?.

It is generally admitted that the defects arise at the point where the silicon freezes last, i.e., for example at the re – entrant corners of a faceted front or at the cusps of a cellular front. Cellular growth has generally been ruled out as the impurity content seems too low. On the other hand, a purely faceted model has been simulated. However, it cannot explain all the experimental effects observed. As suggested recently, the breakdown of the solidification front is probably a convolution of the effects described below.

2.3.1 Silicon reflectivity change: Due to the reflectivity increase of Si on melting, solid and liquid can coexist throughout a range of incident radiation power. This is the "lamella" effect first described by Bosch and al /5/. This effect will be dominant at low scanning speed where the morphology of the liquid/solid interface is qualitatively the same in the solidification front as inside the molten zone.

2.3.2 Impurity rejection: Since the theoretical values for constitutional supercooling can be achieved in Graphite – and Lamp – ZMR, impurity rejection cannot be ruled out even for low scanning speeds. Moreover, this can be observed in – situ; the solidification front exhibits subgrains which are more or less faceted but are separated by deep cusps of liquid Si. In this case, if the scanning speed is increased, the deep cusps become unstable and pockets of liquid Si can be trapped by the solidification front.

2.3.3 Effect of thickness fluctuations: After ZMR, the Si surface always presents undulations running roughly parallel to the scan direction. The amplitude of these fluctuations increases as the thickness of the SiO2 cap is reduced and is dependent on the thermal gradient existing at the Si/SiO2 interface. On the other hand, it must be remembered that the melting temperature of a thin SOI film depends on its thickness. Thus, the following circular scheme was proposed where one effect increases the other until the front breaks down.

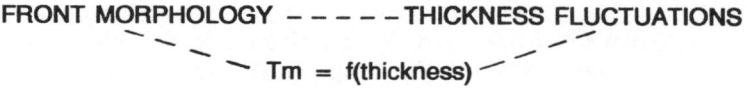

FRONT MORPHOLOGY – – – – –**THICKNESS FLUCTUATIONS**

$Tm = f(thickness)$

2.4 Defect localization techniques

Since recrystallized materials generally present defects, several studies have been devoted towards their localization. Taking into account the results presented above, different techniques have been presented.

2.4.1 Graphite and Lamp ZMR – The patterned structures: At the beginning of these kinds of studies, the aim was to use the optical properties of the thin layers, i.e., reflective and anti – reflective conditions /6/. Moreover, this should have the advantage of patterning the upper layer (above the SOI layer), which can be removed after recrystallization. Nitride and nitride plus oxide layers have been successfully used in laser – ZMR for their anti – reflective properties, while carbon and silicide have been used in graphite and lamp – ZMR for their reflective properties /3/. The choice between one effect or the opposite effect was correlated with the molten dwelling time.

However, in graphite and lamp – ZMR, numerous studies have been devoted towards the heat – sink effect. This effect is due to differential heat dissipation and occurs when steps exist in the SOI structure prior to recrytallization. The easiest pattern is a network of stripes, regularly spaced /7/. In a typical application, with a 40 um periodic network, 4 um in width are devoted to the defect zone, leaving 36 um of defect – free stripes. Indeed, since this network can be included in the mask set of the circuits, we can design the elementary elements to be located inside the future defect – free zones. By patterning any layer of the SOI by ZMR structure, reliable results can be obtained.

– For cap and polysilicon modulations, it is the 4 um wide stripes which are thinned, leaving defects roughly localized in the middle of the 36 um stripes, or in the 4 um stripes for cap and polysilicon modulations respectively. However, the efficiency of localization is not as good as required for device applications.

– The best result is obtained with the patterning of the underlying insulating oxide. In this pattern, it is the 4 um stripes which are thicker than the 36 um stripes. The defects are precisely and accurately localized on the 4 um steps. However, since this latter technique implies the patterning of the underlying substrate, this network follows what happens in the substrate. Unfortunately, the Si substrate undergoes large stress due to the thermal gradients which generate glide planes in the <100> directions. For a 4" wafer, it leads to a shrink of roughly 20 um on the diameter perpendicular to the scan direction. The localization network must therefore be designed to take this shrink into account. Note that the SOI layer itself is not concerned by this effect and the recrystallized film is free of stress.

From a practical point of view, even if the defect localization techniques described above can eliminate the effects of the remaining defects, the designers are very reticent to use them as they imply that the period network dictates the location of the circuits. In fact, most future circuits will be designed through large, sophisticated libraries which will not be modified for the sake of time and money !!! Thus, unless the requirements is for specific applications in which need for absolutely defect – free areas is essential or for circuits of which the design is genetically periodic (Gate Array...), the need for SOI wafers by ZMR will be directed to plain wafers with the lowest defect density as possible. This is now being achieved in several laboratories and factories /8/.

2.4.2 E – beam and cw laser ZMR – The seed problem

In e – beam or laser ZMR, the scanning speed must compensate for the small molten zone extension resulting from the small spots used. Since the thermal gradients are very high in these cases, ZMR is essentially dominated by the solidification front which is faceted and tends to follow the geometrical shape of the molten line or spot. Two constraints are then encountered.

– Since crystallization proceeds at high speed (several cm/sec.), the orientation cannot practically be kept after 60 – 100 um (stability and reproducibility of the spots and the scanning systems).

– A defined crystalline orientation must be given at the beginning of recrystallization as the interfacial effects cannot be taken into account.

Usually, a seed is open in the underlying oxide. In this case, the problem again exists of the presence of a molten zone (SOI area) with a solid zone (substrate bulk seed area) resulting in the existence of voids (see above). It is therefore necessary to carefully study the thermal behaviour of the spot area which covers both the seed and the SOI area. Moreover, the seed area must not be too large if we want to seed the SOI zone, i.e., too high a power will melt the seed but will be too high for the SOI adjacent zone. Several studies /9/ have been developed in this way and the conclusions are the following: (i) the best spot must be either elliptical (laser case) or such as to induce a pseudo – line (e – beam case), (ii) the smaller the seed surface (either continuous (1 – 2 um wide) or discontinuous (1.5 um wide spaced by 1.5 um)), the larger the power window, (iii) if the seed is filled by a Selective Epitaxial Process, the window power is increased, (iv) the best results are obtained with a periodic network structure, (v) the scan strategy must minimize the ratio between the seed and the SOI areas, (vi) as in the preceding cases, the flowing properties of the cap are important especially as they are related to the scan strategy: in some cases, a non – flowing cap (thin nitride plus oxide layer) will behave better than a flowing cap (thick undoped or doped oxide).

2.5 Conclusions

ZMR has been the pioneer technique for obtaining SOI structures. 4" and 5" wafers are now available in all the techniques which have been initiated. These results have been obtained as a lot of fundamental studies have been carried out in order to understand the physical phenomena involved in the recrystallization of thin films whatever the scanning speeds, thermal gradients, sample structures... Below, we will discuss how this technique has to compete now with the other techniques which have been developed in the shadow of this ZMR technique.

3. OXIDATION OF POROUS – Si

Porous – Si is a very sophisticated material which has over the last five years been of considerable interest /10/. Porous – Si is obtained by an electrochemical technique: the Si substrate is made like an anode inside a cell whose electrolyte is HF + H2O. A continuous current is established between Si and a cathode. Depending on the current and voltage, several regimes can be observed. The most well – known is the electropolishing and arises for low HF concentrations. For a given set of parameters, small pores are formed inside bulk Si. All the studies have been concerned with the size and distribution of the pores according to the starting material (N or P type), doping level, and electrolyte concentration.

The main advantage of porous – Si is its apparent high rate of oxidation due to the high surface/volume ratio (in the 300 m2/cm3 range). This allows oxidation of thick porous layers in a short time. On a micronic scale, porous – Si is a homogeneous material, made by crystalline silicon and voids. The best characterization tools are therefore Transmission Electron Microscopy and gas adsorption isotherms. Several kinds of pore networks can be found: either they can be very homogeneously distributed across the porous layer with an average diameter, or on the contrary, they can be entangled with varying diameters across the porous layer.

3.1 pore behaviour as a function of substrate doping The starting bulk – Si material is the principal parameter controlling the nature of the pore network. Porous layers are easily obtained on a P type Si substrate of any resistivity. It has been indeed demonstrated that the carriers involved during the anodic dissolution were holes: consequently, porous material can be obtained by tunneling effects on N + materials, whereas for N type Si, porous layers can be obtained only under illumination or under high voltages. However, the pore network is different in each case of substrate doping:

– In P + and N + substrates, the pore network is equivalent and characterized by an anisotropy following the current lines, the average pore diameter being in the 20 – 100 A range, the porosity varying between 30 and 70% according to the preparation conditions, where porosity is the ratio between the vacuum volume and the silicon (i.e., the reverse of density).

– In P substrates, the pores are very small (less than 20 A), with a very fine texture. The specific surface is high (600 m2/cm3) with a porosity between 40 and 60%.

– In N substrates under illumination or high voltages, the pores are large (1000 A), but the porosity is generally lower than 10%.

3.2 Other material characteristics: The HF concentration and the current density are the other parameters which define the porosity of the porous – Si /11/. Briefly, (i) a decrease in the HF concentration leads to the formation of a higher porosity and (ii) at a fixed HF concentration, the porosity increases when the current density is increased.

The crystalline quality depends first on the doping level of the substrate and secondly on the porosity. On P + and N + materials, when the porosity is of the order of 30%, a monocrystalline behaviour is obtained. On the other hand, porous – Si on P type is no longer monocrystalline. In the latter case, the specific surface is large and it is supposed that oxygen is trapped on this surface and involves an amorphous contribution of SiO2 in the X – ray diagrams.

3.3 Mechanism formation of the pores: No definite interpretation has yet been given. It has only been established that the electrochemical reaction exchanges 2 holes for one silicon dissolved atom according to:
Si + 2 (h+) – – > Si+ +. Since H2 has been detected at the beginning of the porous formation and after stopping of the electrochemical reaction, it has been proposed that a chemical reaction and the electrochemical reaction occur altogether. However, the question is: How can the anodic dissolution proceed while leaving the silicon unattacked ? A model has been proposed which takes into account the charge transfer at the semiconductor/electrolyte interface. Roughly speaking, the thickness between two remaining Si walls is related to the depletion layer of the silicon. This depletion layer is due to the Schottky contact between Si and the electrolyte and it is thus supposed that the pore formation stops for lack of carriers. This explains certain observed trends but still cannot account for the evolution of pore size according to formation parameters such as current and HF concentration.

3.4 Oxidation of porous – Si: Si migration along the surfaces of the pore walls helps bring out the origin of the thermal instability of the as – prepared material. When heated at moderate temperatures (>400 °C), the original microstructure of porous – Si coalesces leading to the formation of large cavities and thick Si blocks. This prevents any thermal treatment and therefore limits the usefulness of the material. However, structuring can be avoided by a mild "pre – oxidation" at 300 °C in dry oxygen before any restructuring takes place.

Preoxidized layers of porous – Si can be completely oxidized at relatively low temperatures (800 °C) as there is a high surface/volume ratio in the porous layers. However even if the porosity is 56% (ideal porosity needed to accommodate the volume change resulting from the Si – >SiO2 transformation), the quality of the oxidized material at 800 °C is still very different from that of standard SiO2. A densification step at a higher temperature is necessary to induce the thermal viscous flow of silica and obtain an oxide equivalent to thermal SiO2. Generally, very good results are obtained at 1050 – 1090 °C in wet oxygen followed by a nitrogen annealing.

The densification of the oxide does not involve extra stress in the structure. The residual stress is that of the thermal constraints existing between silica and silicon. However, since these constraints increase rapidly with the oxide thickness, this effect can limit the maximum oxide thickness that can be formed by this technique.

3.5 SOI structures tested:
– 3.5.1 Epitaxy on porous – Si: Since the surface of a porous – Si structure prepared on highly doped materials is monocrystalline, it is possible to grow an epitaxial layer on it in order to prepare a SOI structure. However, this epitaxial step must be made at a relatively low temperature in order to avoid coalescence of the pores. Although this technique seems the easiest, it suffers

from the surface preparation prior to epitaxy (surface cleaning). In the past, this epi step was performed by Molecular Beam Epitaxy /12/, but in future, the new developments observed in low temperature cleaning and epitaxy procedures make this technique very attractive.

   – 3.5.2 The N/N+/N structure: Since the formation of porous – Si is selective, an N+ buried layer can be transformed into porous – Si without affecting the upper layer. Originally, this layer was made directly by implantation. Now, a better way /13/ is (i) first, to make an N+ implant in an N wafer, (ii) secondly, to grow epitaxially an undoped layer which will be the SOI layer. One drawback of this method is the possible doping of the SOI layer by N+ impurities from the buried layer during the oxidation steps. However, it has recently been proven that this effect can be minimized.

   In the porous approach, plain wafers of SOI cannot be obtained as accesses to the buried N+ layer are needed. This process is therefore an in – line process, i.e., it must be included in the process flow. However, it should be noted that it does not involve any design constraints for the circuits provided that maximum size in one direction is less than 150 um. This distance is around the maximum that can be obtained in order to join two adjacent accesses to the buried N+ layer which is transformed into porous – Si and oxidized.

3.6 Conclusion

   Porous – Si is the technique which has been the most difficult to handle. However, although this technique has been the latest to develop, requiring several fundamental studies, the knowledge accumulated is being now rapidly transferred to promising new technological routes.

4. OXYGEN IMPLANTATION

   For people familiar with silicon processes, SOI by oxygen implantation is just one particular technological step in a well known technique. This SOI technique in fact requires only an oxygen implant plus annealing. However, even if the basic physics involved is very classic, these two steps are not common: the implant step is at doses in the E18 At/cm2 range, and the annealing is at temperatures higher than 1300 °C.

4.1 Implantation step: The implanter required for such high oxygen doses is not a classical implanter. It is a huge machine which has to date been developed in only one factory as several sophisticated problems must be solved: absence of contamination, homogeneity of the flux, reasonable implantation time, erosion of the source... Moreover, the wafer must be maintained at above 500 °C during the implant in order to avoid amorphization of the substrate which after annealing would lead to the formation of a polycrystalline top Si layer /14/. This temperature can be obtained either by beam or external heating but the best results have been obtained with external heating by lamps. The energy of the implant is limited to less than 200 keV, and this limits the thickness of the SOI layer.

4.2 Annealing step: The annealing is used (i) to eliminate the defects created by the implantation, and (ii) to allow the dissolution of all SiO2 precipitates and the segregation of the implanted oxygen into the buried layer.

Two types of defects are found in the so–called SIMOX structures: dislocations and SiO2 precipitates. Total dissolution of the oxide precipitates is obtained by high temperature annealing while the formation of SIMOX substrates with very low dislocation density is dependent on the implantation conditions.

– 4.2.1 Formation of the oxide buried layer: When oxygen is supersaturated in Si, internal oxidation occurs, giving SiO2. The change in free energy is (classical theory) mainly a function of the diameter of the oxide precipitates and the Si/SiO2 interface energy. Two major consequences follow: the existence of a critical radius and the coalescence process. Below the critical radius, precipitates dissolve. Above this value, they remain stable and increase in size with increasing temperature. By coalesence, oxygen migrates towards the precipitate which has an infinite radius and which has become the buried oxide layer. Hence, in time, we obtain the complete segregation of oxygen into the buried layer.

– 4.2.2 Low defect density SIMOX structures

Two sources of dislocations are present in the SOI layer: defects created by the implantation process, and interstitial atoms created during the oxidation of Si. Several methods /15/ are under development in order to eliminate these defects. One technique uses a channeling implantation at a very low dose rate. In a second technique, several implantation and annealing steps are sequentially performed. In this method, the formation of defects is avoided by the sequential dissolution of SiO2 precipitates before they reach a critical concentration. The third method proposes the formation of cavities in the upper Si layer during implantation. These cavities will act as sinks for the interstitial Si atoms and will prevent the dislocation formation. The remaining dislocation densities in these three techniques is in the E4–E5 /cm2 range.

4.3 Conclusion

No specific physics has to be developed in the SIMOX approach. This appears a very attractive technique for the industrial community. Moreover, it is an off–line process, i.e., plain wafers are available. However, whether or not the process conditions defined above can be met in an industrial context represents the main drawback with this technique.

5. POTENTIAL OR PROVEN ADVANTAGES OF SOI

Advances in new SOI materials have been coupled with device characteristics whose performances are inherently better than bulk–Si CMOS circuits. This has opened several possible windows for new SOI techniques such as a commercial mainstream ULSI contender. Many advantages can be envisaged with SOI technology. Briefly, they include:

5.1 Higher performance:

With SOI, there is a reduction in parasitic capacitance surrounding the intrinsic transistors and a complete removal of the latch–up phenomenon (parasitic thyristor effect between two adjacent transistors). It is generally admitted that with SOI, the gain is equivalent to a generation in process evolution.

5.2 Thin film devices: Since thin films (< 1500 A) of SOI can be obtained, the electrical field distribution in such a device can have some beneficial aspects. A totally depleted device can now be made with the following advantages which will not be detailed here: reduction in short channel effects, improved sub – threshold slopes, increased saturation current, reduction in hot electron degradation, elimination of kink effect, reduction in crosstalk and susceptibility to Single Event Upset...

5.3 Simpler processing: With SOI, isolation techniques (which are sophisticated in bulk and increase the cost) are avoided. Other steps are simpler or are eliminated: mesa etching, contact to active areas and gates, planarization... Another great advantage is that SOI technology will use standard processing equipment and will follow the trend towards large wafer diameters.

5.4 Higher packing density: This is mainly due to the absence of LOCOS bird's beak. But other area savings are due to the availability of overlapping contacts. However, we must recall that ultimately, any area saving will be dictated by the interconnect levels as in bulk – Si technology.

5.5 Ease of design: This is a new field since the design community is not used to thinking with an SOI mind. The gain is still to be scaled, but some considerable effects can be expected: with no – latch – up, the circuits can be placed anywhere, non nested contacts can be used... Simulation will also be easier with the lowering of the parasitic diodes and transistors.

All the advantages listed above will promote synergy and consequently improvements in turnaround time of the tested circuits and technologies.

5.6 Devices options: Although the advantages listed above rely mostly on extrapolation from bulk technology, SOI opens up the possibility of new kinds of circuits and concepts.

5.6.1 The 3D circuits: 3D or stacked circuits were proposed during the earlier stages of the laser – ZMR experiments and some impressive results have been obtained: for example, a 3 layer stacked structure in which each level has a specific function was fabricated and successfully tested. However, the main drawback of this technique does not stem from the material aspect (several stacked layers of SOI), but from the very high number of technological steps which lowers the yield and increases the cost of such circuits. This technique must therefore be restricted to very specific circuits which cannot be made in bulk – Si.

5.6.2 Mixed functions: This chapter will be short as everything still has to be achieved. The idea is to mix circuits of a different nature on the same chip: one of the possibilities is to mix analog and digital circuits in order to have "intelligent devices" such as smart power circuits, lateral bipolar devices mixed with CMOS ones...

6. SOI PROSPECTS

Up to now, SOI technology has been restricted to its specific and native military and spatial applications, which are niche markets. With the development of the new techniques described above, SOI can now create several civilian applications and may represent a serious competitor to bulk – Si in the ULSI challenge. The SOI material techniques described are in their development stages. Each of them has unique advantages. Unfortunately, each technique also has its drawbacks, and no technique is expected to be THE technique in the near future. According to the final use of SOI, one SOI technique will be better adapted.

To conclude, we propose below to list the advantages and drawbacks of each technique presented and their potential fields of application.

6.1 Zone melting recrystallization: In ZMR, the different layers of the SOI structure can be adjusted throughout a wide range. This technique will provide the highest thicknesses of underlying oxide and silicon. It is also the cheapest technique, and can be either in – line or off – line. In the latter case, the residual defect density is in the E4 – E5 /cm2 range. At present, the main drawback is the silicon thickness fluctuation which in fact renders this technique only slightly suitable for thin film applications (< 1500 A). The principal fields of application of this technique are the power, ASICs, and ULSI applications /16/. The 3D field will be that of laser – and e – beam – ZMR, which are the only suitable techniques for this range of applications.

6.2 Oxidation of porous – Si: In this technology, the thickness of the silicon layer can be as thin as desired, and the underlying oxide thickness is in the 0.3 – 3 um range. Its cost is moderate. Its main drawback is that it is, to date, an in – line process, which means that any industrialist would have to buy the fabrication of a porous – Si machine with its process parameters. The principal fields of application are (i) the smart power circuits as it is the only technique which can use the silicon substrate in a mixed process, and (ii) the VLSI and ULSI applications.

6.3 Oxygen implantation: In this technique, the thicknesses of oxide and silicon layers are limited (< 3000 A). It is an off – line process with a residual defect density in the E4 – E5 /cm2 range. The drawback is the high cost of this technique for public market applications. Its main application will be the ULSI field /17/.

REFERENCES

1. : Sheftal NN, Bouzynin NA, Bull. Moscow Univ. 27/3, 102 (1972).
2. : Gat A, Gerzberg, L, Gibbons JF, Magee TJ, Peng P, Hong JD, Appl. Phys. Lett., 33, 8 (1978).
3. : Fan JCC, Tsaur B – Y, Geis MW, J. Cryst. Growth, 63, 453 (1983).
4. : Dutartre D, MRS meeting, Boston, H3 – 1, Dec. 1987. To appear in the proceedings 1988.

5. : Bosh MA, Lemons RA, Phys. Rev. Lett., 47, 1151 (1981).
6. : Colinge JP, Demoulin E, Bensahel D, Auvert G, Appl. Phys. Lett., 41, 346 (1982).
7. : Dutartre D, Bensahel D, Haond M, Mat. Res. Soc. Symp. Proc., MRS, 74, 561 (1987).
8. : Vu DP, Allen LTP, Henderson WR, Zavracky PM, Fan JCC, Eur. SOI Workshop, Meylan (France), Ed. SEE/CNET, A – 03, March 1988.
9. : Regolini JL, Bensahel D, Perio A, Karapiperis L, E – MRS, Strasbourg 1987, Ed. de Physique 15, 623 (1987).
10. : Bomchil G, Hérino R, Barla K, E – MRS, Strasbourg 1985, Ed. de Physique, 463 (1985).
11. : Halimaoui O, Ronga I, Hérino R, Bomchil G, id. Ref. 8, C – 06.
12. : Konaka S, Tabe M, Sakai T, Appl. Phys. Lett., 41, 86 (1982).
13. : Barla K, Bomchil G, Hérino R, Monroy A, Gris Y, Electron. Lett., 22, 1291 (1986).
14. : Reeson KJ, IBMM 86, Catania, Nucl. Instr. Meth. Phys. Res., North Holand, B19/20, 269 (1987).
15. : Bruel M, Margail J, Jaussaud C, Papon AM, id. Ref. 8, D – 01.
16. : Gris Y, id. Ref. 8, D – 01.
17. : Colinge JP, id. Ref. 8, D – 06.

COMPLETE EXPERIMENTAL AND THEORETICAL ANALYSIS OF ELECTRICAL
TRANSPORT OF S.O.S. FILMS : THE PARTICULARITY OF HEAVILY
DOPED SAMPLES

G. KAMARINOS (1), G. GHIBAUDO (1), D. TSAMAKIS (2),
C. PAPATRIANTAFILLOU (2), E. ROKOFILLOU (2)

(1) LPCS-ENSERG 23, Rue des Martyrs
    38031 Grenoble Cedex, France
(2) CEN-"Democritos" Aghia Paraskevi
    Athens, Greece

1. INTRODUCTION
   The silicon On Insulator (SOI) techniques are now considered as very
promising for rapid VLSI Electronics. Indeed SOI allows : (a) high density
of elementary devices in I.C. (b) high tolerance to ionizing radiations
(α particles for example) (c) a drastic reduction of parasitic capacitances
(increase of I.C. speed and decrease of power dissipation) (d) absence of
latch-up.
   Among the different SOI techniques the Silicon On Sapphire (SOS) is now
used as a mature technique for fabricating fast and/or hardened Integrated
Circuits. S.O.S. has been studied for more than 20 years and sophisticated
circuits, such as microprocessors, fabricated.
   Nevertheless only after a very dense research work done during the last
ten years (1977-1987) a clear knowledge of the electrical transport properties in this material is achieved. ([1]...[5]). We present here a study of
theelectrical transport effects (conductivity σ, Hall effect R_H, and thermoelectric Power S) in a wide range of temperatures (4,2K ≤ T ≤ 400K).

2. SAMPLES
   The choice of sapphire (Al_2O_3) is due to the acceptable mismatch of its
thermal and crystallographic properties with those of the Silicon ([1],
[2], [3]).
   The studied S.O.S. samples have been prepared in LETI/CENG and Thomson
laboratories by Silane (SiH_4) pyrolysis at 1000°C ([6], [7]). Indeed in
this temperature (950°C ≤ T ≤ 1150°C), the autodoping of the Silicon film
by substrate reaction is acceptable, so that carrier concentrations as low
as $10^{15}cm^{-3}$ can be adjusted by including dopants in the gas. The rate of
deposition is about 2 μm min^{-1} and the usual thickness of Si film is on
the order 0,6 μm [2] . The doping of films is generally obtained by ion implantation; the heavy doping by phosphorus corresponds to implantation
doses of the order of $5x10^{14}cm^{-2}$.
   Several physico-chemical studies (see [2] and [3] for references) show
that :
(a) the films is seriously stressed (up to 10^9 dyn/cm^2 for thicker
than 500Å film ; 10^{12}dyn/cm^2 for thiner [8])
b) the films are frequently of a columnar structure (mosaic structure) consisting of large blocks with a minimum size of about 6000Å
c) a very perturbated transi tion layer in the interface Al_2O_3-Si exists.
Point and linear defects and contaminants (Al, C ...) are present. The
thickness of this layer is estimated being of the order of some 500Å

303

Y. I. Nissim and E. Rosencher (eds.), Heterostructures on Silicon: One Step Further with Silicon, 303–309.
© 1989 by Kluwer Academic Publishers.

(average value) or less.

3. MODERATE DOPING ($N \leq 10^{18} cm^{-3}$)

A large number of papers is dedicated to the electrical properties of moderately doped SOS films. (See reviews [2] and [3]). Summarizing here we can give the most relevant results :

The Influence of internal mechanical stress is important : The drift and Hall mobilities are decreased in 25% and 50% percentages in regard of their bulk Silicon values ([9], [10]). The magnetoresistance also is very alterated. ([2], [3], [10]). The lifetimes of injected carriers are low, particularly near the $Si-Al_2O_3$ interface, where, more precisely, the surface recombination velocity is very large ([11]).

Some sophisticated methods of characterization have given the profiles of mobilities, carriers concentrations, and lifetimes for moderately doped SOS films ([2], [3]).

For the theoretical analysis of electrical measurements a two-layer (N^+N) simple model is chosen : the main silicon film "1" is grown on a thin very perturbated silicon film "2" which can be indentified with the transition layer in the interface Al_2O_3-Si.

The total conductivity σ of the sample can therefore be expressed by [12] :

$$\sigma = \frac{\lambda}{1+\lambda} \sigma_1 + \frac{1}{1+\lambda} \sigma_2 \qquad (1)$$

where $\lambda = t_1/t_2$ the ratio of the thichnesses of the two layers. The total Hall mobility is found [12] :

$$\mu_H = \frac{\lambda\sigma_1\mu_{H1} + \sigma_2\mu_{H2}}{\lambda\sigma_1 + \sigma_2} \qquad (2)$$ and the Hall constant R_H :

$$R_H = \frac{\lambda\sigma_1^2 R_{H1} + \sigma_2^2 R_{H2}}{(\lambda\sigma_1+\sigma_2)^2} \qquad (3)$$

Concerning the total thermoelectric Power S ($\vec{E} = \frac{\vec{j}}{\sigma} + S\nabla T$) we found [12]

$$S = \frac{\lambda\sigma_1 S_1 + \sigma_2 S_2}{\lambda\sigma_1 + \sigma_2} \qquad (4)$$

In a previous paper [12] we show from (2), (3), and (4) that for a two-layered film the thermoelectric Power measurements are more sensitive to variations of global transport coefficients than the Hall effect (see also S vs T in Fig. 1 : The curve corresponding to the bulk silicon is radically different from the curves corresponding to different two-layered films; compare with μ vs T curves in Fig. 2).

For the expressions of S we take : in the supposed homogenous and degenerated (interfacial) layer 2 :

$$S_2 = \frac{\pi^2}{3} \frac{k}{q} \frac{kT}{E_{F2}} (\alpha + \frac{3}{2}) \qquad (5) \qquad ([13, 14, 15])$$

in the supposed homogeneous and main (non degenerated) layer 1 :

$$S_1 = \frac{k}{q} (-\frac{E_{F1}}{kT} + \frac{\Delta E_c}{kT}) \qquad (6) \qquad ([13, 14, 15, 16])$$

where : E_{F1} and E_{F2} are the Fermi Levels, α is defined in the relation connecting the relaxation time to the energy of free carriers $\tau \sim E^\alpha$;

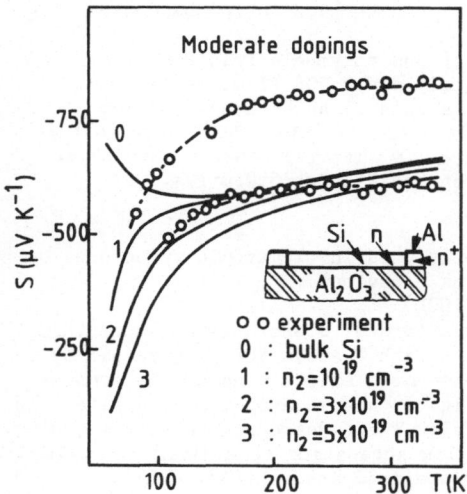

FIGURE 1. Typical experimental and theoretical dependance of S vs T

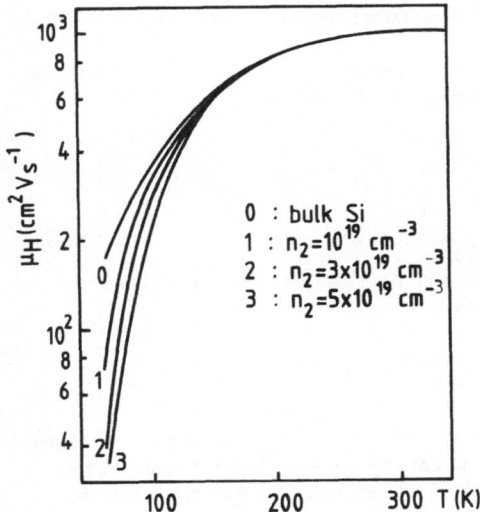

FIGURE 2. Hall mobility vs. temperature for a two-layers N-N$^+$ Si sample and for the bulk Si

ΔE_c is the kinetic energy of an electron in the conduction band depending on the mode of scattering.

Using so the conduction, Hall effect and thermoelectric Power detailed electrical characterization of moderately doped SOS films is performed ([2], [3], [12]). For example : for a sample of an "average" doping $N \cong 10^{17} cm^{-3}$ we find a lower N^+($N \cong 8 \times 10^{19} cm^{-3}$) thin (140Å) very perturbated layer near the $Si-Al_2O_3$ interface ($\mu \simeq 2 cm^2/Vs$) under a main layer ($\simeq 0,6 \mu m$; $N \simeq 10^{17} cm^{-3}$) with an electron mobility $\mu \simeq 550 cm^2/Vs$.

4. HEAVY DOPING ($N > 10^{18} cm^{-3}$)

These dopings are now more frequently used in electronic structures on SOS. The samples exhibit a more complex behaviour :

4.1. In low temperatures (4,2K < T < 100K)

We observe dramatically anomalous variations of σ and S with temperature (Fig. 3, 4). The linear increasing of S with T above 80K is preceded by a peak at 20K which is not due to a phonon-drag effect. Indeed the conductivity is strongly correlated to S (Fig. 4) and so a phonon-drag effect is excluded. Moreover the conductivity below T=20K decreases very slowly with T (Fig. 5). The log σ follows a $T^{-1/4}$ law suggesting so a Mott' law relative to a hopping conduction between localised states ([4], [16]).

We interpret these results by assuming a mobility edge in the band structure ([4], [16]).

Two conduction mechanisms are present relating to carriers above and below the mobility edge each of them contributing to σ and S ([4], [16], [17], [18], [19]). The mechanism "1" gives the hopping conductivity σ_1 ($\sigma_1 \sim T^{-1/4}$ at low T) and the TEP S_1 which depends strongly on the form of the density of states $N(E)$ around the Fermi Level E_F and is zero for a constant $N(E)$ around E_F.

The mechanism "2" due the extended states gives σ_2 and S_2 which can be obtained by the Mott [20] and Fritzsche [18] formulas :

$$\sigma_2 = \int_{E_c}^{\infty} \sigma(E) \left(- \frac{\partial f}{\partial E}\right) dE \quad (5); \quad S_2 = \frac{k}{q\sigma} \int_{E_c}^{\infty} \frac{E-E_F}{KT} \sigma(E) \left(- \frac{\partial f}{\partial E}\right) dE \quad (6)$$

where f is the Fermi-Dirac distribution.

The total transport coefficients σ and S, that result from the simultaneous presence of the two mechanisms are :

$$\sigma = \sigma_1 + \sigma_2 \quad (7) \quad \text{and} \quad S = \frac{\sigma_1 S_1 + \sigma_2 S_2}{\sigma} \quad (8)$$

Fitting $\sigma(T)$ for T > 22K we obtain from (5) (if the hypothesis of a linear $\sigma(E)$ form is adopted [4]) the relation for E_F versus T. With the values of parameters extracted from the above fitting and the E_F (T) relation, we calculate from (6) and (8) the total TEP S; we find that the experimental data are very satisfactorily modellized. Besides in [4] we show how it is possible to deduce the approximate form of the density of states $N(E)$ around the mobility edge (Fig. 6).

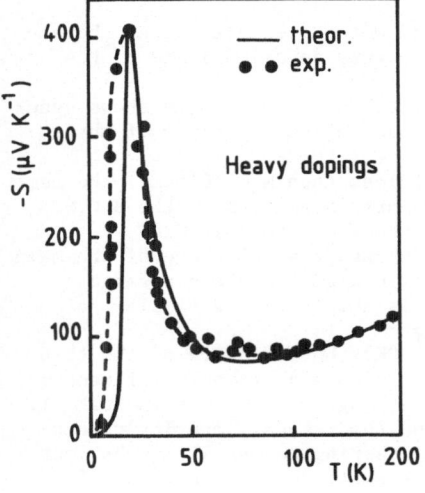

FIGURE 3. TEP vs T

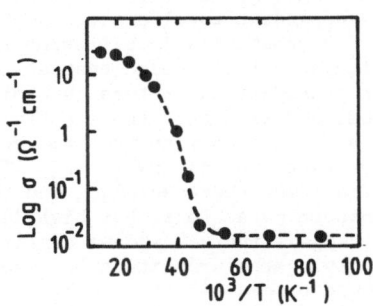

FIGURE 4. Measured σ vs T⁻¹

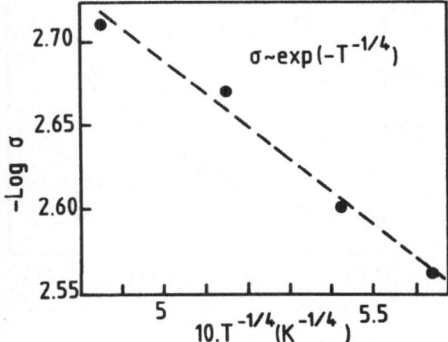

FIGURE 5. Conductivity vs temperature for heavily doped SOS

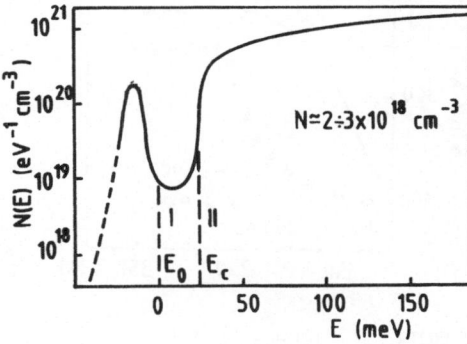

FIGURE 6. Energy density of States
vs E. E_0 is the mobility edge; E_c
lower end of conduction band

4.2. In high temperatures (100K < T < 400K)

The analysis of the experimental results (Fig. 7,8) led us to conclude
that the heavily doped SOS (until $5 \times 10^{18} cm^{-3}$) samples behave practically
the same way as bulk Si ones having a doping level just below the one
causing the Metal-Insulator Transition ($n_{dc} \simeq 3 \times 10^{18} cm^{-3}$; [14], [21],
[5].)

5. CONCLUSION

A variety of classical or sophisticated methods of electrical characterization of SOS films exists and gives a very detailed image of the
material.

The combined experimental and theoretical analysis clearly shows generally that the SOS films are non-homogeneous in the direction of their
growth.

For average impurity concentrations of less than $N = 10^{18} cm^{-3}$ the temperature dependent total electrical transport parameters of the S.O.S.
samples are substantially different from those of the bulk Si.

Particularly it is found that at low dopings the influence of the main
layer on transport parameters is dominant in high temperature range
(T > 100K) and also that their behaviour is qualitatively similar to
that of bulk Silicon with the same doping.

In contrast, in the low T range (T < 100K) the influence of the thin
transition layer (between Al_2O_3 and the main layer) strongly influences
the transport parameters. For high level dopings we find, as in bulk Si,
mobility edge structure and we confirm the theoretical impurity concentration giving the transition from semiconducting to metallic behaviour.

It would be exciting to compare these results to the behaviour of
new SOI films (like SIMOX) and which can face the possibility to work
in relatively low temperature [22].

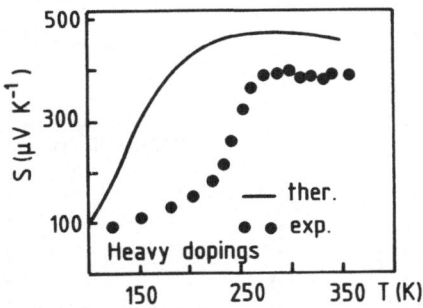

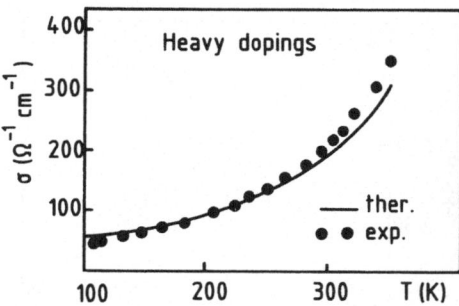

FIGURE 7. TEP S vs T   FIGURE 8. σ vs T

REFERENCES

1 . Cullen GW, Wang CC: Heteroepitaxial Semiconductors for Electrical
    Devices, New-York, Spinger, 1978.
2 . Cristoloveanu S,Ghibaudo G, Kamarinos G: Rev. Phys. Appl.19, 161, 1984.
3 . Cristoloveanu S: Rep. Progr. Phys.50, 327, 1987.
4 . Ghibaudo G, Tsamakis D, Papatriantafillou C, Kamarinos G, Rokofillou E:
    J. Phys. C: Sol. St. Phys.16, 4479, 1983.
5 . Tsamakis D, Rocofillou E, Papatriantafillou E, Ghibaudo G, Kamarinos G:
    J. Phys. C: Sol. St. Phys.20, 1285, 1987.
6 . Gonchond JP: Thesis USM Grenoble, 1978.
7 . Carfagnini ML: Thesis INP Grenoble, 1981.
8 . Hamar-Thibault S: Rev. Phys. Appl.16, 317, 1981.

9 . Lee JH: Thesis INP Grenoble, 1981.
10. Lee JH, Cristoloveanu S, Chovet A: Sol. St. Electr.$\underline{25}$, 947, 1982.
11. Cristoloveanu S, Chovet A, Kamarinos G: Sol. St. Electr.$\underline{21}$, 1563, 1978.
12. Ghibaudo G, Kamarinos G: Rev. Phys. Appl.$\underline{17}$, 133, 1982.
13. Smith RA: Semiconductors; Cambr. Univ. Press, 1978.
14. Morin F, Maita J: Phys. Rev.$\underline{96}$, 28, 1954.
15. Mac Donald DK: Thermoelectricity: An Introduction to the principles
    (Wiley, New-York), 1962.
16. Mott N, Davis E: Electronic Processes in non crystalline materials
    (Oxford: Clarendon Press, London), 1979.
17. Overhof H: Phys. St. Sol.$\underline{67}$, 709,1975.
18. Fritzsche H: Sol. St. Comm.$\underline{9}$, 1813, 1971.
19. Pollak M: Phil. Mag.$\underline{36}$, 1157, 1977.
20. Mott. N: J. Non-Cryst. Sol.$\underline{1}$, 1, 1968.
21. Paalanen MA, Ruckenstein AE, Thomas GA: Phys. Rev. Lett.$\underline{54}$, 1295, 1985.
22. 1988 European Silicon On Insulator Workshop: From Material to devices
    Meylan, France, 15-17 March 1988, Ed. D. Bensahel and G. Bomchil
    CNET/CNS Grenoble.

HETEROEPITAXIAL GROWTH OF SIC ON SI AND ITS APPLICATION

HIROYUKI MATSUNAMI

Department of Electrical Engineering, Kyoto University,
Kyoto 606, Japan

ABSTRACT
 Single crystals of SiC were reproducibly grown on Si by chemical vapor
deposition(CVD). A carbonized layer on single crystalline Si was used to
overcome the large lattice mismatch of 20 % between SiC and Si. Structures
and roles of the carbonized layer are discussed. Crystal growth of SiC was
examined on different crystal planes of Si substrates. Antiphase domains
on Si(100) are originated from atomic steps of the surface. Single-phase
SiC was grown on Si(100) substrates with off-axis orientation towards
[011]. Electrical properties of the grown layers were measured, and
anisotropy in electron mobility was found in the grown layer on Si(100)
with off-axis orientation.   Ion implantation and fabrication of active
electronic devices are described.

1. INTRODUCTION
   Electronic devices using silicon carbide(SiC) are strongly expected for
the use at high temperatures and/or under harsh environments, since SiC has
mechanically hard, refractory, chemically inactive and radiation-resistant
properties.   Beta(cubic) SiC, only one cubic structure among various SiC
polytypes, is promising owing to a high electron mobility of 1000 cm^2/Vs[1]
with an energy gap of 2.3 eV. The calculated saturation drift velocity of
2.7×10^7 cm/s for electrons[2] with high breakdown field is attractive for
high power microwave devices. However, only tiny plate-like crystals of
cubic SiC have been obtained by ordinary sublimation methods above 2000 $^{\circ}$C
or by solution growth using a Si melt in a graphite crucible. In order to
obtain cubic SiC crystals with large area for electronic application,
heteroepitaxial growth on foreign substrates is strongly desired. Among
various kinds of substrate materials, silicon(Si) is the most suitable for
the growth of SiC, since it is one component of SiC and wafers of high
purity and large area are easily available.
   Although chemical vapor deposition(CVD) of cubic SiC on Si has been tried
by many researchers, reproducible growth was difficult mainly due to the
large lattice mismatch of 20% between SiC and Si (a_{SiC}=4.358 Å, a_{Si}=5.430
Å)[3].   After our proposal for reproducible growth using a carbonized
layer of a Si substrate prior to CVD[4], epitaxial layers of cubic SiC up
to 2-3 inches in diameter can be grown on Si at present[5].
   In this paper, heteroepitaxial growth of cubic SiC on Si -highly
mismatched case- is described[6,7]. The optimum conditions to make the
carbonized layers and their structure and the role are discussed. Crystal

311

Y. I. Nissim and E. Rosencher (eds.), Heterostructures on Silicon: One Step Further with Silicon, 311–321.
© 1989 by Kluwer Academic Publishers.

growth of cubic SiC on different crystal planes of Si substrates is examined. Antiphase dopmains(APDs) observed on (100) epitaxial layers are discussed together with the generation mechanism using an atomic step model for the initial stage of carbonization. Growth of APD-free epitaxial layers can be realized on Si(100) off-axis oriented towards [011] through growth experiments on spherically polished Si(100) substrates. Anisotropy in electrical properties of the epitaxial layers on off-axis substrates is discussed. Finally some applications to electronic devices are presented.

2. HETEROEPITAXIAL GROWTH

Heteroepitaxial growth of cubic SiC was carried out at 1 atmosphere in a hori-zontally-set quartz reac-tion tube using SiH_4, C_3H_8 and H_2. The reaction tube has a circular cross sec-tion of 5 cm in diameter and a water-cooling jacket. A Si substrate was put on a SiC-coated graphite suscep-tor after etching in HF solution and rinsing in deionized water. The entire reaction system was evac-uated by a rotary pump and then filled with H_2. The Si substrate was heated by an

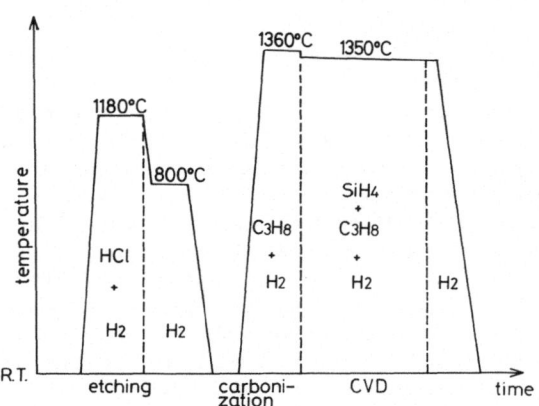

FIGURE 1. Temperature program of heteroepitaxial growth.

rf induction with 200 kHz. Substrate temperature was measured with a pyrometer, which was corrected using the emissivity of Si. Single crystals of cubic SiC was heteroepitaxially grown on Si by the three processes shown in Fig.1[7].

After being etched with HCl gas (HCl: 10 sccm, H_2: 1 SLM) at 1180 °C for 5 minutes, the Si substrate was cooled to near room temperature. The second step is carbonization of the Si substrate, which provides a carbonized layer prior to single crystal growth of cubic SiC by subsequent CVD. A flow of C_3H_8(1.2 sccm) was put into the H_2 carrier gas(1 SLM) just before substrate heating. The substrate temperature was raised to 1360 °C in 2 minutes: the ramp of the heating process is a key to control the quality of the carbonized layer. An optimum flow rate and a temperature program were determined through various experiments[7]. Single crystals were grown by subsequent CVD at around 1330 °C with the standard gas composition of SiH_4: 0.3 sccm and C_3H_8: 0.25 sccm. During the process, the H_2 carrier gas was flowed with a rate of 3 SLM, corresponding to a linear velocity of 5 cm/s. The typical growth rate was about 5 Å/s.

Reflection high energy electron diffraction(RHEED) patterns changed gradually from a spot pattern to a streaked one with increasing layer thickness, which indicates the improvement in the crystal quality. Kikuchi lines and bands are clearly observed in the thick layers as shown in Fig.2

for the layer of 5.4 µm. To confirm
the improvement, channeling technique
by Rutherford back scattering(RBS)
using 1 MeV He$^+$ ions was utilized.
The yield of the backscattered ions
in the channeling spectra decreased
with the thickness.

A cross-sectional transmission electron microscope(XTEM) photograph is
shown in Fig.3, which was observed
from the [110] direction for a CVD-
grown SiC layer with a thickness of
100 nm. Stripes spread along the
[111] directions both in SiC and Si
layers, since they form an angle of
55° with the (100) plane and the
interval of the stripes corresponds
to the lattice spacing of (111)
planes[8]. Dark and white lines along
[111] directions in the SiC layer
indicate plane defects such as stacking faults and microtwins. The interface between SiC and Si is rather
sharp. Since the lattice spacings
differ in SiC and Si, all the stripes
cannot be connected one by one. One
out of 4 or 5 lines of the stripes is
disconnected at the interface showing
the existence of a dislocation line,
which well agrees with the lattice
mismatch of 20%. By detailed XTEM
observations for cubic SiC with va-

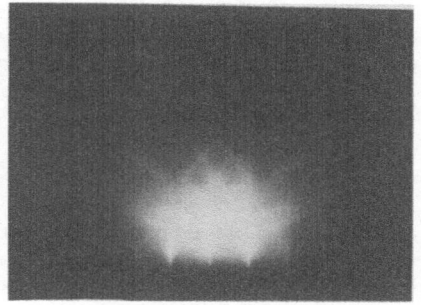

FIGURE 2. [011]-azimuth
RHEED pattern of a grown
layer with 5.4 µm on
Si(100).

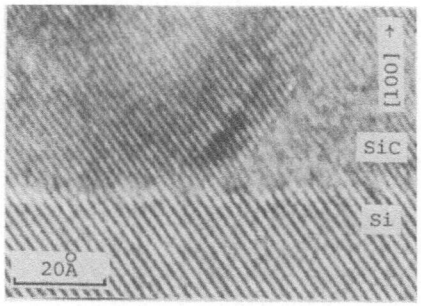

FIGURE 3. XTEM photograph
of a grown layer on Si(100).

rious thicknesses, the plane defects were reported to be decreased with the
thickness up to 3-4 µm of the grown layer[9].

3. CARBONIZED LAYER

The crystal quality of the carbonized layer was examined by RHEED
measurements. The layer prepared with the optimum condition shows a spot
pattern of single crystalline cubic SiC. When the carbonization condition
is far from the optimized one: e.g., too low C_3H_8 flow rate or too long
carbonization time, the RHEED pattern shows polycrystalline feature. Even
the optimized carbonized layer gives extra spots and streaks spreading to
[111] directions in its RHEED pattern, indicating the existence of stacking
faults and microtwins. The thin carbonized layers contain crystal defects
due to the large lattice mismatch and stress induced by the difference in
thermal expansion.

The quality of the carbonized layer is strongly influenced by preparation
conditions such as temperature program, C_3H_8 gas flow rate and
carbonization time[7]. The surface cleaness is also important for the

314

formation of the carbonized layer. In fact, if the Si surface after HCl gas etching was not mirror-like, the RHEED pattern of the carbonized layer contained faint rings, and CVD grown layers on it become polycrystalline.

To determine the optimum condition for the carbonization, values of full width at half maximum(FWHM) of X-ray rocking curves of the CVD layers of cubic SiC were examined[7]. The carbonization time was determined as 2 minutes. For longer or shorter carbonization times, the FWHM increased, indicating a carbonized layer of low quality. The FWHM shows a minimum for a certain value of the C_3H_8 flow rate, e.g., 1.2 sccm for Si(111). For Si(100), the range of the C_3H_8 flow rate giving the narrow FWHM is wider than for Si(111). The minimum FWHM for Si(111) was lower than for Si(100).

In order to study the structure of the carbonized layer, a depth profile analysis using Auger electron spectroscopy(AES) was carried out[7]. For the carbonized layer prepared with the optimum condition, the compositional ratio of Si and C was close to unity in the surface layer of 8 nm. Gradual change in the composition from SiC to Si was observed in the layer of 22 nm under the surface SiC layer. The region with the gradual composition change is hardly understandable, since no crystalline compounds with the form of $Si_{1-x}C_x (x<1)$ except for SiC exist. Also it contradicts the rather sharp interface between SiC and Si shown in Fig.3. The surface of the carbonized layer observed by a Nomarski microscope showed slight roughness. If the roughness comes from nonuniformity in the thickness of SiC, it may give the gradual composition change in the Auger depth profile. The nonuniformity might come from the rough surface of the Si substrate due to high temperature processes such as etching or carbonization.

Recently, detailed analysis of the carbonized layer and the interface region was carried out using medium-energy ion scattering [10]. The following results were deduced, from the two spectra shown in Fig.4. (1) The alignment of atoms in the SiC layer with respect to the Si(100) orientation becomes better (better epitaxy) in the surface region (∿4 nm) than near the interface. (2) The interface between SiC and Si may have an undulation of a-round 2 nm like that of SiO_2-Si. (3) There exists a distort-ed region in SiC of around 2 nm near the interface. (4) The very thin interface region of 4-5 nm extending above and below the real interface of SiC-Si works as a real buffer layer.

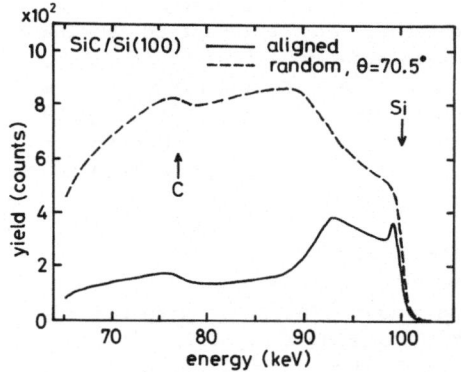

FIGURE 4. Backscattering spectra in medium-energy (121 keV) ion scattering from a carbonized layer on Si(100).

4. GROWTH ON DIFFERENT SI PLANES

Cubic SiC of about 2 μm was grown on Si(111), (110) and (211) substrates by the combination of the carbonization and the CVD growth similarly to the case on Si(100)[11]. The RHEED pattern for Si(111) was streaked, and the crystal quality of the CVD grown layers on it was improved with the thickness. In contrast with the grown layers on Si(100) and (111), the RHEED patterns on Si(110) and (211) gave extra spots and Debye rings. The CVD grown layers showed no significant improvement in the crystal quality, and the surfaces were very rough, even though various trials for optimization were carried out. This is due to the bad crystal quality of the carbonized layers on Si(110) and (211) as determined by the RHEED analysis[11].

As described above, single crystals of cubic SiC were heteroepitaxially grown on Si(100) and (111) by the carbonization and subsequent CVD, whereas polycrystals on Si(110) and (211). To explain this difference, the following mechanism for the formation of SiC-Si interface during the carbonization was proposed[11]. The carbonization was found to be governed by Si diffusion through the SiC layer using the ^{14}C tracer method[12]. The initial stage of the carbonization is assumed to be composed of the following two steps: (1) Adhesion of monoatomic-layer C atoms on the Si surface, and (2) growth of cubic SiC at the surface consuming outdiffused Si atoms and gaseous C_3H_8. The first grown layer consists of only C atoms based on step 1. Figure 5 shows the arrangements of the atoms around the ideal SiC-Si interface for Si(100) and (111). In the figures the lattice mismatch is neglected to make the discussion simple[11].

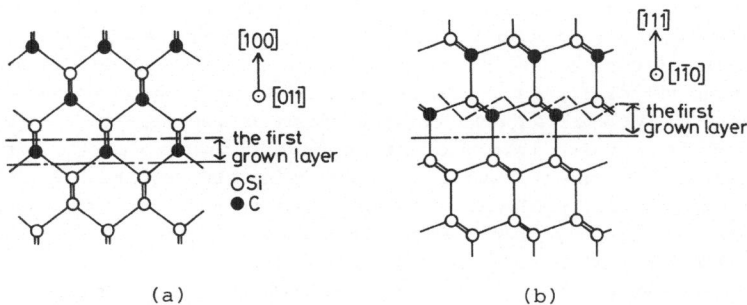

(a)   (b)

FIGURE 5. Ideal atomic arrangements near the SiC-Si
interface. (a) on Si(100) and (b) on Si(111).

The ideal first layer of polar planes like SiC(100) and (111) should have only C atoms. Since the first layer by the carbonization consists of only C atoms, single crystalline cubic SiC can be obtained by subsequent layer-by-layer growth of Si, C, Si and C. In the case of non-polar planes like SiC(110) and (211), however, the ideal first layer should be composed of both C and Si atoms as in Fig.6. The carbonization brings only C atoms in the first layer, therefore single crystalline networks of cubic SiC cannot be obtained on Si(110) and (211) planes[11].

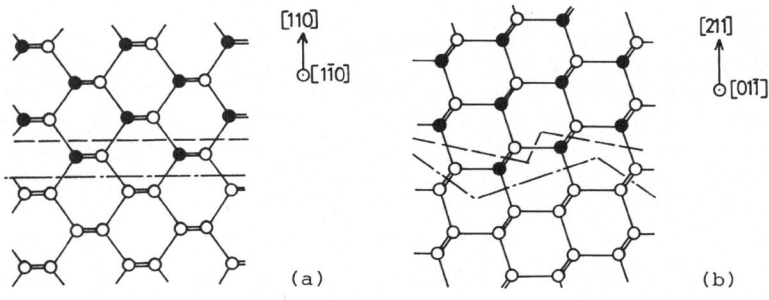

FIGURE 6. Ideal atomic arrangements near the SiC-Si
interface. (a) on Si(110) and (b) on Si(211).

5. ELIMINATION OF ANTIPHASE DOMAINS

Single crystalline cubic SiC can be heteroepitaxially grown on Si (100)
and (111) by the combination of the carbonization and subsequent CVD.
However, probelms remain in the grown layers on these substrates: antiphase
domains(APDs) on Si(100) and warping and cracks on Si(111). Since warping
and cracks are severe problems for device applications, cubic SiC on
Si(111) cannot be used at this moment. Hereafter, the discussion will
concentrate on the elimination of APDs on Si(100).

APDs usually observed in cubic SiC on (100) well-oriented Si substrates
give texture-like morphology consisting of wedge shapes which intersect
perpendicular to each other. The APDs were confirmed as random grooves by
molten KOH etching at about 600°C[11], since the etch pits were put
perpendicularly each other in adjoining regions separated by the grooves.

APDs in epitaxial layers should be eliminated, because they cause surface
roughness and may work as scattering centers or recombination centers. To
obtain APD-free grown layers of zincblende structure on (100) plane of
diamond structure, the following two conditions are required[13]. (1) There
should be a chemical preference in which one of the elements constituting
the zincblende strongly bonds with the element of the substrate than the
other. (2) The substrate should not have steps with mono(or odd number)-
atomic layer height. In the case of cubic SiC on Si(100), since the first
grown layer consists of only C atoms by the carbonization, the condition
(1) is satisfied. Hence, if the Si surface has steps with bi(or even
number)-atomic layer height, APD-free growth is realized.

By the introduction of off-axis orientation, modification of the step
height on Si(100) was carried out[5,14-16]. To study the direction and the
magnitude of the off-axis angle, cubic SiC was grown on a spherically
polished Si(100) substrate. A photograph of the surface of the grown layer
is shown in Fig.7. Two black lines cross at the (100) pole and spread
towards the directions equivalent to [011]. The surface was mirror-like on
the two black lines, whereas that on the other white areas was rough.
Figures 8(a)-8(d) show typical Nomarski microphotographs of the surfaces at
the (100) pole and at other points shown in Fig.7. A texture-like

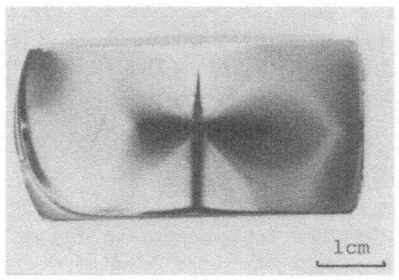 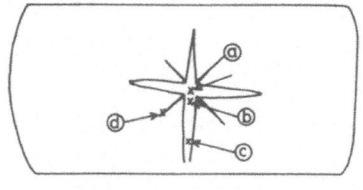

FIGURE 7. Photograph of a grown layer on a spherically polished Si(100).

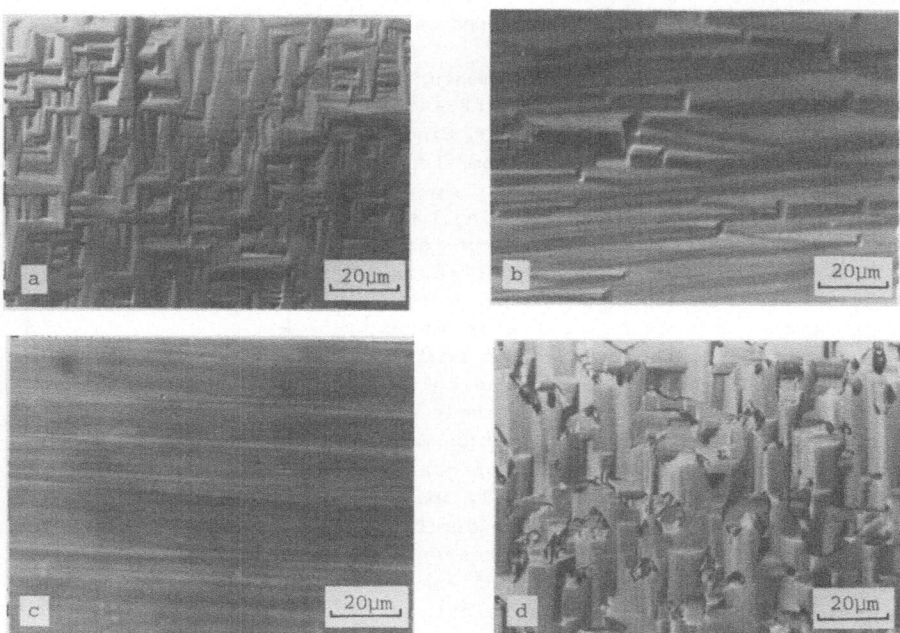

FIGURE 8. Nomarski microphotograph of the grown layer on the spherically polished Si(100). (a) to (d) correspond to the points shown in Fig.7.

morphology indicating the existence of APDs was observed in the grown layer at the (100) pole as shown in Fig.8(a). In the grown layers on the off-axis towards the [011] direction, the wedge-like crystal habit pointing the perpendicular to the off-axis direction becomes dominant. Beyond 2° of the off-axis angle, APDs were not observed as shown in Fig.8(c). Up to 5°, such single-domain morphology was observed. Off-axis orientations except for towards [011] resulted in very rough surfaces with APDs as shown in

318

Fig.8(d). These results were applied to the growth on Si(100) wafers, mainly the wafers oriented 2° off towards [011]. The etched surface with molten KOH showed no grooves, and all etch pits aligned in the same direction, which indicates the elimination of APDs. The existence and elimination of APDs was also confirmed by RHEED observation[8,17].

6. ELECTRICAL PROPERTIES

Undoped layers showed n-type conduction. P-type conduction can be controlled by doping of B or Al during epitaxial growth[8]. B-doping gave highly-resistive epitaxial layers probably due to appropriate compensation by B acceptors with a deep level. Hall measurements were carried out by the van der Pauw method, after Si substrates were removed by a mixture of HNO_3 and HF. The relation between electron mobility and carrier concentration of undoped layers was examined[8,18]. The maximum electron mobility on Si(100) substrates was 488 cm^2/Vs with a carrier concentration of 3.18×10^{16} cm^{-3} for a sample of 5 μm. The undoped carrier concentration was usually in the range of 3.0×10^{16}-2.0×10^{17} cm^{-3}.

By the van der Pauw method, the measurements for the grown layers on off-axis substrates gave asymmetrical results. Since the asymmetrical features seem to come from the anisotropy in the electrical properties, standard Hall measurements using long sheet-shaped samples were carried out parallel and perpendicular to the off-axis direction. Grown layers of 8-24 μm on the 2°, 4° and 6° off-axis substrates were used. In Fig.9, are shown electron mobilities of the parallel and perpendicular directions against the off-axis angle of the substrates. The parallel electron mobility was independent of the off-axis angle, and the values were constantly about 500 cm^2/Vs. However, the perpendicular mobility was reduced with an increase in the off-axis angle. Even for an off-axis angle of 2°, the perpendicular electron mobility was about half of the parallel electron mobility, though the carrier con-

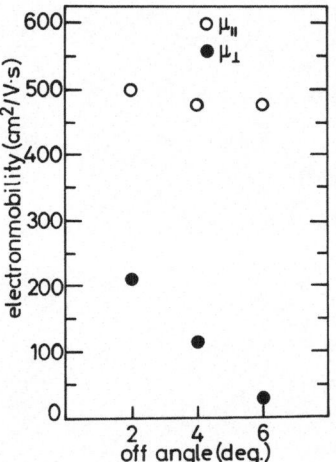

FIGURE 9. Electron mobility vs. off-axis angle.

centrations of these samples were in the same range(3-5×10^{16} cm^{-3}). The difference in the electron mobilities of parallel and perpendicular to the off-axis direction can be explained using the difference in crystal defects, since the grown layers showed anisotropic surface morphology as described in the previous section.

7. APPLICATIONS

Many trials to fabricate electronic devices have been extensively carried out[19-24], based on the recent progress in heteroepitaxial growth of cubic SiC on Si. Fundamental electronic devices such as pn junctions,

Schottky barrier and MOS diodes were described previously[5]. In this part, results on ion implantation and application to MOSFETs are briefly described.

P^+- and N_2^+-implantations into highly-resistive B-doped cubic SiC grown layers were carried out at room temperature using the following total doses: (1) P^+ ions 1.3×10^{15} cm^{-2} (1×10^{15} cm^{-2}, 100 keV; 3×10^{14} cm^{-2}, 25 keV), (2) P^+ ions 3.9×10^{14} cm^{-2} (3×10^{14} cm^{-2}, 100 keV; 9×10^{13} cm^{-2}, 25 keV) and (3) N_2^+ ions 1.95×10^{14} cm^{-2} (1.5×10^{14} cm^{-2}, 100 keV; 4.5×10^{13} cm^{-2}, 25 keV)[23]. After implantation, annealing was carried out in Ar atmosphere by rf induction heating for 30 minutes. In Fig.10, is shown electrical activation as the ratio of the sheet carrier concentration and the total dose. Electrical activation increased with annealing temperature. It should be noticed that the layer with a total dose of 3.9×10^{14} cm^{-2} shows higher electrical activation than that of 1.3×10^{15} cm^{-2} for P^+-implantation. N_2^+-implanted layers showed higher electric activation than P^+-implanted layers.

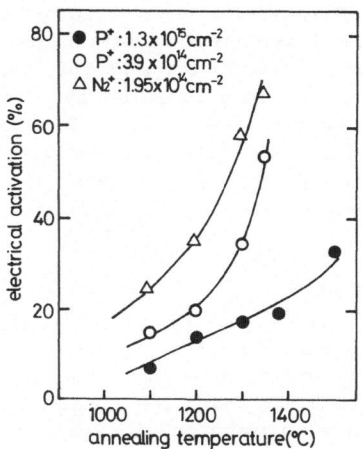

FIGURE 10. Electrical activation rate as a function of annealing temperature.

Planar pn junction diodes were fabricated by implantation of P^+ ions with a total dose of 3.9×10^{14} cm^{-2} or N_2^+ ions of 1.95×10^{14} cm^{-2} into Al-doped p-type cubic SiC on Si(100)[23].

Rectification of the diodes annealed at 1100 $^\circ$C and 1200 $^\circ$C was superior to that at 1300 $^\circ$C. As shown in Fig.10, electrical activation increased abruptly by annealing above 1300 $^\circ$C. Therefore, the change in rectification with annealing temperature indicates the change from n-layer to n$^+$-layer. The capacitance and the magnitude of its variation becomes large with annealing temperature, which means that the diodes annealed at low temperatures have pin structures and the i layer becomes thinner with annealing temperature.

As an application to electronic

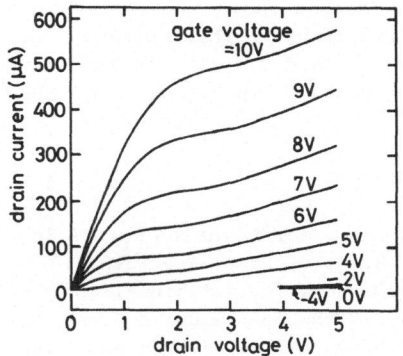

FIGURE 11. Drain characteristics of an inversion-type MOSFET at room temperature.

devices, inversion-type n-channel MOSFETs were fabricated on APD-free cubic SiC on Si(100)[14,21]. The n-channel was made on B-doped p-SiC of $2\,\mu m$ thick grown on n-SiC of $7\,\mu m$. P^+ ions were implanted to make source and drain electrodes. Annealing was carried out at 1080 $^{\circ}$C in Ar for 1 hour in an IR radiative heating furnace. The gate insulator of 60 nm was made by thermal oxidation in dry oxygen at 1050 $^{\circ}$C for 6 hours. The gate length and width were 20 μm and 500 μm, respectively. Ohmic contacts and gate electrode were prepared by Al evaporation. Drain characteristics at room temperature are shown in Fig.11. The threshold voltage was estimated to be 2.8 V and the effective mobility to be about 100 cm^2/Vs. Characteristics of field effects are retained even at 400 $^{\circ}$C except for leaky problems in the gate insulator.

8. SUMMARY

Single crystals of cubic SiC were heteroepitaxially grown on Si by introduction of a carbonized layer on Si to solve the large lattice mismatch of 20 % between SiC and Si. The structure and the role of the carbonized layer were analyzed using various methods. Experiments of crystal growth on Si crystals with different orientations gave single crystals only on Si(100) and (111). Particular texture-like morphology appeared on Si(100) is caused by antiphase domains(APDs). The generation mechanism of APDs was discussed with a model of the carbonization process. Through growth experiments on spherically polished Si(100), elimination of APDs was attained by using Si(100) off-axis orientation towards [011]. Anisotropic electrical properties were observed in SiC grown on Si(100) off-axis substrates. Ion implantation and its application to MOSFET fabrication were described.

ACKNOWLEDGEMENT

The author shows much thanks to Drs. K.Shibahara and S.Nishino for their contributions to this research. This research was partially supported by a Grant-in-Aid for Scientific Research from the Ministry of Education, Science and Culture, Japan and also by a Mazda Foundation Research Grant.

REFERENCES

1. W.E.Nelson, F.A.Halden and A.Rosengreen, J. Appl. Phys. 37 (1986) 333.
2. D.K.Ferry, Phys. Rev. B12 (1975) 2361.
3. H.Matsunami, in "Thin Films from Free Atoms and Particles", edited by K.J.Klabunde (Academic Press, 1985) pp. 301-324.
4. H.Matsunami, S.Nishino and H.Ono, IEEE Trans. Electron Devices ED-28 (1981) 1235.
5. H.Matsunami, in "Novel Refractory Semiconductors", edited by D.Emin, T.L.Aselage and C.Wood (Mater. Res. Soc. Proc. 97, Pittsburgh, PA 1987) pp. 171-182.
6. H.Matsunami, in "Heteroepitaxy on Si", (Mater. Res. Soc. Proc., 1988)

to be published.

7. S.Nishino, H.Suhara, H.Ono and H.Matsunami, J. Appl. Phys. 61 (1987) 4889.

8. K.Shibahara, Doctor thesis in Kyoto University, 1988.

9. C.H.Carter,Jr., R.F.Davis and S.R.Nutt, J. Mater. Res. 1 (1986) 811.

10. M.Iwami, M.Kusaka, M.Hirai, H.Nakamura, T.Koshikawa, K.Shibahara and H.Matsunami, to be published.

11. K.Shibahara, S.Nishino and H.Matsunami, J. Cryst. Growth 78 (1986) 538.

12. J.Graul and E.Wagner, Appl. Phys. Lett. 21 (1972) 67 .

13. H.Kroemer, K.J.Polasko and S.C.Wright, Appl. Phys. Lett. 36 (1980) 763.

14. K.Shibahara, S.Nishino and H.Matsunami, Ext. Abstr. 18th Conf. Solid State Devices and Materials (1986) p.717.

15. K.Shibahara, S.Nishino and H.Matsunami, in "Novel Refractory Semiconductors", edited by D.Emin, T.L.Aselage and C.Wood (Mater. Res. Soc. Proc. 97, Pittsburgh, PA 1987) pp. 183-188.

16. K.Shibahara, S.Nishino and H.Matsunami, Appl. Phys. Lett. 50 (1987) 1888.

17. K.Shibahara and H.Matsunami, to be published.

18. K.Shibahara and H.Matsunami, to be published.

19. Y.Kondo, T.Takahashi, K.Ishii, Y.Hayashi, E.Sakuma, S.Misawa, H.Daimon, M.Yamanaka and S.Yoshida, IEEE Electron Device Lett. EDL-7 (1986) 404.

20. S.Yoshida, H.Daimon, M.Yamanaka, E.Sakuma, S.Misawa and K.Endo, J.Appl. Phys. 60 (1986) 2989.

21. K.Shibahara, T.Saito, S.Nishino and H.Matsunami, IEEE Electron Device Lett. EDL-7 (1986) 692.

22. K.Furukawa, A.Hatano. A.Uemoto, Y.Fujii, K.Nakanishi, M.Shigeta, A.Suzuki and S.Nakajima, IEEE Electron Device Lett. EDL-8 (1987) 48.

23. K.Shibahara, T.Takeuchi, T.Saitoh, S.Nishino and H.Matsunami in "Novel Refractory Semiconductors", edited by D.Emin, T.L.Aselage and C.Wood (Mater. Res. Soc. Proc. 97, Pittsburgh, PA 1987) pp. 247-252.

24. H.S.Kong, J.W.Palmour, J.T.Glass and R.F.Davis, Appl. Phys. Lett. 51 (1987) 442.

NUCLEATION STEP OF GaAs/Si AND GaAs/(Ca,Sr)F$_2$/Si : AES AND RHEED STUDIES

C. FONTAINE, J. CASTAGNE, E. BEDEL and A. MUNOZ-YAGUE

Laboratoire d'Automatique et d'Analyse des Systèmes du CNRS, 7 Avenue du Colonel Roche, 31077 TOULOUSE, FRANCE

1. INTRODUCTION

Over the last years, the growth of heteroepitaxial GaAs layers has received considerable attention. Two principal areas have been investigated : (i) direct growth of GaAs on Si with a view to replacing GaAs substrates and fabricating two-level integrated circuits ; (ii) growth of GaAs on monocrystalline fluoride layers previously grown on semiconductor substrates, the intermediate fluoride allowing for lattice-matching and electric insulation between the two semiconductors.

Although some interesting demonstrations have already been reported, the characteristics and reliability of the devices realized with these heteroepitaxial structures have yet to be improved. This necessitates to ameliorate the crystallinity of the heteroepitaxial material grown on Si and fluoride, which contains a high density of crystalline defects, -twins and dislocations-, the latter being known to affect adversely the semiconducting properties.

As to heteroepitaxial growth, it is possible to propose some parameters which play a preeminent part in defect formation : namely, nucleation step, growth and thermal stresses.

In this paper, the emphasis is placed on the early stages of GaAs growth. A three-dimensional (3D) growth mechanism is known to occur in the two heteroepitaxial systems considered, when standard heteroepitaxial GaAs conditions are used. This mechanism should lead to defect formation upon island coalescence. Here, a technique is proposed, which forces this mode mode of growth towards a two-dimensional (2D)-like mechanism. The coverage of the substrate surface is obtained by growing a thin, amorphous GaAs layer at reduced temperature, whose crystallinity and stoichiometry are restored by annealing at a moderate temperature. In addition, the technique proposed is found to suppress the crystalline defects formed during the initiation of GaAs growth.

2. EXPERIMENTAL

The study was done in an ultrahigh vacuum RIBER molecular beam epitaxy (MBE) system, equipped with Auger electron spectrometry (AES) and reflection high energy diffraction (RHEED) apparatuses.

The (100) oriented Si substrates used were prepared according to a procedure described elsewhere [1], in order to obtain an oxide -and carbon- free surface at the time of loading into the vacuum chamber. Once in the growth chamber, the substrate was subjected to an atomic flux of Si, provided by the doping effusion cell available in the MBE system. Under the Si flux, the substrate surface was found to undergo a (2 x 1) +

Y. I. Nissim and E. Rosencher (eds.), Heterostructures on Silicon: One Step Further with Silicon, 323–328.

(1 x 2) reconstruction ; the layer growth was then initiated on this surface.

The thin $(Ca,Sr)F_2$ layers were grown at a substrate temperature of 600 C and a growth rate of 0.1 μm/h. The fluoride layers grown were monocrystalline, as was evidenced by RHEED observations.

The growth conditions of GaAs/Si and $GaAs/CaF_2$ are detailed in the next section.

3. 3D NUCLEATION UNDER STANDARD CONDITIONS

Heteroepitaxial GaAs/Si growth conditions usually consist of two steps : first, a moderate substrate temperature ($T_s \sim$ 400-450 C) together with a low growth rate ($V_d \sim$ 0.1 μm/h) enable the growth of a 200 A° thick GaAs layer ; then, homoepitaxial GaAs conditions ($T_s \sim$ 550-600 C ; $V_d \sim$ 1 um/h) are applied. The initial step, which was determined empirically, turned out to be the most appropriate for drastically improving the properties of the resulting GaAs layers [2].

As far as GaAs/fluoride is concerned, a comparable procedure was proposed by Tsutsui et al [3].

When considering epitaxial growth mechanisms, T_s and v_d are coupled parameters. Obtaining good crystallinity while reducing T_s necessitates to decrease simultaneously v_d as shown by Metze and Calawa [4] for GaAs homoepitaxial growth. On the other hand, Kashchiev [5] has demonstrated the relative influence of the two parameters, T_s and v_d, on the mode of growth of a layer ; the transition from 3D to 2D nucleation can be caused by either lowering T_s or increasing v_d.

These general considerations indicate that the conditions used for growing heteroepitaxial GaAs on Si and fluoride, i.e., $T_s \sim$ 400 C and v_d 0.1 μm/h, can be regarded as a good compromise in terms of layer crystallinity and substrate coverage. Nevertheless, under these experimental conditions, 3D nucleation is known to occur for GaAs/Si, despite the reduction of T_s, even if the cluster-size and their spacing are smaller than at higher T_s [6].

An AES study enabled us to compare the nucleation mechanisms involved for GaAs/Si and $GaAs/(Ca,Sr)F_2$ under the standard aforementioned heteroepitaxial GaAs conditions. It was recorded how the Auger signal amplitudes of the different elements evolved as a function of the deposited layer thickness. The information obtained with both systems is identical, as shown in figure 1. The variation of the Ga (55 eV) Auger peak amplitude and of those corresponding to an element of the substrate considered, namely Si (92 eV) and Ca (292 eV) for $(Ca,Sr)F_2$, evidences that the mode of growth of $GaAs/(Ca,Sr)F_2$ is similar to that of GaAs/Si, i.e., three-dimensional.

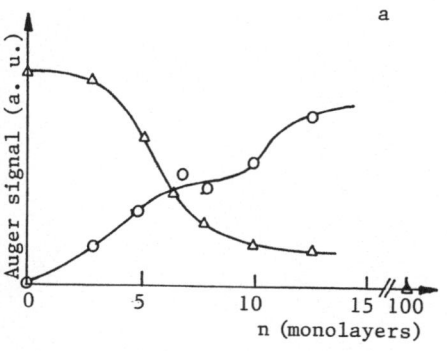
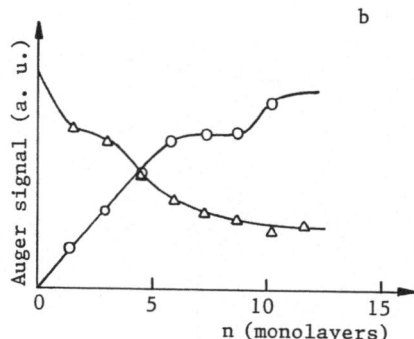

n (monolayers)　　　　　　　　　　n (monolayers)

FIGURE 1. GaAs growth at 400 C : variation of the Auger peak amplitudes as a function of the grown thickness for some of the elements involved : a) For GaAs/Si, o Ga (55 eV), Δ Si (92 eV) ; b) For GaAs/(Ca,Sr)F$_2$, o Ga (55 eV), Δ Ca (232 eV).

4. 2D-LIKE NUCLEATION

Because the 3D mechanism is thought to be partly responsible for the defect formation in the growing layer, a 2D mechanism should be preferred. It therefore appears that GaAs growth would be improved further if experimental conditions leading to the occurrence of the latter mechanism could be found.

This could be obtained by taking into account Kashchiev's theory, T_s being decreased and v_s remaining constant.

Some reported experimental results [7] show that when GaAs is grown at 225 C and 0.4 μm/h, coverage of the Si surface is effectively achieved for a 200 A° thick layer. Unfortunately, these low substrate temperatures involve a degradation of the deposit crystallinity ; to overcome this drawback, the grown layer can be annealed, providing the thickness grown is sufficiently small. Procedures taking advantage of these features require further investigation.

The technique proposed in this context consists of two steps :
(i) deposit of a thin (~ 15 A°) amorphous GaAs layer at low temperature (<100 C) and standard growth rate (0.15 μm/h) ; (ii) annealing of the amorphous layer at a moderate temperature. This technique was studied for both types of heteroepitaxial structures considered, i.e., GaAs/Si and GaAs/fluoride.

At this low growth temperature, the deposit is non-stoichiometric, since the As sticking coefficient value becomes more elevated as temperature decreases. The V/III ratio was fixed as the minimal ratio which leads to a reconstructed (2 x 4) GaAs surface at 550 C. In such conditions, the thickness grown should be at least twice as high as that expected for a stoichiometric deposit.

This is evidenced by the present AES study. Indeed, by raising the substrate temperature to about 400 C, the As excess can be desorbed, as shown in figure 2 for GaAs/Si. A similar observation was made for GaAs/(Ca,Sr)F$_2$.

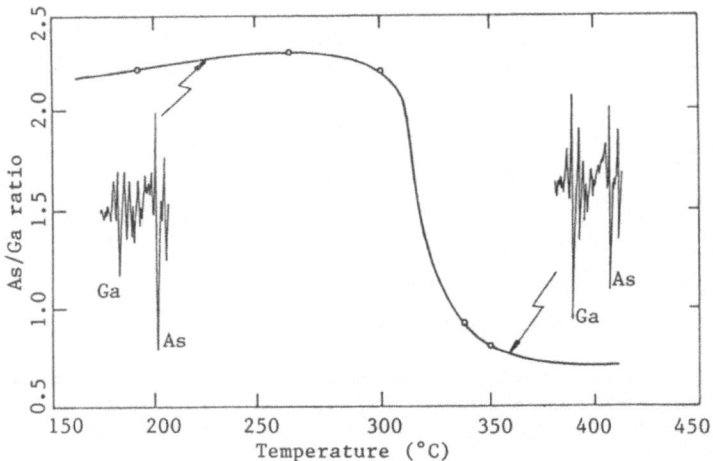

FIGURE 2. Variation of the Auger signal ratio As (1210 eV)/Ga (1065 eV) for a deposit at room temperature when the temperature is raised.

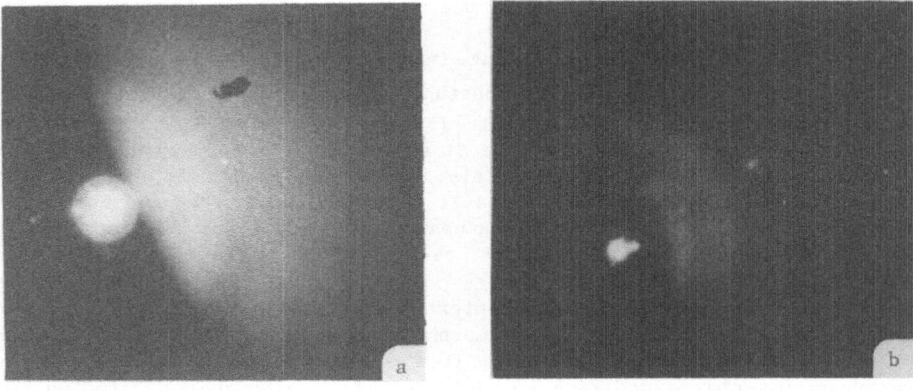

FIGURE 3. RHEED diagrams obtained at room temperature following growth of a 15 A° thick GaAs layer on a) Si and b) (Ca,Sr)F$_2$.

The coverage of the substrates by the thin GaAs layers was effective, as evidenced by RHEED observations. The RHEED diagrams obtained in both cases (figure 3) are composed of two diffuse halos , associated with the short distance order of the amorphous state. AES analyses indicate the same result.

These AES studies were carried out taking into account the non-stoichiometry of the layer grown. Following growth of two monolayers, T_s was increased to 400 C for desorbing the excess As and then lowered for recording the AES spectrum. This procedure was repeated several times. The evolution of the substrate element (Ca or Si) peak amplitude was then plotted as a function of the GaAs thickness. The result obtained in the case of GaAs/(Ca,Sr)F$_2$ is presented in figure 4. The same was found for GaAs/Si [8]. In figure 4, the variation of the substrate signal for a GaAs growth temperature of 400 C is also indicated for comparison purposes : here, as previously discussed, the substrate coverage is indeed achieved for a higher layer thickness (200 A°).

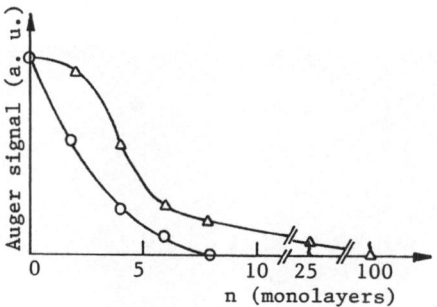

FIGURE 4. Variation of the Auger signal amplitude for Ca (292 eV) as a function of GaAs thickness grown on (Ca,Sr)F$_2$; o : T$_s$ < 100 C, Δ : T$_s$ = 400 C.

Improvement of the crystallinity of the thin GaAs amorphous layers by thermal annealing was next considered. RHEED observations were made, while T$_s$ was progressively increased after growth of the 15 A° thick amorphous GaAs layer. As already noticed in the case of layer stoichiometry, a temperature of 400 C was found to be sufficient to restore the layer crystallinity ; indeed, at about this temperature, the RHEED diagrams change rapidly and finally indicate monocrystallinity of the annealed layers. Twin spots are sometimes noted during heating, which disappear quickly, leaving a defect-free streaky pattern. Surface reconstructions are often detectable, as seen in figure 5, indicating ordering and smoothing of the layer surface. This result points to the effectiveness of thermal treatments on such thin amorphous heteroepitaxial layers.

The absence of twinning shows that the procedure employed would be well suited for heteroepitaxial growth, where twin formation occurs easily at the heteroepitaxial interface as a response to mismatch stress. The role played by such a procedure in stress relaxation has yet to be determined.

5. CONCLUSION

The nucleation step of heteroepitaxial GaAs growth on Si and (Ca,Sr)F$_2$ surfaces was studied using AES and RHEED.

Under the experimental conditions generally used the growth mode is 3D in both cases, and this is known to be a source of crystal defects at the coalescence step. On the basis of the kinematic analysis of epitaxial growth, we propose a procedure to convert this growth mode to a 2D-like mechanism. This procedure consists of growing first an amorphous, thin (15 Å) GaAs layer at T$_s$ < 100 C, whose stoichiometry and crystallinity are

328

restored by raising the substrate temperature to 400 C. This has been shown to provide a complete coverage of the starting surface (Si or $(Ca,Sr)F_2$) for very low film thicknesses thus reducing the formation of defects -namely twins- observed at this nucleation step of GaAs heteroepitaxial growth, and improving the surface morphology of the grown layer.

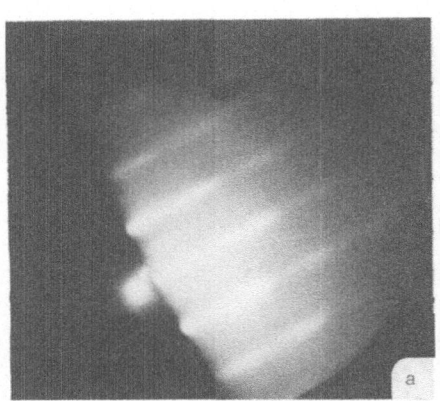 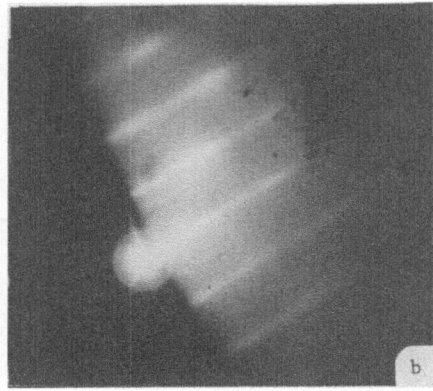

Figure 5. RHEED diagrams obtained after annealing of the 15 A° amorphous GaAs layer at 400 C : a) GaAs/Si, b) GaAs/(Ca,Sr)F_2.

REFERENCES

1. Castagné J., Fontaine C., Bedel E. and Munoz-Yague A., J. Appl. Phys., to be published, in June 1988.
2. Wang W.I., Appl. Phys. Lett., 44, p. 1149, 1984.
3. Tsutsui K., Ishiwara H. and Furukawa S., Appl. Phys. Lett., p. 587, 1986.
4. Metze G.M. and Calawa A.R., Appl. Phys. Lett., 42, p. 818, 1983.
5. Kashchiev D., Thin Solid Films, 55, p. 399, 1978.
6. Rosner S.J., Koch S.M. and Harris J.S., Jr, Appl. Phys. Lett., 49, p. 1764, 1986.
7. Biegelsen D.K., Ponce F.A., Smith A.J. and Tramontana J.C., J. Appl. Phys., 61, P. 1856, 1987.
8. Castagné J., Fontaine C., Bedel E. and Munoz-Yague A., J. Appl. Phys., to be published.

EPITAXIAL CaF$_2$–SrF$_2$–BaF$_2$ STACKS ON Si(111) AND Si(100)

S. BLUNIER, H. ZOGG, H.WEIBEL
AFIF at Swiss Federal Institute of Technology
ETH-Hönggerberg, CH-8093 Zürich, Switzerland

ABSTRACT

Stacks of epitaxial CaF$_2$, SrF$_2$ and BaF$_2$ layers have been grown by MBE onto Si(111), and, for the first time, onto Si(100). The layers are applied as epitaxial buffers for heteroepitaxy of lattice mismatched II-VI- and IV-VI-compound semiconductors on Si. On CaF$_2$ or SrF$_2$ covered Si(111) surfaces, growth of BaF$_2$ is 2-d after formation of the first monolayers, despite a lattice mismatch of 14%. On Si(100), BaF$_2$ grows with the same lattice orientation as the underlying substrate if a thin ($\approx$100 Å) intermediate CaF$_2$- or SrF$_2$-layer is deposited first, despite its preferred (111)-growth mode. Growth is 3-d on (100)-surfaces because of the large (100)-surface free energy of the group IIa-fluorides. In situ short anneal cycles increase the crystalline quality for (111)- as well as for (100)-orientation. – The lattice orientation of BaF$_2$ on Si(100) can be switched from (100) to twinned (111) by a short in situ anneal at very early stages of growth.

1. INTRODUCTION

Heteroepitaxy of II-VI- and IV-VI compound semiconductors on silicon would open the door to many interesting device applications. E.g. infrared focal plane arrays could be constructed with the infrared sensors fabricated in a narrow gap compound semiconductor top layer, and multiplexing of the high number of electrical signals performed in the silicon substrate. Such a design would lead to a true monolithic device, much simpler than present hybrid approaches. However, it is not possible in most cases to grow these materials directly onto Si even with MBE or related techniques. This is partly due to the high lattice mismatch, e.g. near 20% between CdTe or PbTe and Si. In addition, thermal expansion mismatch is another very severe problem leading to high stresses or even cracks in the epitaxial layers.

Most of the work on epitaxial fluoride growth on Si has been performed with CaF$_2$ [1–3]. Epitaxial CaF$_2$ on Si was overgrown with Si, Ge or GaAs layers with 4% lattice mismatch for Ge and GaAs [4–6]. These studies were performed on (111)- or (100)-oriented substrates.

With stacked or compositionally graded intermediate fluoride buffers, we found that BaF$_2$ with 14% lattice mismatch compared to Si can be easily grown on Si(111) [7,8]. In the most simple growth procedure, untwinned BaF$_2$ layers on Si(111) were obtained by growing first some CaF$_2$ (mismatch 0.6% at RT) on Si, eventually followed by SrF$_2$ (mismatch 7%). These layers were successfully applied as intermediate buffers on Si(111) substrates for further growth of high quality CdTe [9] as well as for PbSe [7], PbTe and Pb$_{1-x}$Sn$_x$Se [10], and background noise limited infrared sensors have been fabricated in the narrow gap lead chalcogenide layers. One very important fact is that thermal mismatch strain is relieved in such structures containing fluoride buffers. This is due to movement of misfit dislocations, most

329

Y. I. Nissim and E. Rosencher (eds.), Heterostructures on Silicon: One Step Further with Silicon, 329–334.
© 1989 by Kluwer Academic Publishers.

probably at or near the fluoride-Si interface [11]. The relief is possible under clean conditions only. E. g., when low purity material is deposited, or the substrate surfaces are contaminated, movement of dislocations is hindered and cracks often form.

Fluoride growth is easier with (111)- than with (100)-surface orientation. While the lattice orientation of CaF_2 on Si(100) is the same as that of the substrate, it was found in an early work [1] that BaF_2 grown directly on Si(100) does not have the (100) orientation, but consists of crystallites with a strong (111)-texture with the mean [111]-axis aligned perpendicular to the Si(100)-surface (this work was performed under non UHV-conditions). It is therefore not obvious that BaF_2 with its strongly preferred (111)-growth mode can be grown epitaxially on a largely mismatched (100)-surface.

In this presentation, we will discuss our results of BaF_2 growth on Si with intermediate stacked CaF_2 and/or SrF_2, and the effects of short in situ anneals. It will be shown that in situ annealing improves the crystalline quality of layers grown on Si(100) as well as on Si(111).

2. EXPERIMENTAL

A MBE apparatus equipped with a direct radiative sample heater and RHEED- and AUGER-facilities was used for the present work. Si wafers were cleaned with a modified Shiraki method. The protective oxide which forms in the last chemical cleaning step is removed by heating up to 900°C immediately prior to growth. The RHEED-pattern of the Si-surface then showed the 7 x 7- or 2 x 1- surface reconstruction for (111)- and (100)-oriented samples, respectively. C-contaminations at this stage were less than 0.01 monolayers, and background pressures were in the 10^{-9} mbar range during depositions. Growth temperatures were around 700°C for (111)-, and 550-600°C for (100)-oriented wafers, and growth rates in the low Å/sec range. Some CaF_2 ($\approx$100 Å) was deposited first in most cases, followed by 1000-4000 Å BaF_2. In some experiments, a thin ($\approx$100 Å) additional intermediate SrF_2 layer was deposited, too, and in some other experiments, $\approx$100Å of SrF_2 was grown directly onto Si and followed by BaF_2, thus omitting the CaF_2 step.

3. EFFECT OF ANNEALING ON (111)-ORIENTED Si-WAFERS

Growth of CaF_2 on Si(111) is 2-dimensional, streaked RHEED patterns are observed after formation of the first monolayers [2,3]. However, we sometimes observed a spotty RHEED-pattern in our system, or even rings as characteristic for polycrystalline growth became visible after starting CaF_2 growth [12]. This occurred even if the substrate exhibited a sharp 7 x 7 reconstructed surface RHEED-pattern prior to growth, and all other growth parameters were judged to be optimal.

Sometimes, the patterns transferred to streaks in such cases during further growth after the layer thickness reached some 10 Å, or if growth was interrupted for a few minutes. Solid state recrystallization of the thin layers is therefore effective already at 700°C, the growth temperature we used for most experiments. By short annealing (a few seconds at 1000°C) at this stage of growth, still improved RHEED-patterns were obtainable.

The same growth mode (streaks, or rings, or both at very early stages of growth) occurred when depositing SrF_2 directly onto Si(111), and in situ anneals improved the quality of the layer in the same way as for CaF_2 on Si(111).

After starting SrF_2 or BaF_2 deposition on top of the thin CaF_2 layer on Si (or BaF_2 on top of thin SrF_2/Si), the distance of the streaks changed within the first few monolayers to that corresponding to the new equilibrium

a)   b)

Fig. 1. RHEED-patterns of BaF_2 on CaF_2/Si(111) before (a) and after (b) an in situ anneal at ≈1000°C for a few seconds. Note that the rings in (a) caused by misoriented grains have disappeared after the anneal.

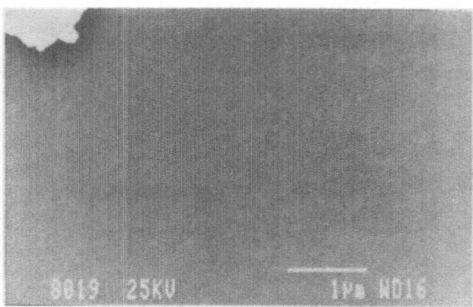 

a)   b)

Fig. 2. (a) Surface morphology of a 3800 Å thick BaF_2(111) layer on CaF_2 covered Si(111). (b) Corresponding electron channeling pattern.

lattice constant. The quality of the RHEED-pattern again increased after similar short anneals as described above were performed. Fig. 1a) shows an example of a RHEED-pattern where nonoriented BaF_2 grains give rise to rings in addition to the streaks. These rings disappear after anneal (Fig. 1b).

The surfaces of these layers were perfectly smooth, no structures were discernible with Nomarksi interference contrast or scanning electron microscopy (Fig 2a). Electron channeling patterns indicate untwinned layers with type B orientation compared to the Si-substrate (Fig. 2b). A possible amount of twinned type A grains in the type B matrix is below 2% volume fraction, the sensitivity limit of the X-ray technique used for these determinations [8]. Rutherford backscattering minimum channeling yields using 2MeV He^+ ions were below 4%, indicating again a high structural quality of the surface region of the samples. Fig. 3a shows a spectrum of a CaF_2-SrF_2-BaF_2 on Si(111) stack with minimum yield X_{min}=3.5%. Note the increase in intensity in the channeled spectrum towards the Si-interface which reflects the increase in defect density due to the discommensurate growth mode.

4. GROWTH ON (100) ORIENTED WAFERS

Spotty RHEED-patterns are observed for all fluorides when grown in a (100)-orientation. The 3-d growth mode is due to the high surface free

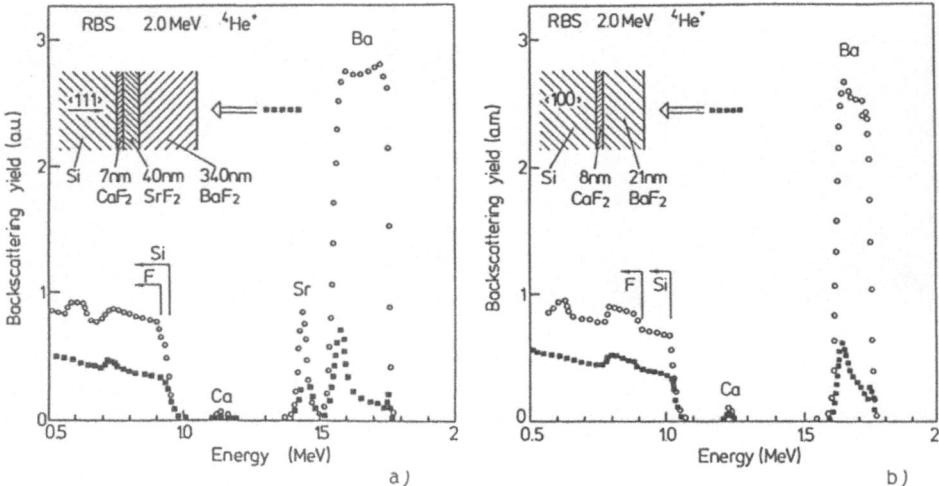

Fig. 3. Random and aligned Rutherford backscattering spectra for
(a) CaF_2-SrF_2-BaF_2 on Si(111) with $X_{min}=3.5\%$,
(b) CaF_2-BaF_2 on Si(100) with $X_{min}=8\%$.

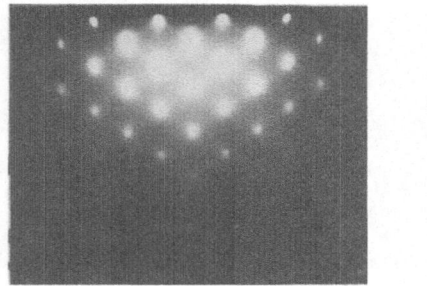

a)

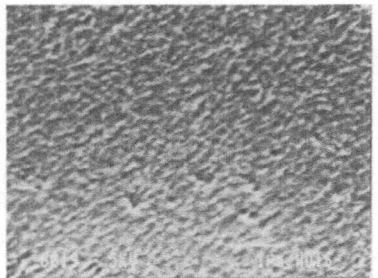

b)

Fig. 4. (a) RHEED-pattern and (b) surface morphology of a 3000 Å thick
BaF_2(100) film on CaF_2 covered Si(100).

energy of the (100)-surfaces of fluorides. Best quality of CaF_2 is obtained
at 550 °C growth temperature [1-3]. The surface is not perfectly smooth at
this growth temperature, but consists of columnar structures on a sub-micron
scale [3]. Very recently, it has been shown that perfectly smooth CaF_2(100)
surfaces with a streaky RHEED-pattern are obtained if the substrate temperature is increased to 850°C after growth of some hundred Å [13].

In our experiments, we deposited BaF_2 on top of thin (mean thicknesses
≈100 Å) CaF_2, SrF_2, or CaF_2/SrF_2 bilayers on Si(100). In all cases with
growth temperatures around 550-600°C, epitaxial BaF_2 with (100)-orientation
was obtained. RHEED-patterns were spotty, indicating 3-d growth (Fig. 4a),
and the surfaces were not smooth on a microscopic scale, they closely
resemble those of CaF_2 on Si(100). In situ rapid thermal anneal increased
the quality, and rings originating from misoriented grains at the very early

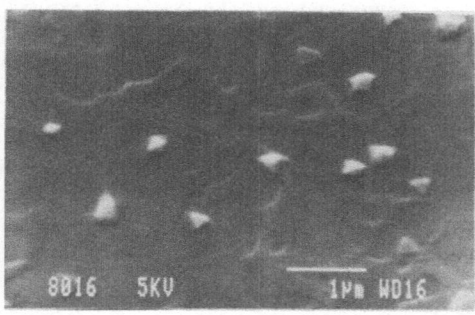

a) b)

Fig. 5. (a) Surface morphology and (b) electron channeling pattern of twinned (111)-oriented BaF$_2$ grains on CaF$_2$ covered Si(100). The orientation switched to (111) by annealing during growth before the layer thickness reached ≈30 Å.

stages of growth disappeared on proper annealing [14]. Minimum Rutherford backscattering channeling yields were below 10% for BaF$_2$ layers of 2000Å thickness (Fig. 3b). Values below 4% were obtained with thicker layers. Even a 4800Å thick BaF$_2$ layer on intermediate SrF$_2$ (but no CaF$_2$) showed a X_{min} value of 4.2%. The layers were free of cracks, and thermal mismatch strain was relaxed similar to the Si(111) case.

If in situ anneals were performed at very early stages of growth (mean layer thickness below 30Å), it was possible to switch the orientation of the BaF$_2$ from (100) to (111). These layers were of low quality. Minimum channeling yield was 30%, the surfaces were structured (Fig. 5b), and the layers were twinned with 12-fold symmetry. This twinning can be explained with the 4-fold (100)-substrate symmetry combined with the 6-fold symmetry of (111)-surface lattices.

5. CONCLUSIONS

In conclusion, we have shown that high quality BaF$_2$ can be grown epitaxially onto Si(100) with (100)-lattice orientation despite the high lattice mismatch (14%) and its preferred (111)-growth mode. (100)-orientation on Si(100) was obtained reproducibly if a thin (≈100 Å) intermediate CaF$_2$, SrF$_2$, or stacked CaF$_2$/SrF$_2$ layer was deposited first. Twinned (111)-oriented BaF$_2$ grows if the sample is shortly annealed in situ before the mean film thickness reaches 30 Å. This change of orientation is related with the lower (111)- free surface energy of fluorides and the enhanced surface diffusivity at higher temperatures.

In situ thermal anneal cycles during growth improve the crystallographic quality both for (100)- and (111)-substrate orientation. These cycles can be performed at an early stage of layer growth (≈100 Å). They are effective with lower temperatures and/or less duration compared to rapid thermal anneals (RTA) on completed several 1000 Å thick layers [3].

For applications as buffer layers for growth of compound semiconductors and fabrication of sensor devices in these layers, (100)-orientation is preferred if CMOS circuits have to be fabricated in the Si-substrate for signal processing purposes.

ACKNOWLEDGEMENTS

The authors would like to thank P. Waegli, ETH Zürich, for performing the SEM investigations, and Prof. V. Meyer and Dr. R. Pixley, University of Zürich, for the kind permission and help to use their Van-de-Graaf apparatus for the RBS investigations. The work is sponsored by the Swiss National Science Foundation, and by the Swiss Defense Technology and Procurement Agency.

REFERENCES

[1] T. Asano, H. Ishiwara, N. Kaifu, Jap. J. Appl. Phys. 22, 1474, 1983.
[2] L.J. Schowalter, R.W. Fathauer, R.P. Goehner, L.G. Turner, R.W. DeBlois, S. Hashimoto, J.-L. Peng, W.M. Gibson, J.P. Krusius, J. Appl. Phys. 58, 302, 1985.
[3] L. Pfeiffer, J.M. Phillips, T.P. Smith, III, W.M. Augustyniak, K.W. West, Appl. Phys. Lett. 46, 947, 1985.
[4] H. Onoda, T. Katoh, N. Hirashita, M. Sasaki, Techn. Digest Int. Electron Devices Meeting IEDM, Washington D.C. Dec. 1985, p. 680.
[5] R.W. Fathauer, N. Levis, E.L. Hall, L.J. Schowalter, J. Appl. Phys. 60, 3886, 1986.
[6] H. Ishiwara, T. Asano, H.C. Lee, Y. Kuriyama, K. Seki, S. Furukawa, Mat. Res. Soc. Symp. Proc. 67, 105, 1986.
[7] H. Zogg, M. Hüppi, Appl. Phys. Lett. 47, 133, 1985.
[8] H. Zogg, P. Maier, H. Melchior, J. Crystal Growth 80, 408, 1987.
[9] H. Zogg, S. Blunier, Appl. Phys. Lett. 49, 1531, 1986.
[10] H. Zogg, P. Norton, Techn. Digest Int. Electron Devices Meeting IEDM, Washington D.C. Dec. 1985, p. 121.
[11] H. Zogg, Appl. Phys. Lett. 49, 933, 1986.
[12] H. Zogg, S. Blunier, J. Masek, Proc. 2nd Int. Symp. on Silicon Molecular Beam Epitaxy, Honolulu, Hawaii, Oct. 1987, Electrochem. Soc., Proc. Vol. 88-8, p.321
[13] Y. Morimoto, S. Sudo, K. Yoneda, presented at the MRS spring meeting Reno NV, April 1988; to appear in Mat. Res. Soc. Symp. Proc. Vol. 116.
[14] S. Blunier, H. Zogg, H. Weibel, ibid.

YTTRIA STABILISED ZIRCONIA HETEROEPITAXY ON SILICON BY ION BEAM SPUTTERING

C. PELLET, C. SCHWEBEL
Institut d'Electronique Fondamentale (UA22) CNRS–Université Paris Sud,
91405 Orsay Cedex.

P. LEGAGNEUX
Thomson CSF/LCR Domaine de Corbeville,
B.P. 10, 91401.Orsay.

J. SIEJKA
Groupe de Physique des Solides de l'Ecole Normale Supérieure
Tour 23, 2 Place Jussieu, 75221 PARIS CEDEX.

1. INTRODUCTION

Among the different substrates for silicon on insulator structures, yttria stabilised zirconia (YSZ) offers several advantages as high chemical stability, good resistance to cosmic ray and the possibility of silicon oxydation through the zirconia at the Si/YSZ interface (1). However the silicon on zirconia (SOZ) technology is presently restrained by the size of available zirconia monocrystal (namely 2 inch in diameter) which is inconsistent with usual silicon processes. To overcome this problem, a solution consists in the use of zirconia heteroepitaxial thin film deposited on silicon wafers, as substrate, instead of bulk zirconia.In this way, we have deposited YSZ thin film on Si by ion beam sputter deposition (IBSD) at the Institut d'Electronique Fondamentale.This technique is used in this laboratory since few years, for the deposition of dielectric (2) metalliC (3,4) or epitaxial semiconducting films (5). Especially we have demonstrated the feasability of silicon homoepitaxial thin films with good electrical properties (6) . The IBSD technique consists in the sputtering of a target by an energetic ion beam issued from an ion source. In this configuration the plasma is isolated from the deposition chamber where ultra high vacuum condition could be achieved. Comparatively to conventionnal R.F. sputtering, this allow purer layers and *in situ* characterization. This paper present our first results in the study of YSZ heteroepitaxy on Si.

Y. I. Nissim and E. Rosencher (eds.), Heterostructures on Silicon: One Step Further with Silicon, 335–340.
© *1989 by Kluwer Academic Publishers.*

2. EXPERIMENTAL SET UP

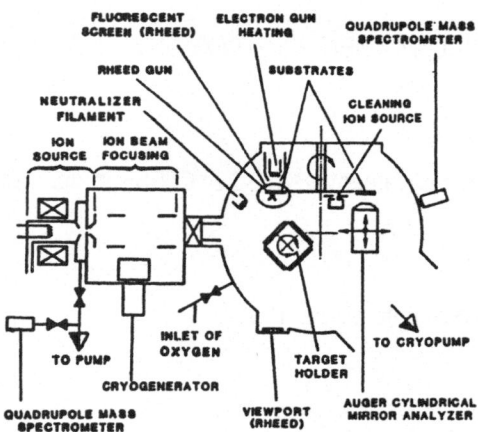

FIGURE 1 :Schematic drawing of the experimental set-up

The experimental set up (see Fig. 1) previously described (6) consists of three mains parts : the duoplasmatron source, the intermediate chamber (focusing chamber) and the deposition chamber. This later is equipped with three diagnostics technique, reflection high energy electron diffraction (RHEED), Auger electron spectrometry (AES) and quadrupole masss spectrometer (QMS). The ultimate pressure in this chamber is $5 \cdot 10^{-8}$ Pa. For this work, it has been necessary to put an electron gun in front of the target in the way to avoid the target's charge due to the ion beam.

3. DEPOSITION PARAMETERS

The target used in these experiments is a commercial sintered YSZ target with a typical composition $(Zr\,O_2)_{0.77}\,(Y_2O_3)_{0.23}$. When an ion strikes a solid surface it can be either trapped or backscattered. In the way to minimize the second process which could induce defects in the growing film (7) all of our experiments are carried out with primary ion heavier than zirconium or yttrium (i.e. xenon ion). The ion energy was 20 keV and the ion beam current was 1,7 mA corresponding to 100 $\mu A/cm^2$ maximum current density.For beam current of 1,7 mA, the working pressure, which is nearly proportionnal to the ion flux emitted by the source, was 3.10^{-5} Pa. This atmosphere consists mainly of xenon (93 %). The gaseous impurity (CO, H_2O, CO_2 : 7 %) are due principally to the target's out gazing.In addition we are able to introduce oxygen with a partial pressure up to 10^{-3} Pa in the deposition chamber during the thin film's growth.

Before their introduction in the deposition chamber, the silicon substrates ((100) oriented 3-6 Ω cm) are chemically cleaned. The last stage of this treatment consists in the growth of a protective oxide which will be sublimated in vacuum prior to the film deposition. This is achieved by flash heating at 1000°C for 2 minutes. The surface cleanliness is then checked *in situ* by means of Auger electron spectrometry. During the growth the substrate temperature has been adjust from 20°C up to 900°C.

4. RESULTS

4.1. Deposition rate

The deposition rate for IBSD is proportionnal to the ion beam current. With a xenon ion beam of 1,7 mA we have obtained a growth rate of $8 \ 10^{-2}$ nm/s. All the films described here are 150 nm thick.

4.2 - Composition of the films

The film composition has been studied *in situ* using Auger Electron Spectroscopy (AES) and *ex situ* at the Groupe de Physique des Solides by Proton induced X-ray emission (PIXE), Rutherford Backscattering (RBS) and nuclear microanalysis. The Y/Zr ratio, estimated by PIXE, is found equal to 0.5 corresponding to a $(ZrO_2)_{0.8} \ (Y_2O_3)_{0.2}$ composition close to the target one. The AES Y_{MNN}/Zr_{MNN} peak to peak height ratio of 0.67 (8), for a YSZ film deposited at 700°C under an $1.2 \ 10^{-4}$ Pa oxygen pressure, confirms this composition. This result can be explained by the fact that :

a/ At the steady state the composition of sputtered flux is the same than the target's one.

b/ For the two compound yttrium and zirconium, the vapor pressure is very low leading to a sticking coefficient of one.

Then the Y/Zr ratio is maintained from the target of the source. As shown in the figure 2, the O/Y+Zr ratio is greatly dependant of the oxygen partial pressure. It varies form 1.2 for deposition without additionnal pressure up to 1.6 for an oxygen partial pressure of 10^{-4} Pa. In all cases the value is below 1.83 corresponding to a stoechiometric film $(Zr \ O_2)_{0.8} \ (Y_2O_3)_{0.2}$.

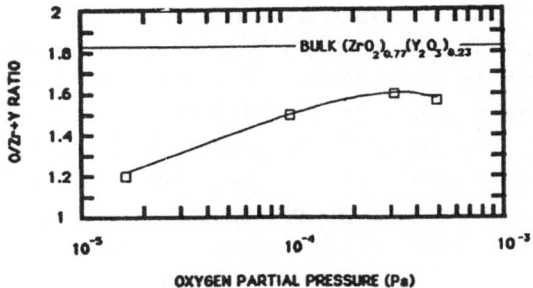

FIGURE 2: O/Zr+Y ratio as a function of oxygen pressure
(films deposited at 750°C)

4.3. Structure

The crystalline structure was checked using *in situ* RHEED measurement and *ex situ* X-ray diffraction at the Thomson-CSF and channeling at the Groupe de Physique des Solides. YSZ epitaxial growth could not be achieved without additionnal oxygen pressure. Within the range 700°-800°C and under an oxygen partial pressure of the order of 2.10^{-4} Pa the layers exhibited monocrystalline cubic phase. This can be illustrated by figure 3 showing a RHEED pattern of a YSZ film grown at 750°C under an $1.2\ 10^{-4}$ Pa oxygen pressure.

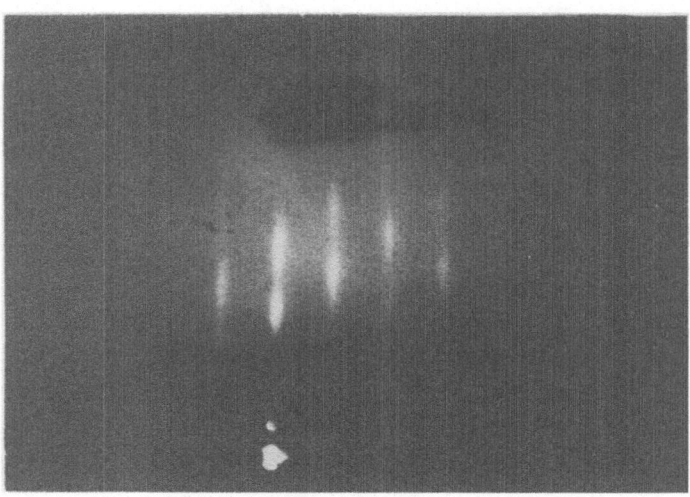

FIGURE 3: RHEED pattern of a YSZ film deposited at 750°C under an
$1.2\ 10^{-4}$ Pa oxygen pressure

The figure 4 shows the X ray diffraction pattern of a YSZ film deposited at 700°C under an 1.2 10^{-4} Pa oxygen pressure. For both YSZ and Si, one can observe exclusively the peaks of the (100) family planes (e.g. YSZ peaks (200) and (400) and Si peaks (400)). The lattice parameter deduced from the X ray spectrum was found to be 0.518 nm. This value corroborate the Y/Zr ratio of 0.5.

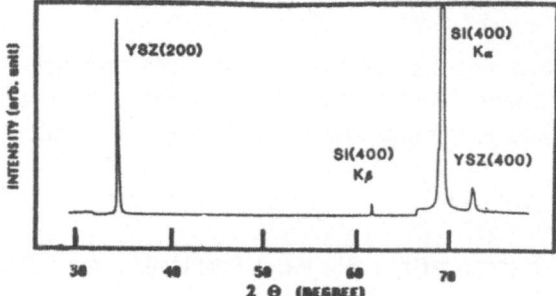

FIGURE 4: X-ray diffraction pattern of a YSZ film deposited at 750°C under 1.2 10^{-4} Pa oxygen pressure

The good cristallographic continuity between the silicon substrate and the YSZ film has been estimated by channelling of 0.75 MeV He ions. The figure 4 shows a random and aligned RBS spectra of a thin YSZ heteroepitaxial film. The resulting χ_{min} value of 17 % is to compare with a bulk value of 3.5 %. The rather low thickness of the film could explain the difference. Indeed, as currently observed in the case of heteroepitaxy (9), one can hope, the χ ratio decrease when the film thickness increase. Nevertheless, our films probably contains twins and dislocations.

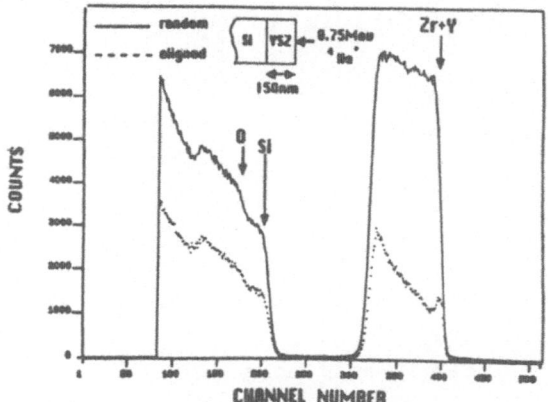

FIGURE 5: Random and (100) aligned energy spectra of 0.75 Mev ^{4}He$^+$ ions scattered from a 150 nm thick YSZ film

5. CONCLUSION

We have reported the characteristics of thin YSZ heteroepitaxial films on Si deposited using IBSD.The mains parameters, in the way to obtain epitaxy, are the substrate temperature and the oxygen partial pressure. Within the temperature range 700°-800°C and for oxygen pressure of the order of 10^{-4} Pa YSZ hetero- epitaxy on Si is achieved.

Electrical properties of the YSZ films are currently studied, by C(V) and I(V) measurement. These characterization will allow the optimisation of the parameters and so the improvement of these first results.

REFERENCES

1. D. Pribat, L.M. Mercandalli, J. Siejka, J. Perrière, J. Appl. Phys. 58(1), 1985, P. 313.
2. D. Bouchier, A. Bosseboeuf,Thin Solid Films, 139 (1986), p. 95.
3. F. Meyer, C. Pellet, C. Schwebel,Le Vide Les couches minces, 42, 236 (1987).
4. F. Meyer, E. Velu, C. Pellet, C. Schwebel, C. Dupas, Revue de Physique Appliquée (Mai 88).
5. C. Schwebel, G. Gautherin, 34th A.V.S. Symposium, Nov. 1987.
6. C. Schwebel, F. Meyer, G. Gautherin, C. Pellet,J. Vac. Sci. Technol. B4(5), Sept.-Oct. 1986.
7. C. Schwebel, C. Pellet, G. Gautherin,Nucl. Instr. and Meth. B18, 525 (1987).
8. J.S. Solomon, J.T. Grant,J. Vac. Sci. Technol. A3(2), 373 (1985).
9. M.S. Abrahams, J.L. Hutchinson, G.R. Booker,Phys. Stat. Sol. (a) 63, K3, (1981).

HETEROEPITAXY OF SEMICONDUCTOR/FLUORIDE/Si STRUCTURES

Tanemasa ASANO, Hiroshi ISHIWARA, and Seijiro FURUKAWA

Graduate School of Science and Engineering, Tokyo Institute of Technology, 4259 Nagatsuda, Midoriku, Yokohama 227, Japan

ABSTRACT
        Heteroepitaxial growth of alkaline earth fluoride films on Si substrates and Si, Ge, and GaAs films on the fluoride/Si structures, is reviewed. Growth of single crystalline fluoride films on Si and CaF_2/Si interface characteristics are first discussed. Then the usefulness of novel heteroepitaxial technologies, the thin amorphous layer predeposition method and the electron beam irradiation method, is demonstrated in the growth of semiconductor films on fluoride layers. Finally, use of vicinal (100) substrates in conjunction with rapid thermal annealing treatments is shown to be effective in growing high quality GaAs(100) films on fluoride layers.

1. INTRODUCTION
        Heteroepitaxial structures composed of semiconductors and insulators have a variety of potential applications ranging from high mobility surface channel transistors to semiconductor-on-insulator structures, with emphasis on very high speed integrated circuits(ICs), 3 dimensional ICs, optoelectronic ICs and intelligent sensors. In particular those prepared on Si substrates are very attractive for various applications. One of the most promising approaches to realize the semiconductor/insulator layered structures is the epitaxial growth of alkaline earth fluorides such as CaF_2, SrF_2, BaF_2 and mixtures thereof. This is because 1) the lattice constants of the fluorides can be continuously varied in the range between 0.546nm and 0.620nm by forming mixed fluorides, which cover the lattice constants of semiconductors of interest as shown in Fig. 1, and 2)thin films of these fluorides can be formed by vacuum evaporation or molecular beam epitaxy (MBE) method.
        Since the success of growing epitaxial fluoride films on semiconductors[1] and Si/CaF_2/Si double heteroepitaxial structures[2,3], considerable progress has been made on the growth of fluoride/Si and semiconductor/ fluoride/Si structures composed of Si[4], Ge[5-7], GaAs[8], PbTe[9], CdTe[10] overlayers. Besides there have been many extensive studies on the growth of fluoride films on semiconductor substrates other than Si for the purpose of forming stable surface passivation films and semiconductor-on-insulator structures[11-15]. Feasibility of such semiconductor/fluoride layered structures for device applications has been demonstrated by fabricating metal-oxide-semiconductor(MOS)[16] and metal-semiconductor(MES) field effect transistors (FETs)[17], metal-epitaxial insulator-semiconductor FETs(MEISFETs)[18], inverter chains[19], and infrared sensors[20].
        However the growth of semiconductor/fluoride/Si structures has not been straightforward. Owing to the large difference in material properties between the fluoride and semiconductors, many difficulties have arisen. Some examples are as follows. The thermal expansion coeficients of the fluorides are about 20×10^{-6}/deg at room temperature, which is much

Y. I. Nissim and E. Rosencher (eds.), Heterostructures on Silicon: One Step Further with Silicon, 341–357.
© 1989 by Kluwer Academic Publishers.

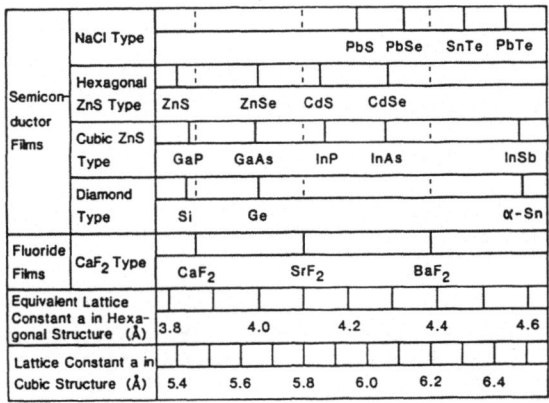

Semiconductor Films	NaCl Type						PbS	PbSe	SnTe	PbTe
	Hexagonal ZnS Type	ZnS		ZnSe	CdS	CdSe				
	Cubic ZnS Type	GaP	GaAs		InP	InAs				InSb
	Diamond Type	Si	Ge							α-Sn
Fluoride Films	CaF₂ Type	CaF₂			SrF₂			BaF₂		
Equivalent Lattice Constant a in Hexagonal Structure (Å)		3.8		4.0		4.2		4.4		4.6
Lattice Constant a in Cubic Structure (Å)		5.4		5.6		5.8		6.0		6.2	6.4

Fig. 1. Lattice constants of fluorides and various semiconductors.

higher than those of Si(2.5×10^{-6})/deg and GaAs(6×10^{-6}/deg). This difference can induce cracks and crystalline defects in the fluoride and semiconductor films[21]. The fluorides are ionic bonding material and the surface free energy of the (111) of the fluorides is particularly low. This results in poor wettability of semiconductors to the fluoride surface, appearance of the microfacets at the (100) fluoride films[22,23], and so on. Moreover, as is in the growth of other heterostructures, the interfacial reaction takes place between deposited semiconductor and fluorides at high growth temperatures.

In order to reduce or eliminate defects owing to these differences in material properties, several new heteroepitaxial growth techniques have been developed. That is, the thin amorphous layer predeposition method[4], the electron beam exposure epitaxy[7,24], rapid thermal annealing(RTA) treatments[21,25], etc. In this paper, mainly focusing on such new growth processes, we review recent progress in the research of growing fluoride films on Si and Si, Ge, and GaAs films on fluoride/Si structures.

2. GROWTH OF SINGLE CRYSTAL FLUORIDE FILMS ON Si

Fluoride films having almost stoichiometric composition can be formed by simply evaporating the fluorides in vacuum, because fluorides evaporate as molecule. Thus, so far, epitaxial fluoride films have been grown by vacuum evaporation or molecular beam epitaxy(MBE). The crystalline quality of fluoride films grown on Si substrates has been found to depend strongly on the lattice mismatch. CaF₂ films can be grown epitaxially on various crystal planes of Si substrates. Figure 2 shows the substrate-temperature dependence of the channeling minimum yield X_{min} in the Rutherford backscattering spectroscopy(RBS) for CaF₂ films grown on (100), (311), (111), and (110) oriented Si substrates. Precise analyses with ion channeling measurements and transmission electron microscopy(TEM) have shown that single crystal CaF₂ films can be grown on (111) and (100) oriented substrates at 600–800°C and 500–600°C, respectively[26].

It has been pointed out that the cryatal orientation of the single crystal CaF₂ films grown on Si(111) substrates rotates by 180° about the surface normal <111> axis of the substrate (rotationally twinned,or so-called type B orientation)[27]. Recently ultrathin(1nm) CaF₂ films grown on

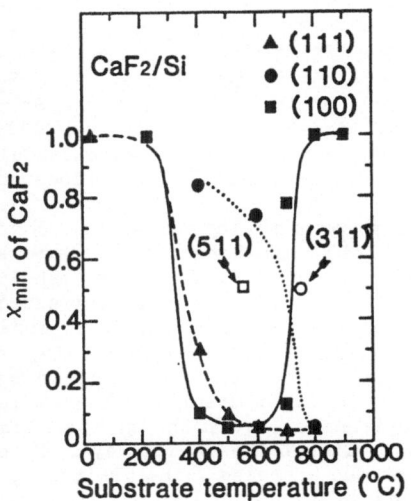

Fig. 2. Dependence of the substrate temperature of the channeling minimum yield for CaF$_2$ films grown on (100),(511),(311),(111), and (110) oriented Si substrates.

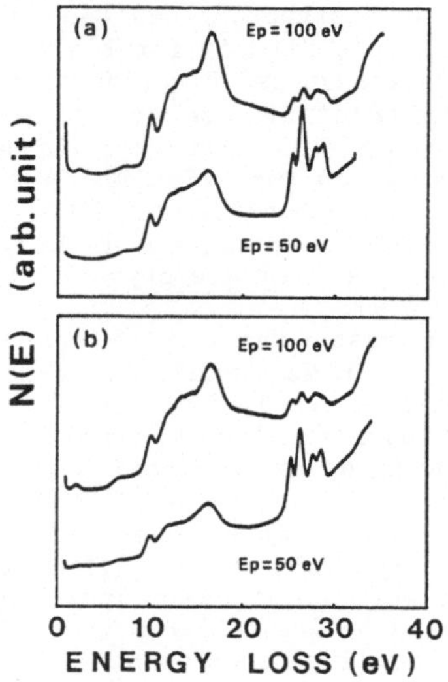

Fig. 3. LEELS spectra taken from (a) a bulk CaF$_2$ single crystal and (b) a 1nm-thick CaF$_2$ film grown on Si(111).

(111)Si have been analyzed by Auger electron spectroscopy(AES) and low-energy electron energy loss spectroscopy(LEELS)[28]. Figure 3 shows LEELS spectra taken from a 1nm thick CaF$_2$ film on (111)Si and from a bulk (111)CaF$_2$. It has been found that such ultrathin CaF$_2$ films grown on (111)Si are uniform and have the same electronic structure as that of a bulk CaF$_2$.

On Si(100) substrates, since the optimum temperature range is relatively narrow, misoriented grains appear in CaF$_2$ films when the substrate temperature is deviated from the desired one. Pfeiffer et al have shown that such misoriented grains are recrystallized by rapid thermal annealing at about 1000°C and CaF$_2$ films having excellent crystalline quality can be formed[25]. But there was a problem that a cracks were generated in the CaF$_2$ films. They have used an ex situ RTA system. That is, the growth system and the annealing system are independent. On the other hand, it has recently been found that in situ RTA treatment(RTA in a growth system) is effective to improve the quality of CaF$_2$ films without generating cracks[23]. This result suggests that absorption of contaminants such as oxygen from the ambient into the film reduces the mechanical strength of CaF$_2$ films.

On the contrary to the CaF$_2$ films, lattice-mismatched fluoride films such as SrF$_2$, BaF$_2$, and (Ca,Sr)F$_2$ do not grow on Si substrates as single crystal. On Si(100) substrates, they become polycrystalline films containing (111) oriented crystallites. On Si(111) substrates, they are composed of two types of epitaxial crystallites; e.g. one has orientations identical to those of the substrate(so-called type A orientation) and the other has the orientations of type B mentioned above. In order to grow single crystal films of the lattice-mismatched fluorides, the intermediate

344

CaF$_2$ layer structure has been developed[29]. In this structure, a single crystal CaF$_2$ film is first grown on Si substrates followed by growth of a lattice-mismatched fluoride film on top of the CaF$_2$ film. Figure 4 shows X-ray diffraction spectra taken from a SrF$_2$/Si(100) structure and a SrF$_2$/CaF$_2$/Si(100) structure. As can be seen clearly in the figure, the SrF$_2$ film grown on the CaF$_2$/Si(100) structure grows epitaxially while the SrF$_2$ film grown directly on Si(100) shows strong signals diffracted from (111) oriented SrF$_2$. The intermediate CaF$_2$ layer has been found to be also useful to grow type B single crystal films of the lattice-mismatched fluorides on Si(111)[29].

Besides, there is a problem in the growth of lattice-mismatched fluoride films. Cracks are generated in thick, mismatched films. Cracks appear more easily in mixed fluoride films such as (Ca,Sr)F$_2$ films than in pure fluoride films. We have found that crack generation can be supressed by growing a compositionally graded fluoride layer[30]. In Fig. 5, typical surface morphologies of (Ca,Sr)F$_2$/CaF$_2$/Si(111) and (Ca,Sr)F$_2$/graded-layer /CaF$_2$/Si(111) structures are compared. These samples were prepared by evaporating CaF$_2$ and SrF$_2$ from separate crucibles. No crack is observable in the surface of the sample with the graded layer, even though its film thickness is greater than that of the sample without the graded layer. From ion channeling measurements with ions at various energies , the dislocation density in the top (Ca,Sr)F$_2$ film was found to be reduced by growing the compositionally graded layer beneath[30]. We speculate that the suppression of crack generation is the result of the reduced dislocation density.

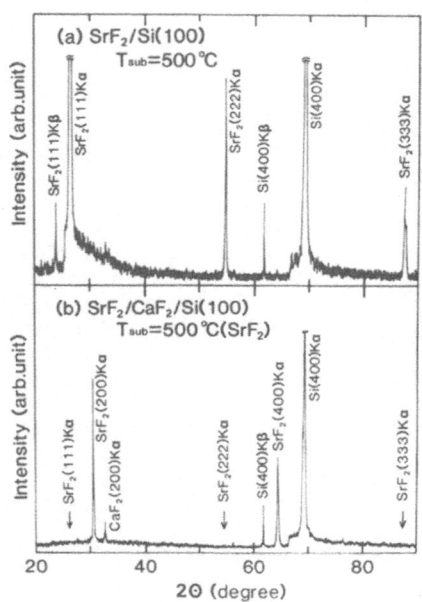

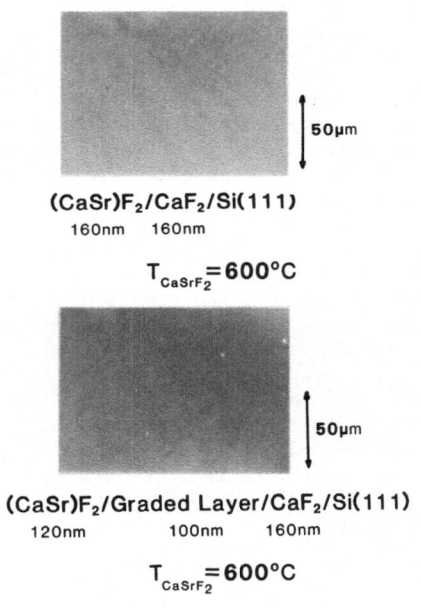

Fig. 4. X-ray diffraction spectra for (a)SrF$_2$/Si(100) and (b)SrF$_2$/CaF$_2$/Si(100) structures.

Fig. 5. Optical micrographs of surfaces of (a)(Ca,Sr)F$_2$/CaF$_2$/Si(111) and (b)(Ca,Sr)F$_2$/compositionally-graded-layer/CaF$_2$/Si(111) structures.

3. CaF$_2$/Si INTERFACE CHARACTERISTICS

Smith et al have demonstrated the successful operation of MEISFETs fabricated by using epitaxial CaF$_2$ films on (100)Si[18]. The observed channel electron mobility was about 400cm^2/Vs, which is lower than the value obtained from usual MOSFETs. This has been attributed to the rather high density of the interface states; about 5x10^{11}/cm^{2}eV. In order to investigate factors which determine the interface states, we have measured the interface characteristics of CaF$_2$/Si(100) structures prepared under various conditions. Figure 6 shows capacitance-voltage characteristics for two CaF$_2$/Si structures. One was prepared by growing a CaF$_2$ film at 600°C. The other was prepared by growing a CaF$_2$ film at 400°C followed by in situ RTA at 900°C for 30sec. The effectiveness of RTA treatments has been demonstrated for CaF$_2$/Si(111) structures[31]. The former sample shows the Fermi level pinning near the valence band edge, while the later sample shows a large variation of capacitance. The maximum and minimum capacitance values almost corresponds to the values for accumulation and inversion states. The interface state density of the RTA processed sample was about 1x10^{12}/cm^{2}eV as measured by the conductance method. The crystalline quality of the RTA processed CaF$_2$ film was slightly better than that of the film grown at 600°C. However, another sample whose CaF$_2$ film was grown at 400°C showed an unpinned interface characteristic, although the film quality of this sample was worse than that of the film grown at 600°C. These results suggest that the bonding structure at the interface changes with the growth temperature and it is an important factor influencing the electrical characteristic at the CaF$_2$/Si interface. Structural characterization of the interface[32,33] will be useful for further improvement of the interface characteristics.

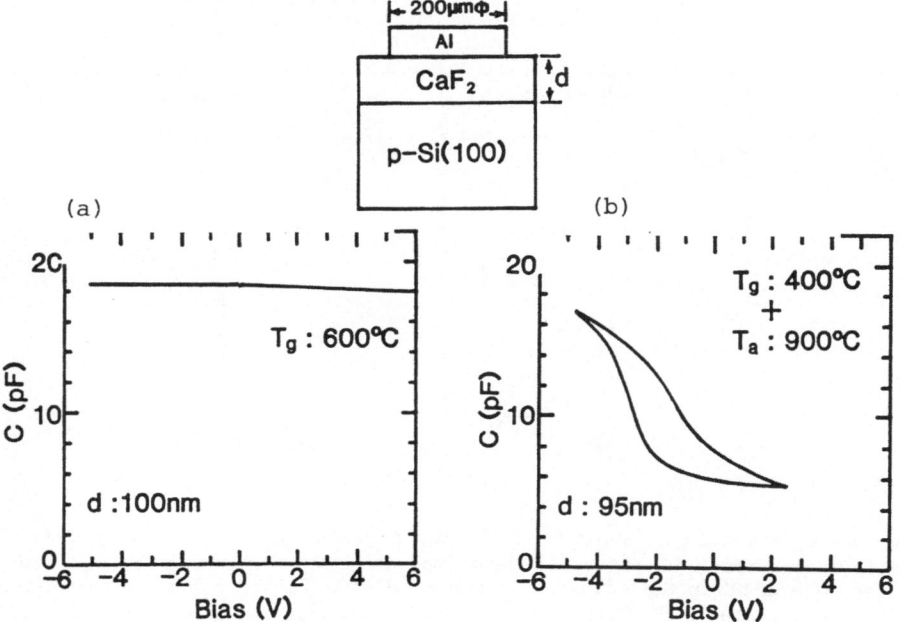

Fig. 6. Capacitanece-Voltage characteristics for CaF$_2$/Si(100) structures grown at 600°C(a) and 400°C followed by RTA at 900°C(b).

346

4. Si OR Ge OVERGROWTH ON FLUORIDE/Si STRUCTURES
4.1 The thin amorphous layer predeposition method

Growth of Si or Ge films on fluoride/Si structures was carried out by using vacuum deposition or MBE. Use of $(Ca,Sr)F_2$ matched in lattice constant to Ge has been found to be useful for growing Ge films having better crystalline quality[34]. But, in order to investigate the factors other than lattice matching, growth of Ge films on CaF_2/Si structures have been mostly investigated as well as growth of Si on CaF_2/Si structures. If these semiconductor films are deposited onto heated fluoride/Si structures, the following problems appear when the substrate temperature is raised.

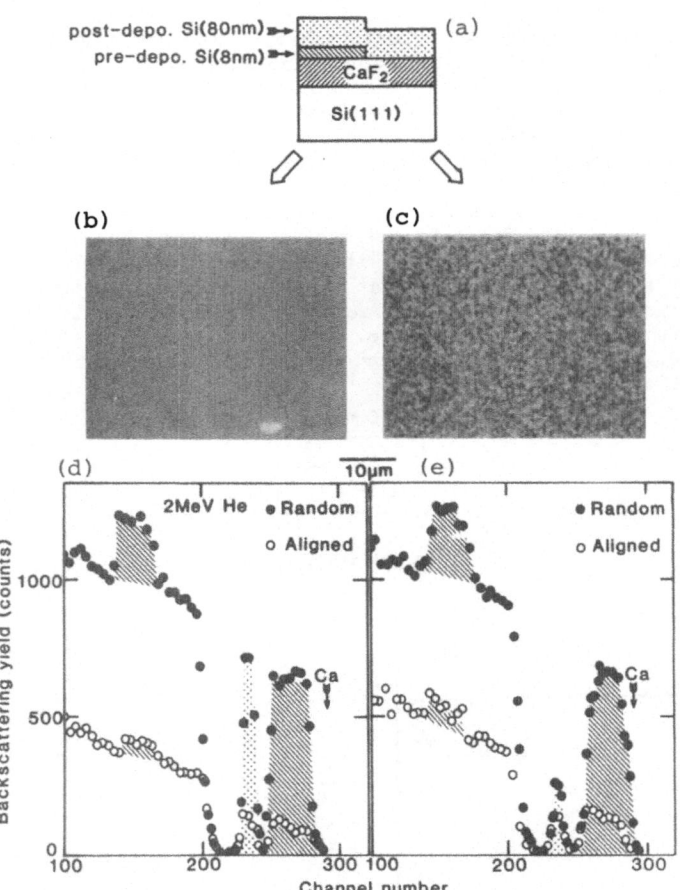

Fig. 7. (a)Schematic illustration of sample geometry. In one region(left side), an 8nm thick Si layer was deposited onto the CaF_2 surface followed by deposition of an 80nm thick Si film at 800°C. In the other region(right side), an 80nm thick Si film was deposited directly on top of the CaF_2 film at 800°C. (b) and (c): Optical micrographs of the surfaces of the respective regions. (d) and (e): Random and aligned backscattering spectra taken from the respective regions.

1) For the growth of Si films on CaF_2/Si structures, interfacial reaction takes place between deposited Si and the underlying CaF_2 at substrate temperatures above 750°C[3].

2) For the growth of Ge films, the uniformity of Ge films become very poor due to island growth of Ge on fluorides[5].

In order to overcome these problems, the method of predepositing a thin amorphous layer has been developed[4,6]. In this method, a thin amorphous Si or Ge layer is deposited onto the CaF_2 surface at room temperature, prior to the deposition of Si or Ge films at elevated temperatures.

Figure 7 demonstrates effects of the thin amorphous Si layer for preventing the interfacial reaction. The sample consists of two regions. In one region, an 8nm thick Si layer was predeposited on CaF_2 at room temperature, followed by deposition of an 80nm thick Si film at 800°C. In the other region, an 80nm thick Si film was directly deposited on CaF_2 at 800°C. The surface morphology of the region with the predeposited layer is rather smooth, while that of the region without the predeposited layer is very rugged. From the Rutherford backscattering spectra shown in the figure, we can see that a uniform Si film grows epitaxially in the region with the predeposited layer.

Figure 8 shows the variation of the channeling minimum yield χ_{min} near the surface of approximately 400nm thick Si films on (111) and (100) substrates with thickness of the predeposited Si layer. The χ_{min}'s of Si films on both (111) and (100) substrates are less than 8% when the predeposited Si layer is about 10nm or less in thickness.

The epitaxial growth mechanism of the predepositon method has been investigated by analyzing the initial stage of the growth using transmission electron microscopy(TEM) and ion channeling measurements[4,35]. The results of these analyses have shown that the predeposited Si is almost aligned epitaxially to the substrate in solid phase when the substrate tempeature is raised to the growth temperature and it acts as precursor for

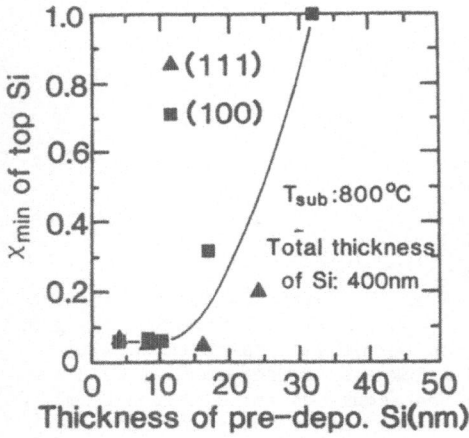

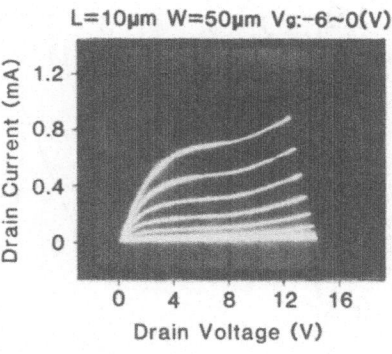

Fig. 8. Variation of near surface χ_{min}'s of 400nm thick Si films grown on CaF_2/Si structures with thickness of predeposited Si layers.

Fig. 9. Drain characteristics of a MOSFET fabricated on a Si/CaF_2/Si(100) structure.

the epitaxial growth of Si deposited at elevated temperatures.

In order to electrically characterize the quality of Si films grown by this method, MOSFETs were fabricated on the Si/CaF$_2$/Si(100) structures[16]. The process employed was a conventional one except the fact that the gate oxide was formed by the wet oxidation. Figure 9 shows typical drain characteristics of a MOSFET. The maximum field effect electron mobility of about 580cm^2/Vs is obtained, which is close to the mobility observed in MOSFETs on bulk Si. But, the leakage current of the MOSFETs on the Si/CaF$_2$/Si structure was about one order higher than that of MOSFETs on bulk Si. Cross-sectional TEM analyses have shown that microtwins are generated in the top Si film and their density is rather high near the Si/CaF$_2$ interface[36]. We speculate that these microtwins are responsible for the higher leakage current observed in MOSFETs. The partial amorphization of the Si film by ion implantation and subsequent solid phase epitaxy(SPE), which was originally developed to improve the quality of Si on sapphire, has been found to be effective to improve the quality of the Si/CaF$_2$/Si(100) structure[37]. The SPE technique is also useful to suppress the auto-doping of Ca in the Si film[19]. Fabrication of inverter chains with 100 stages has been also reported[19]. The minimum propagation dalay at the channel length of 2μm was about 360ps per gate at 5V.

The predeposition method of thin amorphous layer is also effective to improve crystalline quality and film uniformity of Ge films grown on CaF$_2$/Si structures[6]. Figure 10 shows variation of the χ_{min} near the surface of approximately 300nm thick Ge films grown on CaF$_2$/Si(111) structures with thickness of the predeposited Ge layers. Ge films having crystalline quality close to bulk Ge(χ_{min} 3%) is obtained when the thickness of the predeposited layer is about 1nm. The surface of the Ge films grown on the predeposited layer was smooth while that of the Ge film grown directly on the CaF$_2$ surface was very rough. It has been confirmed that interfacial reaction as observed in the Si/CaF$_2$ system does not occur in the Ge/CaF$_2$ system. These results suggest that the predeposited amorphous layer is capable not only to prevent the interfacial reaction but also to change the growth mode of films from the three dimensional growth to the two dimensional growth at the initial stage.

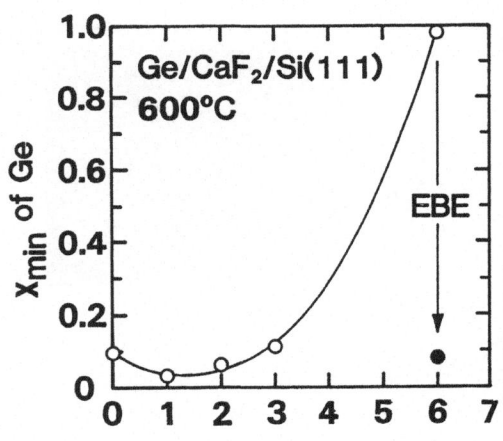

Fig. 10. Variation of near surface χ_{min}'s of 300nm thick Ge films grown on CaF$_2$/Si(111) structures with thickness of predeposited Ge layer. The result of the electron beam epitaxy method was also plotted.

Methods similar to the predeposition technique have been applied successfully to chemical vapor depositon of Si on sapphire[38], MBE of Si on $NiSi_2$/Si[39], Si on SrO/Si[40] and GaAs on Si[41]. In the growth of the Si on sapphire, improvement in surface morphology of the Si film was observed. In the formation of the Si/$NiSi_2$/Si structure, it has been shown that diffusion of Ni in the top Si film can be prevented and uniform Si films having high crystalline quality quality can be formed by a similar process described above. In the Si/SrO/Si system, it has also been shown that the deposition of an amorphous Si layer is useful to suppress autodoping of Sr in the top Si film. These results coincide with the results obtained from the growth of Si or Ge on CaF_2/Si. In the growth of GaAs on Si, there has been no discussion of the amorphous GaAs layer from the view point of interfacial reaction or autodoping. But an interesting aspect of the role of the amorphous layer has been pointed out. That is, the predeposited amorphous GaAs layer has a effect to relax the lattice mismatch between GaAs and Si[41].

4.2 The Electron Beam Exposure Epitaxy

Although the predeposition method is effective for the growth of Ge films, the optimum thickness of the predeposited layer is as thin as 1nm. TEM analyses have shown that polycrystalline nucleation takes place when the thickness of the predeposited layer is as large as 6nm. We presume that this result is due to poor wettability of Ge to the CaF_2(111) surface. Recently we have found that the optimum thickness of the predeposited layer can be drastically increased by electron beam irradiation either directly or through the predeposited layer to the CaF_2 surface[7]. Moreover it has been found that this method can drastically improve the surface morphology of the Ge film. We call this growth method electron beam exposure(EBE) epitaxy. For this growth method, an electrically scanned 3 keV electron beam was irradiated to the sample surface from the direction 3° off from the plane parallel to the sample surface. The temperature of the sample was kept at 600°C during the irradiation. The result of the elctron beam irradiation was also plotted in Fig. 10. The optimum dose range has been found to be of the order of $10^2 \mu C/cm^2$. It is noteworthy that the number of incident electrons almost corresponds to that of atoms at the CaF_2 surface. When the dose was over $10^3 \mu C/cm^2$, both the crystallinity and the surface morphology became worse.

Figure 11 shows TEM cross-sectional view of a Ge/CaF_2/Si(111) structure grown by the EBE epitaxy. Although defects such as dislocations and stacking faults are present, the crystallinity is rather good. Moreover the film uniformity is excellent. In fact, the surfaces of Ge films grown by this method were featureless as observed by scanning electron microscopy.

From several related experiments, the mechanism of the epitaxial growth in this method is understood as follows[7]. Electrons irradiated to the CaF_2 film directly or through the predeposited layer decompose the outermost F atoms of the CaF_2 surface. In fact, deficiency of F at electron irradiated CaF_2 surfaces and appearance of defects related to the F loss have been observed[42]. Then Ge atoms in the predeposited layer take the missing F sites. Because of this substitution, strong bonds between Ge and Ca atoms are formed, in other words, the wettability of Ge to the CaF_2 surface is much improved. Therefore generation of randomly oriented nuclei, which takes place during cohesion of the predeposited Ge, is avoided. Consequently a thick Ge film with a good crystallinity and a flat surface can be grown on the thin Ge layer just in the same manner as the homoepitaxy on bulk Ge. The resultant structure of the interface is schematically

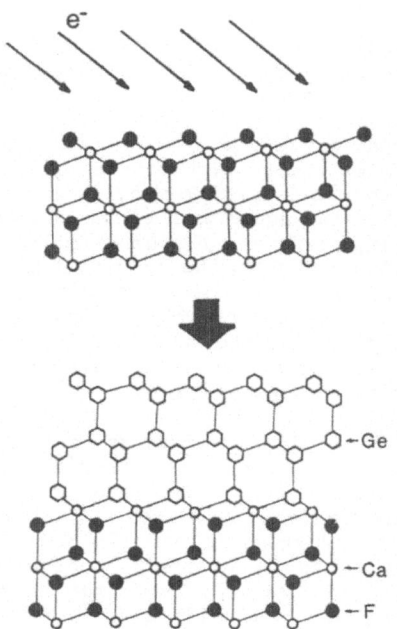

Fig. 11. Cross-sectional TEM bright field micrograph for a Ge/CaF$_2$/Si(111) structure. The Ge film was grown by the EBE epitaxy.

Fig. 12. A growth model of a Ge film on electron beam exposed CaF$_2$.

shown in Fig. 12.

Such modification of the fluoride surface can be also made by irradiation of photons[43]. Therefore use of photons will also be effective in growing high quality semiconductor films on fluoride layers.

5. GaAs OVERGROWTH ON FLUORIDE/Si STRUCTURES

GaAs films can be epitaxially grown on the fluoride/Si structures by MBE. However, as far as the GaAs films were grown by the conventional MBE method, there was a problem that film uniformity as well as the crystallinity is poor. For this problem it has been found that the modification of the fluoride surface by electron beam irradiation is efffective in growing uniform GaAs films having good crystallinity on (111) oriented substrates [24]. For the growth of GaAs films on (100) oriented substrates, it has been found that use of RTA and vicinal (100) substrates is effective. The RTA process has the ability to flatten the fluoride film surface as well as to improve the crystallinity of the fluoride film[23].

5.1 The electron beam exposure epitaxy of GaAs(111) films

In earlier attempts to use the electron beam exposure technique for the growth of GaAs films, electron beam irradiation was carried out through an amorphous GaAs layer predeposited on the fluoride surface as in the growth of Ge films. However this method gave no pronounced improvement of the film quality. On the other hand, it has been found that electron beam irradiation to the fluoride surface under As molecular beam exposure and

MBE growth onto thus irradiated fluoride surface is very effective to grow high quality GaAs films[24].

Figure 13 shows cross-sectional TEM micrographs of GaAs/CaF$_2$/Si(111) structures grown by the conventional MBE method and by the EBE epitaxy. The uniformity of the GaAs film grown by the EBE epitaxy is excellent while that of the film grown by the conventional method is bad having a number of facets at the surface. Furthermore the dislocation density is much lower in the GaAs film grown by the EBE epitaxy, though a small amount of microtwins appear .

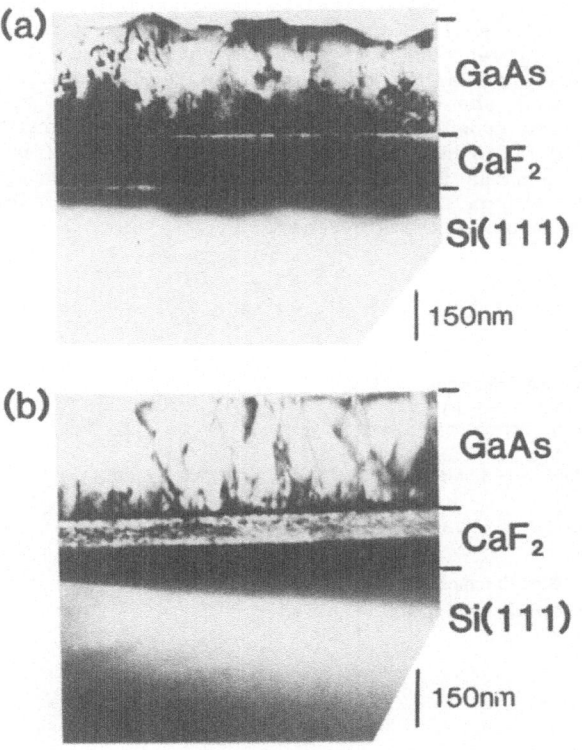

Fig. 13. Cross-sectional TEM bright field micrographs of GaAs/CaF$_2$/Si(111) structures grown by the conventional MBE method(a) and the EBE epitaxy technique.

Figure 14 shows dependence on the electron dose of the channeling minimum yield of GaAs films grown by the EBE epitaxy. The optimum dose range is several hundreds $\mu C/cm^2$. The best X_{min} is 4.6% which is close to the value obtained from a bulk GaAs crystal (3.5%). It has also been found that the electron beam irradiation under As molecular beam impingement is effective to grow single crystal GaAs films without rotationally twined crystallites, e.g., the type A growth with respect to underlying CaF$_2$ films[24]. The optimum dose almost corresponds to that obtained from the growth of Ge films.

The growth mechanism of the GaAs films is considered to be the same as that of Ge films described above. That is, electrons dissociate the outermost F atoms of CaF$_2$ and As atoms occupy the missing F sites, since electrons are irradiated in As atmosphere. The uniform arrangement of As atoms at the CaF$_2$ surface is considered to act as a template for the epitaxial growth of GaAs films[44].

5.2 Use of RTA and vicinal (100) substrates for GaAs(100) films

In the growth of GaAs films on fluoride/Si(100) structures, the EBE technique has not yet given such a dramatic improvement in film quality as observed in the growth of GaAs(111) films. This fact suggests that there are reasons other than the wettability for appearance of roughness at surfaces of GaAs films grown on (100) substrates. The appearance of rough surfaces of GaAs(100) films has been related to such crystalline defects as twins and antiphase domains in the GaAs films[45]. Two reasons, which are peculiar to the growth on fluoride/semiconductor(100) structures, are considered for generation of these defects.

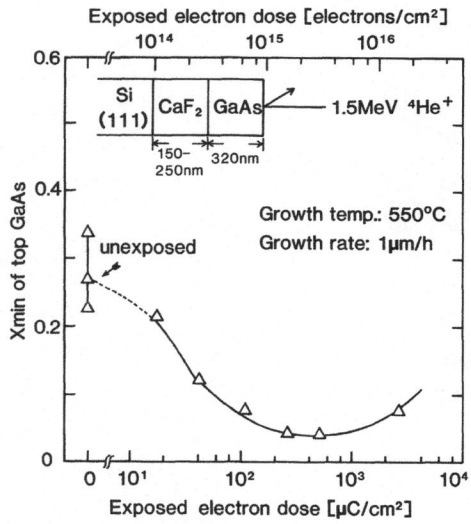

Fig. 14. Dependence on the electron dose of the X_{min} of GaAs films grown on CaF$_2$/Si(111) structures by the EBE epitaxy technique.

The first one is the fact that the surfaces of fluoride films grown on (100) substrates are not flat but faceted[22,23]. The facet plane is {111} and its dimension is a few tens of nm, so that the fluoride surface consists of a number of micro-pyramids as shown in Fig. 15(a). A replica TEM which demonstrates the faceted surface of a CaF_2 film grown on Si(100) is shown in Fig. 15(b). If epitaxial nucleations take place on each facet plane of the pyramid with identical film/substrate orientation, then domains which are rotated 90° to each other (i.e. antiphase domains) will appear in the GaAs film. Also if rotationally twined nucleations take place on the facet, it will result in appearance of microtwins. For this problem, we have recently found that the facets at the surfaces of CaF_2 films grown on Si(100) are extirpated by either ex situ RTA or in situ RTA[23]. An example of the flattened CaF_2 surface is shown in Fig. 16. But the surface flattening accomplished by RTA was insufficient to grow uniform GaAs films.

(a)   (b)

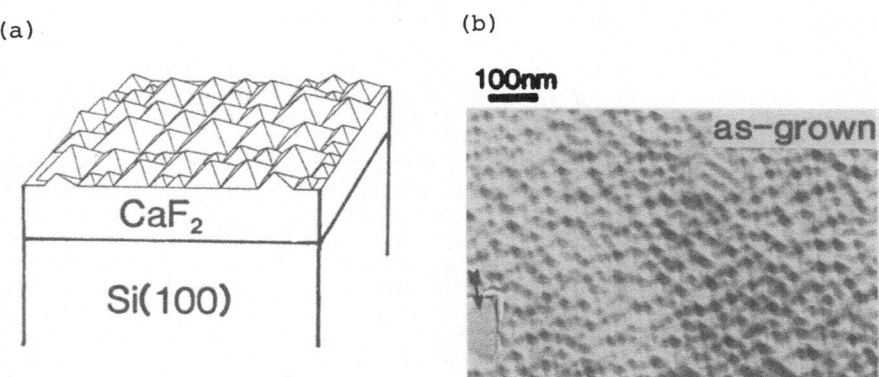

Fig. 15. Schematic illustration(a) and replica TEM micrograph(b) for the surface of a CaF_2/Si(100) structure grown at 600°C.

(a)   (b)

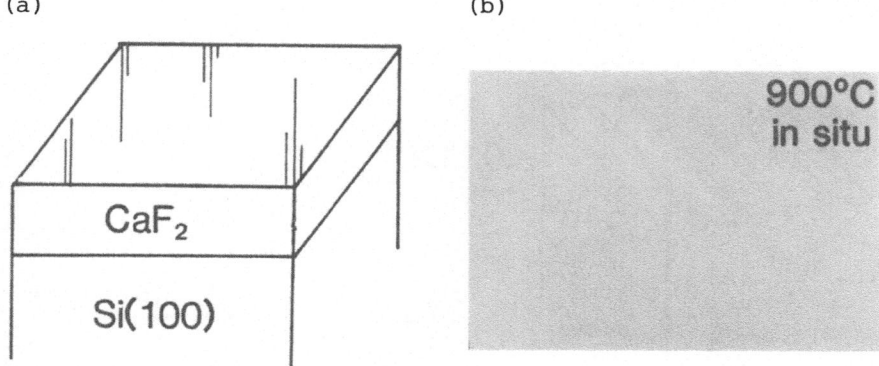

Fig. 16. Schematic illustration(a) and replica TEM micrograph(b) for the surface of a CaF_2/Si(100) structure subjected to in situ RTA at 900°C after the growth at 600°C.

354

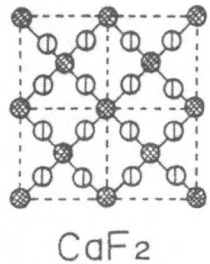
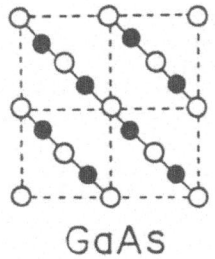

Fig. 17. Projection of the surface monolayer atoms on the (100) plane in GaAs and CaF₂ crystals.

The reason for this is considered as follows. As is shown in Fig. 17, the fluorides have four-fold symmetry in both lattice structure and bonding configuration, so that it can produce 90° rotated domains in overgrown GaAs which has two-fold symmetry in bonding configuration. This is the second reason for the appearance of defects which is responsible for the surface roughness.

In order to overcome this problem owing to the mismatch in symmetry of bonding configuration between fluoride and GaAs, use of vicinal (100) Si substrates is expected to be effective. The reason for this is as follows. As is shown in Fig. 18, the surface of a vicinal (100) such as Si(511) surface usually consists of terraces and steps. The terraces have (100) plane. The steps compensate the angular deviation between (100) and (511) planes. Since (511) is the plane inclined about 15° from (100) toward the [011] axis , the steps will lie along the [011] axis on the average. In other words, a two fold symmetry structure can be produced at the surface. If GaAs is overgrown on fluoride having such surface structure, then the epitaxial nucleation may take place at the steps avoiding generation of nuclei rotated 90° each other.

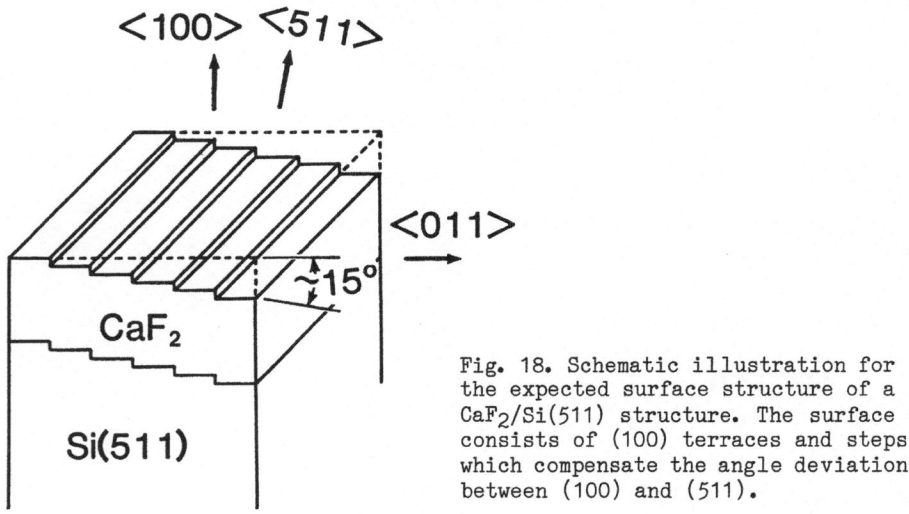

Fig. 18. Schematic illustration for the expected surface structure of a CaF₂/Si(511) structure. The surface consists of (100) terraces and steps which compensate the angle deviation between (100) and (511).

We have recently succeeded in growing uniform GaAs films by using Si(511) as substrates in conjunction with in situ RTA[46]. In the experiments, CaF$_2$ films were grown on Si(511) at 550°C. Some samples were in situ annealed at 900°C for 30sec. Then GaAs films were grown by the two step (400°C +550°C) growth method. It has been found that both the surface smoothness and the crystallinity of the CaF$_2$ films are drastically improved by the RTA process. For example, the canneling minimum yield along the <100> axis of the RTA processed CaF$_2$ was about 5%, while that of the as-grown CaF$_2$ film was about 50%.

Figure 19 shows surface morphologies of 2.2μm thick GaAs films grown on the as-grown and RTA processed CaF$_2$/Si(511) structures. We can see that a GaAs film having a rather smooth surface can be grown on the RTA processed CaF$_2$/Si structure. The channeling minimum yield of this GaAs film was about 6%. It is noteworthy that the surface morphology of the GaAs film on the as-grown CaF$_2$ film is similar to that usually observed in GaAs films grown on (100) substrates.

It is well known that doping of impurities can be made on GaAs(511) surfaces without the compensation problem[47]. Therefore the use of this kind of substrate will be useful for preparation of GaAs-on-insulator structures on Si substrates.

(a)   (b)   10μm

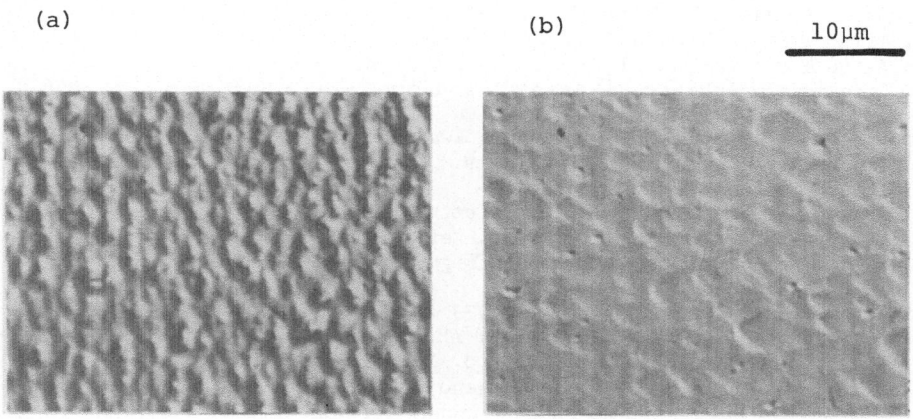

Fig. 19. Optical micrographs showing the surfaces of GaAs films grown on as-grown(a) and 900°C annealed(b) CaF$_2$/Si(511) structures. GaAs films were about 2.2μm in thickness.

6. CONCLUSION

Owing to the developments of the new heteroepitaxial growth processes such as the thin amorphous layer predeposition method, the EBE epitaxy and RTA treatments, the growth of fluoride/Si and semiconductor/fluoride/Si structures has been rapidly progressed and the potentiality of these structures has been greatly enhanced. Further advances toward the exotic devices mentioned in the first section will be made by solving remaining problems, for example;

356

1) Bonding structure at the fluoride/semiconductor interface and related generation mechanisms of the interface states.
2) Electrical and optical characterization of Ge and GaAs films grown on the fluoride/Si structures.
3) Generation mechanisms of defects in semiconductor films on fluoride layers and developments of very thin semiconductor films.
4) Advantage and disadvantage of the use of insulating films having high thermal expansion coefficients in layered structures.
5) Thermal stability of the semiconductor/fluoride layered strucutres.

ACKNOWLEDGMENTS
     The authors are thankful to Prof. A. Koma(Univ. Tokyo), Dr. K. Saiki (Univ. Tokyo), Dr. K. Tsutsui, Dr. S. Kanemaru(presently with Electrotechnical Lab.) and H. C. Lee for their invaluable collaborations. This work was supported in part by the 1987 grant-in-aid for the special distinguished research(No. 59060002) from the Ministry of Education, Science and Culture of Japan.

REFERENCES
  1. Farrow RFC, Sullivan PW, Williams GM, Jones GR, and Cameron DC: J. Vac. Sci. Technol. 19, 415 (1981).
  2. Ishiwara H and Asano T: Appl. Phys. Lett. 40, 66 (1982).
  3. Asano T and Ishiwara H: Thin Solid Films 93, 143 (1982).
  4. Asano T and Ishiwara H: J. Appl. Phys. 55, 3566 (1984)
  5. Asano T and Ishiwara H: Jpn. J. Appl. Phys. 21, L630 (1982).
  6. Kanemaru S, Ishiwara H, Asano T, and Furukawa S: Surf. Sci. 174, 666 (1986).
  7. Kanemaru S, Ishiwara H, and Furukawa S: J. Appl. Phys. 63, 1060 (1988).
  8. Asano T, Ishiwara H, Lee HC, Tsutsui K, and Furukawa S: Jpn. J. Appl. Phys. 25, L595 (1986).
  9. Zogg H and Huppi M: Appl. Phys. Lett. 47, 1133 (1985).
 10. Zogg H and Blunier S: Appl. Phys. Lett. 49, 1531 (1986).
 11. Siskos S, Fontaine C, and Munoz-Yague A: J. Appl. Phys. 56, 1642 (1984).
 12. Sugiyama K: J. Appl. Phys. 56, 1733 (1986).
 13. Tu CW, Sheng TT, Read MH, Schlier AR, Johnson JG, Johostron, Jr. WD, and Bonner WA: J. Electrochem. Soc. 130, 2081 (1983).
 14. Ishiwara H, Kim KH, Tsutsui K, Asano T, and Furukawa S: Proc. Dielectric Films on Compound Semiconductors, Honolulu, 1987, (The Electrochem. Soc.).
 15. Waho T and Yanagawa F: Abs. State-of-the-Art Program on Compound Semiconductors, The Electrochem Soc. Fall meeting, No.1792 SOA (1987).
 16. Asano T, Kuriyama Y, and Ishiwara H: Electron. Lett. 21, 386 (1985).
 17. Tsutsui K, Nakazawa T, Asano T, Ishiwara H, and Furukawa S: IEEE Electron Devices Lett. EDL-8, 277 (1987).
 18. Smith, III TP, Phillips JM, Augustyniak WM, and Stiles PJ: Appl. Phys. Lett. 45, 907 (1984).
 19. Onoda H, Katoh T, Hirashita N, and M. Sasaki: Tech. Dig. Int. Electron Devices Meeting, No.28.3 p.680 (1985).
 20. Zogg H and Norton P: Tech. Dig. Int. Electron Devices Meeting, No.5.4, p.121 (1985).
 21. Phillips JM, Pfeiffer L, Joy DC, Smith, III TP, Gibson JM, Augustyniak WM, and West KW: Proc. 1st Int. Symp. Silicon Molecular Beam Epitaxy, Bean JC(ed.)(The Electrochem. Soc., Inc., Pennington, 1985) p. 296.
 22. Schowalter LJ, Fathaur RW, Turner LG, Robertson CD: in Layered Structures, Epitaxy and Interfaces, Gibson JM and Dawson LR(eds.)(Mat. Res.

Soc., Pittsburgh, 1985) p.151.
23. Asano T, Ishiwara H, and Furukawa S: Jpn. J. Appl. Phys. in press.
24. Lee HC, Ishiwara H, Kanemaru S, and Furukawa S: Jpn. J. Appl. Phys. 26, L1834 (1987).
25. Pfeiffer L, Phillips JM, Smith, III TP, Augustyniak WM, and West KW: Appl. Phys. Lett. 46 947 (1985).
26. Asano T, Ishiwara H, and Kaifu N: Jpn. J. Appl. Phys. 22, 1474 (1983).
27. Asano T and Ishiwara H: Appl. Phys. Lett. 42, 517 (1983).
28. Ando K, Saiki K, Sato Y, Koma A, Asano T, Ishiwara H, and Furukawa S: Jpn. J. Appl. Phys. 27, 170 (1988).
29. Ishiwara H, Kanemaru S, Asano T, and Furukawa S: Jpn. J. Appl. Phys. 24, L56 (1985).
30. Kanemaru S, Ishiwara H, Asano T, and Furukawa S: Jpn. J. Appl. Phys. 26, 848 (1987).
31. Phillips JM and Augustyniak WM: in Heteroepitaxy on Si, Fan JCC and Poate JM(eds.)(Mat. Res. Soc., Pittsburgh, 1986)p. 115.
32. Himpsel FJ, Hillebrecht FU, Hughes G, Jordan JL, Karlsson UO, McFeely FR, Morar JF, and Rieger D: Appl. Phys. Lett: 48, 596 (1986).
33. Olmstead MA, Uhsberg RIG, Bringans RD, and Bachrach RZ: J. Vac. Sci. Technol. B4, 1123 (1986).
34. Ishiwara H and Asano T: Jpn. J. Appl. Phys. Suppl. 22-1, 201 (1983).
35. Asano T, Ishiwara H, and Furukawa S: Proc. Topical Conf. Deposition and Growth, American Vac. Soc., Anaheim, 1987, in press
36. Asano T, Ishiwara H, and Furukawa S: in Heteroepitaxy on Si II, Fan JCC, Phillips JM, and Tsaur BY(eds.)(Mat. Res. Soc., Pittsburgh, 1987) p. 337.
37. Ishiwara H, Asano T, Lee HC, Kuriyama Y, Seki K, and Furukawa S: in Heteroepitaxy on Si, Fan JCC and Poate JM(eds.)(Mat. Res. Soc., Pittsburgh, 1986) p. 105.
38. Ishida M, Ohyama H, Sasaki S, Yasuda Y, Nishinaga T, and Nakamura T: Jpn. J. Appl. phys. 20, L541 (1981).
39. Tung RT and Gibson JM: in Heteroepitaxy on Si, Fan JCC and Poate JM (eds.)(Mat. Res. Soc., Pittsburgh, 1986) p. 211.
40. Kado Y and Arita Y: Ext. Abs. 18th Conf. Solid State Devices and Materials (Tokyo, 1986) p. 45.
41. Akiyama M, Kawarada Y, Nishi S, Ueda T, Kaminishi K: in Heteroepitaxy on Si, Fan JCC and Poate JM(eds.)(Mat. Res. Soc., Pittsburgh, 1986)p. 53.
42. Saiki K, Sato Y, Ando K, and Koma A: Surf. Sci. 192, 1 (1987).
43. Rieger D, Himpsel FJ, Karlsson UO, McFeely FR, Morar JF, and Yarmoff JA: Phys. Rev. B34, 1123 (1986).
44. Lee HC, Asano T, Ishiwara H, Kanemaru S, and Furukawa S: Ext. Abs. 19th Conf. Solid State Devices and Materials (Tokyo, 1987) p. 163.
45. Tsutsui K, Asano T, Ishiwara H, Furukawa S: Proc. 14th Int. Symp. GaAs and Related Compounds, Crete, 1987, in press.
46. Asano T, Ishiwara H, and Furukawa S: Ext. Abs. 5th Int. Workshop on Future Electron Devices, 3-dimensional Integration, Zao, 1988, in press.
47. W. I. Wang: Proc. 2nd Int. Conf. Modulated Semiconductor Structures (Kyoto, 1985) p. 257.